Fundamentals of Probability and Stochastic Processes with Applications to Communications

Kun Il Park

Fundamentals of Probability and Stochastic Processes with Applications to Communications

Springer

Kun Il Park
Holmdel, New Jersey
USA

ISBN 978-3-319-68074-3 ISBN 978-3-319-68075-0 (eBook)
https://doi.org/10.1007/978-3-319-68075-0

Library of Congress Control Number: 2017953254

Printed on acid-free paper

This Springer imprint is published by Springer Nature
The registered company is Springer International Publishing AG
The registered company address is: Gewerbestrasse 11, 6330 Cham, Switzerland

To
Sylvie

Contents

Chapter 1
Introduction

Through many years of teaching probability and stochastic processes, the author has observed that students find these subjects difficult because of the difficulty associated with three particular areas of mathematics—complex variables, linear algebra, and set theory, which are used extensively as analyses tools in these subjects. This book devotes a full chapter to explain the essential elements of these three areas of mathematics.

Probabilities and stochastic processes are essential mathematical theories applied in communications systems analyses. In communications systems analyses, electrical signals are often represented by complex random variables (RVs) and complex stochastic processes. If the RVs and stochastic processes are complex rather than real, the complexity of their analyses multiplies greatly. Analyses based on real RVs and real stochastic processes are not adequate to deal with systems involving complex signals. This book provides comprehensive treatment of complex RVs and complex stochastic processes including extensive analysis and derivation of the parameters such as the mean, variance, autocorrelation, autocovariance, cross-correlation and cross-covariance of complex RVs, and complex stochastic processes as well as the stationarity of complex stochastic processes.

This book draws examples of applications of probability and stochastic processes from various areas of communications such as the modeling of the additive white Gaussian noise (AWGN) communications channel, the estimation theory including the Kalman filtering, and the queueing theory. The Gaussian RV and the Gaussian stochastic process are essential as applied to communications channels, e.g., AWGN model. This book devotes a full chapter to the Gaussian distribution and its properties and presents detailed and complete derivation of the characteristic function of the vector Gaussian RV.

Derivation of the general form of the Kalman filter involves extensive matrix and vector operations. Using the basic linear algebra reviewed in the prerequisite chapter, this book derives and proves all the matrix identities required in the derivation of the Kalman filter.

K.I. Park, *Fundamentals of Probability and Stochastic Processes with Applications to Communications*, https://doi.org/10.1007/978-3-319-68075-0_1

Some of the key topics and concepts discussed in this book include probability axioms, Bernoulli trials; conditional probability; total probability; Bayes' theorem; independence of events; combined experiments; Cartesian product; Cumulative Distribution Function (CDF); probability density function (pdf); mean, variance, and autocorrelation function; stationarity; ergodicity; Gaussian process; estimation theory; minimum mean square error (MMSE) estimation; Kalman filtering; counting process; point process; Poisson process; queueing theory; and Little's law.

The remainder of this book is organized as follows. Chapter 2 deals with the prerequisite mathematical concepts of complex variables, matrix and vector operations, and set theory, and Chap. 3 deals with probability theory with a focus on the axiomatic approach to probability formulation. Chapters 4 and 5 deal with applications of RVs, definition of RVs, the CDFs, the *pdf*s, and other general topics on RVs. Chapter 5 deals with the parameters of RVs such as the mean, variance, and covariance of both real and complex RVs; Chap. 6, stochastic processes, both real and complex with a significant portion of this chapter devoted to the stationarity of a stochastic process, both real and complex; Chap. 7, the Gaussian distribution; and, finally, Chap. 8, examples of application of RVs and stochastic processes drawn from the area of communications such as the AWGN channel modeling, the estimation theory, the queueing theory, and properties of the Gaussian distribution. Included at the end of the book are a bibliography, an index of the terms used in this book, and a brief write-up about the author.

Chapter 2
Basic Mathematical Preliminaries

In this chapter, we review essential prerequisite concepts of complex variables, linear algebra, and set theory required in this book. A reader familiar with these subjects may skip this chapter without losing the logical flow of the material treated in this book.

Complex variable operations are used in analyzing complex random variables and complex stochastic processes. Basic understanding of linear algebra including vector and matrix operations such as matrix multiplication, matrix inversion, and matrix diagonalization is needed for understanding vector random variables, multivariate distributions, and estimation theory. Finally, fundamental concepts of the set theory are needed for the discussion and formulation of probability and random variables.

2.1 Complex Variables

This section defines a complex number and complex conjugate and the four basic algebraic operations performed on complex numbers—addition, subtraction, multiplication, and division. This section also discusses the associative, the commutative, and the distributive properties of the complex variable operations.

2.1.1 Definitions

A complex number x is defined by the following expression:

$$x = x_r + jx_i$$

K.I. Park, *Fundamentals of Probability and Stochastic Processes with Applications to Communications*, https://doi.org/10.1007/978-3-319-68075-0_2

where x_r is called the real component and x_i is called either the imaginary component or the coefficient of the imaginary part. In this book, x_i is called the imaginary component and j the imaginary unit. The imaginary unit is also denoted by i. In this book, j is used. The imaginary unit and the imaginary component put together, i.e., jx_i, is called the imaginary part.

The imaginary unit j denotes

$$j = \sqrt{-1}$$

Therefore, the square of j is given by the following:

$$j^2 = -1$$

2.1.2 Complex Variable Operations

The four basic operations of complex variables—addition, subtraction, multiplication, and division—are defined as follows.

Addition

Consider two complex numbers x_1 and x_2 with the real and imaginary components denoted by the following equations, respectively:

$$x_1 = x_{1_r} + jx_{1_i}$$
$$x_2 = x_{2_r} + jx_{2_i}$$

The addition operation performed on the above two complex numbers produces a new complex number, which is denoted by w with the real and imaginary components w_r and jw_i as follows:

$$x_1 + x_2 = w = w_r + jw_i$$

The addition operation is defined in such a way that the real component of w is equal to the sum of the real components of $x_1$and x_2 and the imaginary component of w, the sum of the imaginary components of $x_1$and x_2 as follows:

$$w_r = x_{1_r} + x_{2_r}$$
$$w_i = x_{1_i} + x_{2_i}$$

Therefore, we have the following addition rule of complex numbers:

$$x_1 + x_2 = x_{1_r} + jx_{1_i} + x_{2_r} + jx_{2_i} = (x_{1_r} + x_{2_r}) + j(x_{1_i} + x_{2_i}) \quad (2.1)$$

Subtraction

In algebraic operations, a number being subtracted is called the "subtrahend" and the number it is subtracted from, the "minuend," and the result of subtraction, the "difference." As in real algebra, a subtraction operation is defined as the inverse operation of the addition operation. Let the minuend and the subtrahend be x_1 and x_2, respectively, and the difference be denoted by w as follows:

$$w = x_1 - x_2$$

The subtraction operation is defined such that the sum of the difference w and the subtrahend x_2 obtained by using the addition operation already defined produce the minuend x_1 as follows:

$$x_2 + w = x_1$$

By an addition operation, the left-hand side of the above equation becomes

$$x_2 + w \quad = x_{2_r} + jx_{2_i} + w_r + jw_i = (x_{2_r} + w_r) + j(x_{2_i} + w_i)$$

which should be equal to x_1 as follows:

$$(x_{2_r} + w_r) + j(x_{2_i} + w_i) = x_{1_r} + jx_{1_i}$$

From the above equation, we derive the following conditions that the real and imaginary components of the difference w_r and w_i must satisfy:

$$\begin{aligned} x_{2_r} + w_r &= x_{1_r} \\ x_{2_i} + w_i &= x_{1_i} \end{aligned}$$

From the above equations, we obtain the following equations:

$$\begin{aligned} w_r &= x_{1_r} - x_{2_r} \\ w_i &= x_{1_i} - x_{2_i} \end{aligned}$$

Therefore, a subtraction operation yields the following expression:

$$x_1 - x_2 = (x_{1_r} + jx_{1_i}) - (x_{2_r} + jx_{2_i}) = (x_{1_r} - x_{2_r}) + j(x_{1_i} - x_{2_i}) \tag{2.2}$$

Multiplication

As in real algebra, a multiplication operation is performed as follows:

$$\begin{aligned} x_1 x_2 &= (x_{1_r} + jx_{1_i})(x_{2_r} + jx_{2_i}) \\ &= x_{1_r}x_{2_r} + jx_{1_r}x_{2_i} + jx_{1_i}x_{2_r} + j^2 x_{1_i}x_{2_i} \end{aligned}$$

Noting that $j^2 = -1$, we rewrite the above equation as follows:

$$x_1x_2 = (x_{1_r}x_{2_r} - x_{1_i}x_{2_i}) + j(x_{1_r}x_{2_i} + x_{1_i}x_{2_r}) \tag{2.3}$$

Division

As with a real variable, a division operation is defined as the inverse operation of the multiplication operation. Therefore, the quotient of dividing a numerator by a denominator must be such that, if the quotient is multiplied by the denominator, the numerator be recovered.

$$w = \frac{x_1}{x_2}$$

or

$$wx_2 = x_1$$

Suppose now that a complex number $x_1 = x_{1_r} + jx_{1_i}$ is divided by a complex number $x_2 = x_{2_r} + jx_{2_i}$ and the quotient is denoted by $w = w_r + jw_i$ as follows:

$$w = w_r + jw_i = \frac{x_1}{x_2} = \frac{x_{1_r} + jx_{1_i}}{x_{2_r} + jx_{2_i}} \tag{2.4}$$

A division operation must produce the quotient $w = w_r + jw_i$ such that w multiplied by the denominator x_2 produces the numerator x_1:

$$wx_2 = x_1 = x_{1_r} + jx_{1_i} \tag{2.5}$$

By the multiplication rule given by (2.3), we obtain the following expression for the left-hand side of the above equation:

$$\begin{aligned} wx_2 &= (w_r + jw_i)(x_{2_r} + jx_{2_i}) \\ &= (w_r x_{2_r} - w_i x_{2_i}) + j(w_r x_{2_i} + x_{2_r} w_i) \end{aligned}$$

Equating the right-hand side of the above equation with the right-hand side of (2.5), we obtain the following equation:

$$(w_r x_{2_r} - w_i x_{2_i}) + j(w_r x_{2_i} + w_i x_{2_r}) = x_{1_r} + jx_{1_i}$$

Equating the real and imaginary components of both sides of the above equation, respectively, we obtain the following equations:

$$\begin{aligned} w_r x_{2_r} - w_i x_{2_i} &= x_{1_r} \\ w_r x_{2_i} + w_i x_{2_r} &= x_{1_i} \end{aligned}$$

Solving the above two simultaneous equations with respect to the real and imaginary components of the quotient to be defined, we obtain the following equations:

$$w_r = \frac{x_{1_r}x_{2_r} + x_{1_i}x_{2_i}}{x_{2_r}^2 + x_{2_i}^2} \tag{2.6a}$$

$$w_i = \frac{x_{2_r}x_{1_i} - x_{1_r}x_{2_i}}{x_{2_r}^2 + x_{2_i}^2} \tag{2.6b}$$

so that the quotient w becomes the following:

$$w = \frac{x_1}{x_2} = \frac{x_{1_r}x_{2_r} + x_{1_i}x_{2_i}}{x_{2_r}^2 + x_{2_i}^2} + j\frac{x_{2_r}x_{1_i} - x_{1_r}x_{2_i}}{x_{2_r}^2 + x_{2_i}^2} \tag{2.7}$$

To perform a division of x_1 by x_2, we apply the same process used in real algebra and confirm if the result agrees with the above definition. First, multiply the numerator and the denominator by the complex conjugate of the denominator as follows:

$$\frac{x_1}{x_2} = \frac{x_{1_r} + jx_{1_i}}{x_{2_r} + jx_{2_i}} = \frac{(x_{1_r} + jx_{1_i})(x_{2_r} - jx_{2_i})}{(x_{2_r} + jx_{2_i})(x_{2_r} - jx_{2_i})}$$

By the multiplication rule, we obtain the following expression for the denominator of the right-hand side of the above equation, where the imaginary unit j has been eliminated:

$$(x_{2_r} + jx_{2_i})(x_{2_r} - jx_{2_i}) = x_{2_r}^2 + x_{2_i}^2$$

By the multiplication rule, we obtain the following expression for the numerator:

$$(x_{1_r} + jx_{1_i})(x_{2_r} - jx_{2_i}) = (x_{1_r}x_{2_r} + x_{1_i}x_{2_i}) + j(x_{1_i}x_{2_r} - x_{2_i}x_{1_r})$$

and, thus, obtain the following expression for the division:

$$\begin{aligned} w - w_r + jw_i \\ &= \frac{(x_{1_r}x_{2_r} + x_{1_i}x_{2_i}) + j(x_{1_i}x_{2_r} - x_{2_i}x_{1_r})}{x_{2_r}^2 + x_{2_i}^2} \\ &= \frac{x_{1_r}x_{2_r} + x_{1_i}x_{2_i}}{x_{2_r}^2 + x_{2_i}^2} + j\frac{x_{1_i}x_{2_r} - x_{2_i}x_{1_r}}{x_{2_r}^2 + x_{2_i}^2} \end{aligned} \tag{2.8}$$

By comparing (2.8) with (2.7), we see that the result of a normal algebraic division operation agrees with the definition of a division operation for the complex number given by (2.7).

2.1.3 Associative, Commutative, and Distributive Laws of Algebra

The addition, subtraction, multiplication, and division operations of complex numbers defined above follow the associative, commutative, and distributive laws of algebra as shown below.

Associative

$$x_1 + (x_2 + x_3) = (x_1 + x_2) + x_3$$
$$x_1(x_2x_3) = (x_1x_2)x_3$$

Commutative

$$x_1 + x_2 = x_2 + x_1$$
$$x_1x_2 = x_2x_1$$

Distributive

$$x_1(x_2 + x_3) = x_1x_2 + x_1x_3$$

2.1.4 Complex Conjugate

The complex conjugate or conjugate of the complex variable $x = x_r + jx_i$, which is denoted by x^*, is defined as follows:

$$x^* = x_r - jx_i \tag{2.9}$$

The following equations hold true for the complex conjugate.

Theorem 2.1.1 The complex conjugate of the sum of two complex variables is equal to the sum of the complex conjugates of the individual complex variables as follows:

$$(x_1 + x_2)^* = x_1{}^* + x_2{}^* \tag{2.10}$$

Proof Let

$$x_1 = x_{1_r} + jx_{1_i}$$
$$x_2 = x_{2_r} + jx_{2_i}$$

Substitute the above two equations into the following operation:

$$\begin{aligned}(x_1 + x_2)^* &= \{(x_{1_r} + jx_{1_i}) + (x_{2_r} + jx_{2_i})\}^* \\ &= (x_{1_r} + x_{2_r}) - j(x_{1_i} + x_{2_i}) = (x_{1_r} - jx_{1_i}) + (x_{2_r} - jx_{2_i}) \\ &= x_1{}^* + x_2{}^*\end{aligned}$$

Q.E.D.

The sum of a complex variable and its complex conjugate is equal to two times its real component as shown below:

$$x + x^* = x_r + jx_i + x_r - jx_i = 2x_r \tag{2.11}$$

Theorem 2.1.2 The complex conjugate of a complex variable is the original complex variable:

$$(x^*)^* = x \tag{2.12}$$

Proof

$$(x^*)^* = \left[(x_r + jx_i)^*\right]^* = [x_r - jx_i]^* = x_r + jx_i = x$$

Q.E.D.

Theorem 2.1.3 If a complex variable is equal to its complex conjugate, the variable is real, that is, if $x = x^*$, then x is real.

Proof If $x = x^*$, we have the following equation:

$$x_r + jx_i = x_r - jx_i$$

Rearranging the terms, we obtain the following equation:

$$2jx_i = 0$$

or

$$x_i = 0$$

Since the imaginary component is zero, the complex variable x is real.

Q.E.D.

Theorem 2.1.4

$$(x_1x_2)^* = x_1^*x_2^* \tag{2.13a}$$

Proof By taking the complex conjugate of both sides of (2.3), we obtain the following result:

$$\begin{aligned}(x_1x_2)^* &= \{(x_{1_r}x_{2_r} - x_{1_i}x_{2_i}) + j(x_{1_r}x_{2_i} + x_{1_i}x_{2_r})\}^* \\ &= (x_{1_r}x_{2_r} - x_{1_i}x_{2_i}) - j(x_{1_r}x_{2_i} + x_{1_i}x_{2_r})\end{aligned}$$

On the other hand, we have the following result:

$$x_1^*x_2^* = \left(x_{1_r} - jx_{1_i}\right)\left(x_{2_r} - jx_{2_i}\right) = (x_{1_r}x_{2_r} - x_{1_i}x_{2_i}) - j(x_{1_r}x_{2_i} + x_{1_i}x_{2_r})$$

By comparing the above two results, we see that

$$(x_1x_2)^* = x_1^*x_2^*$$

Q.E.D.

Theorem 2.1.5

$$\left(\frac{x_1}{x_2}\right)^* = \frac{x_1^*}{x_2^*} \tag{2.13b}$$

Proof By taking the complex conjugate of (2.7), we have the following:

$$\left(\frac{x_1}{x_2}\right)^* = \frac{x_{1_r}x_{2_r} + x_{1_i}x_{2_i}}{x_{2_r}^2 + x_{2_i}^2} - j\frac{x_{2_r}x_{1_i} - x_{1_r}x_{2_i}}{x_{2_r}^2 + x_{2_i}^2}$$

On the other hand, we obtain the right-hand side of the above equation by evaluating the following equation:

$$\frac{x_1^*}{x_2^*} = \frac{x_{1_r} - jx_{1_i}}{x_{2_r} - jx_{2_i}} = \frac{(x_{1_r} - jx_{1_i})(x_{2_r} + jx_{2_i})}{(x_{2_r} - jx_{2_i})(x_{2_r} + jx_{2_i})}$$

$$= \frac{x_{1_r}x_{2_r} + x_{1_i}x_{2_i} - jx_{2_r}x_{1_i} + jx_{1_r}x_{2_i}}{x_{2_r}^2 + x_{2_i}^2}$$

$$= \frac{x_{1_r}x_{2_r} + x_{1_i}x_{2_i}}{x_{2_r}^2 + x_{2_i}^2} - j\frac{x_{2_r}x_{1_i} - x_{1_r}x_{2_i}}{x_{2_r}^2 + x_{2_i}^2}$$

Q.E.D.

The absolute value of a complex variable x, which is denoted by $|x|$, is defined as the square root of the sum of the squares of the real and imaginary components of x as follows:

$$|x| = \sqrt{x_r^2 + x_i^2} \tag{2.14}$$

The absolute value of a complex variable x is called the magnitude of x.

The product of a complex variable x and its complex conjugate is a real variable given by the following equation:

$$xx^* = (x_r + jx_i)(x_r - jx_i) = (x_r x_r + x_i x_i) + j(x_r x_i - x_r x_i) = x_r^2 + x_i^2 \tag{2.15}$$

By comparing (2.15) and (2.14), we obtain the following equation:

$$|x| = \sqrt{xx^*} \qquad \text{or} \qquad xx^* = |x|^2 \tag{2.16}$$

By the Maclaurin series expansion, the following equation holds true:

$$e^{j\omega} = \cos\omega + j\sin\omega \tag{2.17}$$

2.2 Matrix Operations

In analyzing multivariate probability distributions, it is convenient to use the matrix operations. This section reviews the basics of the matrix and vector calculus that will be used in subsequent discussions. Throughout this book, the vectors will be denoted by boldface letters and the matrices by double letters.

2.2.1 *Definitions*

Dimensions of a Matrix
The dimension of a matrix is denoted by $(m \times n)$, where m denotes the number of rows of the matrix and n the number of columns of the matrix. A $(m \times 1)$ matrix is a matrix of one column with m elements, that is, m rows, and is called an m-dimensional column vector. A $(1 \times n)$ matrix is an n-dimensional row vector. Sometimes, the dimension of the matrix may be shown as a subscript as in $\mathbb{A}_{m \times n}$. A matrix may sometimes be denoted by

$$\left[a_{ij}\right]_{i,j=1}^{m,n}$$

which represents a matrix with the ijth element denoted by a_{ij}, where the row number i runs from 1 to m and the column number j from 1 to n.

Square Matrix
If $m = n$, that is, $(n \times n)$, the matrix is called the square matrix. For an $(n \times n)$ square matrix, the elements along the diagonal line of the matrix, that is, a_{ii}, are called the diagonal elements. The elements which are not on the diagonal line, a_{ij}, $i \neq j$, are called the off-diagonal elements.

Diagonal Matrix
A matrix is called a diagonal matrix if all its off-diagonal elements are zero as

$$\begin{bmatrix} a_{11} & \cdot & \cdot & \cdot & 0 \\ \cdot & \cdot & \cdot & \cdot & \cdot \\ \cdot & \cdot & a_{ii} & \cdot & \cdot \\ \cdot & \cdot & \cdot & \cdot & \cdot \\ 0 & \cdot & \cdot & \cdot & a_{nn} \end{bmatrix}$$

Identity Matrix
A matrix $\mathbb{B}$ is defined as the identity matrix, denoted by $\mathbb{I}$, if an arbitrary matrix $\mathbb{A}$ multiplied by $\mathbb{B}$ results in the same matrix $\mathbb{A}$: if $\mathbb{AB} = \mathbb{BA} = \mathbb{A}$, then $\mathbb{B} = \mathbb{I}$. By this definition, it can be seen that the identity matrix is a diagonal matrix in which all diagonal elements are equal to 1:

$$\mathbb{I} = \begin{bmatrix} 1 & . & . & . & 0 \\ . & . & . & . & . \\ . & . & 1 & . & . \\ . & . & . & . & . \\ 0 & . & . & . & 1 \end{bmatrix}$$

Transposed Matrix

An important special type of matrix that is used frequently in this book is a transposed matrix. Given an $(m \times n)$ matrix $\mathbb{A}$, the transposed matrix of $\mathbb{A}$, denoted by $\mathbb{A}^{\mathrm{T}}$, is the $(n \times m)$ matrix whose ijth element is equal to the jith element of $\mathbb{A}$ for $i = 1, \ldots, m$ and $j = 1, \ldots, n$ as follows:

$$\mathbb{A} = \left[a_{ij}\right]_{i,j=1}^{m,n}$$

$$\mathbb{A}^{\mathrm{T}} = \left[a'_{ij}\right]_{i,j=1}^{n,m} = \left[a_{ji}\right]_{i,j=1}^{n,m} \qquad a'_{ij} = a_j i$$

The transposed matrix of a given matrix $\mathbb{A}$ is obtained by taking the ith row of $\mathbb{A}$ as the ith column of the new matrix, for $i = 1, \ldots, m$.

Symmetric Matrix

A square matrix $\mathbb{A}$ is called a symmetric matrix, if all its off-diagonal elements are equal to their mirror image elements across the diagonal line, that is, $a_{ij} = a_{ji}$, $i = 1, \ldots, n, j = 1, \ldots, n$. For a symmetric matrix $\mathbb{A}$, the following holds true:

$$\mathbb{A} = \mathbb{A}^{\mathrm{T}} \tag{2.18}$$

Hermitian Matrix

One special type of matrix, particularly important for complex random variables and stochastic processes, is the Hermitian matrix. A square matrix $\mathbb{A}$ is called a Hermitian, or self-adjoint, matrix if it satisfies the following two conditions:

1. The off-diagonal elements are the complex conjugates of their mirror image elements across the diagonal line of the matrix:

$$a_{ij} = a_{ji}^*, i \neq j, 1 = 1, \ldots, n \tag{2.19}$$

2. The diagonal elements are real:

$$a_{ii} = \text{real number}$$

In fact, the second condition is superfluous because we can simply use the first condition without the restriction $i \neq j$ since, as shown by (2.12), if $a_{ii} = a_{ii}^*$, then, a_{ii} is real, that is, the second condition follows.

The following matrix is an example of the Hermitian matrix.

$$\begin{bmatrix} 1 & 4-j5 & 6 \\ 4+j5 & 2 & j7 \\ 6 & -j7 & 3 \end{bmatrix}$$

For a real matrix, where all its elements are real, the Hermitian matrix and the symmetric matrix are the same because, with real numbers, the following holds true:

$$a_{ij} = a_{ji}^* = a_{ji} \quad i \neq j, 1 = 1, \ldots, n$$

nth Power of a Matrix

A matrix $\mathbb{A}$ raised to the power n, denoted by $\mathbb{A}^n$, is defined as n successive multiplications of $\mathbb{A}$ as follows $\mathbb{A}^n \triangleq \mathbb{A} \times \mathbb{A} \times \ldots. \times \mathbb{A}$.

Submatrix

Given a matrix $\mathbb{A}$, a submatrix of $\mathbb{A}$ is a matrix formed by striking out selected rows and/or selected columns of $\mathbb{A}$. For example, given

$$\mathbb{A} = \begin{bmatrix} 1 & 2 & 3 \\ 4 & 5 & 6 \\ 7 & 8 & 9 \end{bmatrix}$$

the submatrix formed by striking out the second row and the second column, the submatrix formed by striking out the first row, and the submatrix formed by striking out the first and the second rows are, respectively,

$$\begin{bmatrix} 1 & 3 \\ 7 & 9 \end{bmatrix} \qquad \begin{bmatrix} 4 & 5 & 6 \\ 7 & 8 & 9 \end{bmatrix} \qquad [7 \; 8 \; 9]$$

Partitioned Matrix or Block Matrix

A matrix may be divided into submatrices by inserting dividing lines between selected rows and/or columns. For example, a given matrix $\mathbb{A}$ can be partitioned as shown below:

$$\mathbb{A} = \begin{bmatrix} a_{11} & a_{12} & a_{13} & a_{14} & a_{15} \\ a_{21} & a_{22} & a_{23} & a_{24} & a_{25} \\ a_{31} & a_{32} & a_{33} & a_{34} & a_{35} \\ a_{41} & a_{42} & a_{43} & a_{44} & a_{45} \\ a_{51} & a_{52} & a_{53} & a_{54} & a_{55} \end{bmatrix} = \left[\begin{array}{ccc|cc} a_{11} & a_{12} & a_{13} & a_{14} & a_{15} \\ a_{21} & a_{22} & a_{23} & a_{24} & a_{25} \\ a_{31} & a_{32} & a_{33} & a_{34} & a_{35} \\ \hline a_{41} & a_{42} & a_{43} & a_{44} & a_{45} \\ a_{51} & a_{52} & a_{53} & a_{54} & a_{55} \end{array}\right]$$

The dashed lines partition the given matrix $\mathbb{A}$ into four submatrices. The dashed lines are inserted to show the submatrices and do not alter the given matrix $\mathbb{A}$. The

$$
\begin{array}{llll}
(1) \quad \mathbb{A} = \left[a_{ij}\right]_{i,j=1}^{m,n} & \mathbb{A}^{\mathrm{T}} = \left[a_{ij}^{\mathrm{T}}\right]_{i,j=1}^{n,m} = \left[a_{ji}\right]_{i,j=1}^{n,m} & a_{ij}^{\mathrm{T}} = a_{ji} & \\
(2) \quad \mathbb{B} = \left[b_{kj}\right]_{k,j=1}^{p,n} & \mathbb{B}^{\mathrm{T}} = \left[b_{kj}^{\mathrm{T}}\right]_{k,j=1}^{n,p} = \left[b_{jk}\right]_{k,j=1}^{n,p} & b_{kj}^{\mathrm{T}} = b_{jk} & \\
(3) \quad \mathbb{M} = \left[m_{lj}\right]_{l,j=1}^{q,n} & \mathbb{M}^{\mathrm{T}} = \left[m_{lj}^{\mathrm{T}}\right]_{l,j=1}^{n,q} = \left[m_{jl}\right]_{l,j=1}^{n,q} & m_{lj}^{\mathrm{T}} = m_{jl} & q = m + p
\end{array}
$$

As shown by the above three relations, $\mathbb{A}$ has m rows and n columns, $\mathbb{B}$ has p rows and n columns, and $\mathbb{M}$ has $q = m + p$ rows and n columns. The first m rows of $\mathbb{M}$, that is, $l = 1, \ldots, m$, coincide with those of $\mathbb{A}$ and the next p rows, that is, $l = (m + 1), \ldots, (m + p)$, with those of $\mathbb{B}$, and, thus, we have the following relations:

$$
\begin{array}{lll}
(4) \quad m_{lj} = a_{ij} & l = i = 1, \ldots, m; \quad j = 1, \ldots, n \\
(5) \quad m_{lj} = b_{kj} & l = m + k; k = 1, \ldots, p; \quad j = 1, \ldots, n
\end{array}
$$

Now, consider $\mathbb{M}^{\mathrm{T}}$. $\mathbb{M}^{\mathrm{T}}$ has n rows and $m + p$ columns. The first m columns of $\mathbb{M}^{\mathrm{T}}$ coincide with those of $\mathbb{A}^{\mathrm{T}}$ and the next p columns, with those of $\mathbb{B}^{\mathrm{T}}$. By the relations (3), (4), and (5) above, we have the following relations:

$$
\begin{array}{ll}
(6) \quad m_{lj}^{\mathrm{T}} = m_{jl} = a_{il}; & j = i = 1, \ldots, m; \quad l = 1, \ldots, n \\
(7) \quad m_{lj}^{\mathrm{T}} = m_{jl} = b_{kl}; & j = k + m \quad k = 1, \ldots, p; \quad l = 1, \ldots, n
\end{array}
$$

By comparing (6) and (7) with (1) and (2), we obtain the following relations:

$$
\begin{array}{ll}
(8) \quad m_{lj}^{\mathrm{T}} = a_{lj}^{\mathrm{T}} & l = 1, \ldots, n; \quad j = 1, \ldots, m \\
(9) \quad m_{lj}^{\mathrm{T}} = b_{lj}^{\mathrm{T}} & l = 1, \ldots, n; \quad j = (m + 1), \ldots, (m + p)
\end{array}
$$

(8) and (9) above show that the first m columns of $\mathbb{M}^{\mathrm{T}}$ coincide with those of $\mathbb{A}^{\mathrm{T}}$ and the next p columns of $\mathbb{M}^{\mathrm{T}}$, with those of $\mathbb{B}^{\mathrm{T}}$.

Q.E.D.

Proof of (2.25b) The matrix identity (b) follows from (a) as follows. Rewrite $\mathbb{M}$ as follows:

$$
\mathbb{M} = \begin{bmatrix} \mathbb{A} & | & 0 \\ - & + & - \\ 0 & | & \mathbb{B} \end{bmatrix} = \begin{bmatrix} \mathbb{N} \\ - \\ \mathbb{L} \end{bmatrix}
$$

where

$$
\mathbb{N} = [\mathbb{A} \quad | \quad 0] \qquad \mathbb{L} = [0 \quad | \quad \mathbb{B}]
$$

Then, using (a), we have

$$
\mathbb{N}^{\mathrm{T}} = \begin{bmatrix} \mathbb{A}^{\mathrm{T}} \\ - \\ 0 \end{bmatrix} \qquad \mathbb{L}^{\mathrm{T}} = \begin{bmatrix} 0 \\ - \\ \mathbb{B}^{\mathrm{T}} \end{bmatrix} \tag{2.26}
$$

Again, using (a), we have the following matrix identity:

$$\mathbb{M}^{\mathrm{T}} = \begin{bmatrix} \mathbb{N} \\ - \\ \mathbb{L} \end{bmatrix}^{\mathrm{T}} = \left[\mathbb{N}^{\mathrm{T}} | \mathbb{L}^{\mathrm{T}}\right]$$

Substituting (2.26) into the above equation, we obtain the desired result as follows:

$$\mathbb{M}^{\mathrm{T}} = \begin{bmatrix} \mathbb{A}^{\mathrm{T}} & | & 0 \\ - & + & - \\ 0 & | & \mathbb{B}^{\mathrm{T}} \end{bmatrix}$$

Q.E.D.

Illustration of the Matrix Identity (2.25a)
The matrix identity (a) is illustrated below. Let

$$\mathbb{M}_{(m+n)\times l} = \begin{bmatrix} \mathbb{A}_{m\times l} \\ --- \\ \mathbb{B}_{n\times l} \end{bmatrix} = \begin{bmatrix} a_{11} & . & a_{1j} & . & a_{1l} \\ . & . & . & . & . \\ a_{i1} & . & a_{ij} & . & a_{il} \\ . & . & . & . & . \\ a_{m1} & . & a_{mj}\cdot & . & a_{ml} \\ -- & -- & -- & -- & -- \\ b_{11} & . & b_{1j} & . & b_{1l} \\ . & . & . & . & . \\ b_{i1} & . & b_{ij} & . & b_{in} \\ . & . & . & . & . \\ b_{n1} & . & b_{nj} & . & b_{nl} \end{bmatrix}$$

$$\mathbb{A}_{m\times l} = \begin{bmatrix} a_{11} & . & a_{1j} & . & a_{1l} \\ . & . & . & . & . \\ a_{i1} & . & a_{ij} & . & a_{il} \\ . & . & . & . & . \\ a_{m1} & . & a_{mj}\cdot & . & a_{ml} \end{bmatrix}$$

$$\mathbb{B}_{n\times l} = \begin{bmatrix} b_{11} & . & b_{1j} & . & b_{1l} \\ . & . & . & . & . \\ b_{i1} & . & b_{ij} & . & b_{in} \\ . & . & . & . & . \\ b_{n1} & . & b_{nj} & . & b_{nl} \end{bmatrix}$$

Then, by the definition of the matrix transposition, we have the following transpositions of the three matrices:

$$\left\{\mathbb{M}_{(m+n)\times l}\right\}^{\mathrm{T}} = \begin{bmatrix} \mathbb{A}_{m\times l} \\ --- \\ \mathbb{B}_{n\times l} \end{bmatrix}^{\mathrm{T}} = \begin{bmatrix} a_{11} & . & a_{i1} & . & a_{m1} & | & b_{11} & . & b_{i1} & . & b_{n1} \\ . & . & . & . & . & | & . & . & . & . & . \\ a_{1j} & . & a_{ij} & . & a_{mj} & | & b_{1j} & . & b_{ij} & . & b_{nj} \\ . & . & . & . & . & | & . & . & . & . & . \\ a_{1l} & . & a_{il} & . & a_{ml} & | & b_{1l} & . & b_{in} & . & b_{nl} \end{bmatrix}$$

$$\left(\mathbb{A}_{m\times l}\right)^{\mathrm{T}} = \begin{bmatrix} a_{11} & . & a_{i1} & . & a_{m1} \\ . & . & . & . & . \\ a_{1j} & . & a_{ij} & . & a_{mj} \\ . & . & . & . & . \\ a_{1l} & . & a_{il} & . & a_{ml} \end{bmatrix}$$

$$\left(\mathbb{B}_{n\times l}\right)^{\mathrm{T}} = \begin{bmatrix} b_{11} & . & b_{i1} & . & b_{n1} \\ . & . & . & . & . \\ b_{1j} & . & b_{ij} & . & b_{nj} \\ . & . & . & . & . \\ b_{1l} & . & b_{in} & . & b_{nl} \end{bmatrix} \tag{2.27}$$

We see that the two submatrices of $\left\{\mathbb{M}_{(m+n)\times l}\right\}^{\mathrm{T}}$ are $\left(\mathbb{A}_{m\times l}\right)^{\mathrm{T}}$ *and* $\left(\mathbb{B}_{n\times l}\right)^{\mathrm{T}}$.

Example 2.2.1

Consider *a* 5 × 3 matrix $\mathbb{M}$ consisting of 3 × 3 and 2 × 3 submatrices $\mathbb{A}$ and $\mathbb{B}$ as follows:

$$\mathbb{M}_{(3+2)\times 3} = \begin{bmatrix} a_{11} & a_{12} & a_{13} \\ a_{21} & a_{22} & a_{23} \\ a_{31} & a_{32} & a_{33} \\ - & - & - \\ b_{11} & b_{12} & b_{13} \\ b_{21} & b_{21} & b_{23} \end{bmatrix} = \begin{bmatrix} \mathbb{A}_{3\times 3} \\ -- \\ \mathbb{B}_{2\times 3} \end{bmatrix}$$

$$\mathbb{A}_{3\times 3} = \begin{bmatrix} a_{11} & a_{12} & a_{13} \\ a_{21} & a_{22} & a_{23} \\ a_{31} & a_{32} & a_{33} \end{bmatrix} \qquad \mathbb{B}_{2\times 3} = \begin{bmatrix} b_{11} & b_{12} & b_{13} \\ b_{21} & b_{21} & b_{23} \end{bmatrix}$$

Taking the transposition of the above three matrices, we obtain the following matrices:

$$\mathbb{M}^{\mathrm{T}}_{(3+2)\times 3} = \begin{bmatrix} a_{11} & a_{21} & a_{31} & b_{11} & b_{21} \\ a_{12} & a_{22} & a_{32} & b_{12} & b_{21} \\ a_{13} & a_{23} & a_{33} & b_{13} & b_{23} \end{bmatrix}$$

$$\mathbb{A}^{\mathrm{T}}_{3\times 3} = \begin{bmatrix} a_{11} & a_{21} & a_{31} \\ a_{12} & a_{22} & a_{32} \\ a_{13} & a_{23} & a_{33} \end{bmatrix}$$

$$\mathbb{B}^{\mathrm{T}}_{2\times 3} = \begin{bmatrix} b_{11} & b_{21} \\ b_{12} & b_{22} \\ b_{13} & b_{23} \end{bmatrix}$$

By comparing $\mathbb{M}^{\mathrm{T}}_{(3+2)\times 3}$, $\mathbb{A}^{\mathrm{T}}_{3\times 3}$ and $\mathbb{B}^{\mathrm{T}}_{2\times 3}$, we see that

$$\mathbb{M}^{\mathrm{T}}_{(3+2)\times 3} = \left[\mathbb{A}^{\mathrm{T}}_{3\times 3} | \mathbb{B}^{\mathrm{T}}_{2\times 3}\right]$$

2.2.3 *Matrix Multiplication*

If the number of columns of $\mathbb{M}$ is equal to the number of rows of $\mathbb{N}$, the matrix multiplication $\mathbb{M} \times \mathbb{N}$ is defined as follows:

$$\mathbb{L}_{m\times l} = \mathbb{M}_{m\times n} \times \mathbb{N}_{n\times l} = \begin{bmatrix} m_{11} & . & m_{1j} & . & m_{1n} \\ . & . & . & . & . \\ m_{i1} & . & m_{ij} & . & m_{in} \\ . & . & . & . & . \\ m_{m1} & . & m_{mj} & . & m_{mn} \end{bmatrix} \times \begin{bmatrix} n_{11} & . & n_{1k} & . & n_{1l} \\ . & . & . & . & . \\ n_{j1} & . & n_{jk} & . & . \\ . & . & . & . & . \\ . & . & . & . & . \\ n_{n1} & . & n_{nk} & . & n_{nl} \end{bmatrix}$$
$$= \begin{bmatrix} l_{11} & . & l_{1k} & . & l_{1l} \\ . & . & . & . & . \\ l_{i1} & . & l_{ik} & . & . \\ . & . & . & . & . \\ . & . & . & . & . \\ l_{m1} & . & l_{mk} & . & l_{ml} \end{bmatrix} \tag{2.28}$$

where

$$l_{ik} = m_{i1}n_{1k} + \cdots + m_{ij}n_{jk} + \cdots + m_{in}n_{nk} = \sum_{j=1}^{n} m_{ij}n_{jk}, \quad i = 1, \ldots, m; k = 1, \ldots, l \tag{2.29}$$

By comparing the last four matrix products with the four submatrices of the first matrix product $\mathbb{L}_{(m+n)\times(m+n)}$, we see that they are indeed the same so that (2.31) results.

It will be useful later in the covariance analysis to note that, by (2.77), $\mathbb{L}$ is symmetric. This can be confirmed by noting that the two diagonal submatrices are symmetric and the transposition of the upper right submatrix is the same as the lower left submatrix.

Example 2.2.3

Let

$$\mathbb{M} = \begin{bmatrix} 1 & 2 \\ 3 & 4 \\ - & - \\ 1 & 3 \end{bmatrix} \qquad \mathbb{A} = \begin{bmatrix} 1 & 2 \\ 3 & 4 \end{bmatrix} \qquad \mathbb{B} = [\,1 \quad 3\,]$$

$$(1)\quad \mathbb{A}\mathbb{A}^{\mathrm{T}} = \begin{bmatrix} 1 & 2 \\ 3 & 4 \end{bmatrix}\begin{bmatrix} 1 & 3 \\ 2 & 4 \end{bmatrix} = \begin{bmatrix} (1\times 1)+(2\times 2) & (1\times 3)+(2\times 4) \\ (3\times 1)+(4\times 2) & (3\times 3)+(4\times 4) \end{bmatrix} = \begin{bmatrix} 5 & 11 \\ 11 & 25 \end{bmatrix}$$

$$(2)\quad \mathbb{B}\mathbb{B}^{\mathrm{T}} = [1\ \ 3]\begin{bmatrix} 1 \\ 3 \end{bmatrix} = [\,10\,] = 10$$

$$(3)\quad \mathbb{A}\mathbb{B}^{\mathrm{T}} = \begin{bmatrix} 1 & 2 \\ 3 & 4 \end{bmatrix}\begin{bmatrix} 1 \\ 3 \end{bmatrix} = \begin{bmatrix} (1\times 1)+(2\times 3) \\ (3\times 1)+(4\times 3) \end{bmatrix} = \begin{bmatrix} 7 \\ 13 \end{bmatrix}$$

$$(4)\quad \mathbb{B}\mathbb{A}^{\mathrm{T}} = [\,1 \quad 3\,]\begin{bmatrix} 1 & 3 \\ 2 & 4 \end{bmatrix} = [\,(1\times 1)+(3\times 2) \quad (1\times 3)+(3\times 4)\,] = [\,7 \quad 13\,]$$

We now find $\mathbb{M}\mathbb{M}^{\mathrm{T}}$ first by direct multiplication and then by the multiplication of the partitioned $\mathbb{M}$ as follows:

$$\mathbb{M}\mathbb{M}^{\mathrm{T}} = \begin{bmatrix} 1 & 2 \\ 3 & 4 \\ 1 & 3 \end{bmatrix}\begin{bmatrix} 1 & 3 & 1 \\ 2 & 4 & 3 \end{bmatrix} = \begin{bmatrix} 5 & 11 & 7 \\ 11 & 25 & 13 \\ 7 & 13 & 10 \end{bmatrix}$$

Suppose that we have already obtained the matrix products (1)–(4), then $\mathbb{M}\mathbb{M}^{\mathrm{T}}$ can be obtained simply by inserting the products as the submatrices as follows:

$$\begin{bmatrix} 1 & 2 \\ 3 & 4 \\ - & - \\ 1 & 3 \end{bmatrix}\left[\begin{array}{cc|c} 1 & 3 & 1 \\ 2 & 4 & 3 \end{array}\right] = \left[\begin{array}{c|c} \mathbb{A}\mathbb{A}^{\mathrm{T}} & \mathbb{A}\mathbb{B}^{\mathrm{T}} \\ \hline \mathbb{B}\mathbb{A}^{\mathrm{T}} & \mathbb{B}\mathbb{B}^{\mathrm{T}} \end{array}\right] = \left[\begin{array}{cc|c} 5 & 11 & 7 \\ 11 & 25 & 13 \\ \hline 7 & 13 & 10 \end{array}\right]$$

Notice that $\mathbb{M}\mathbb{M}^{\mathrm{T}}$ is a symmetric matrix in which the diagonal submatrices are symmetric and the off-diagonal matrices are the transposition matrix of one another. Note also that $\mathbb{B}\mathbb{A}^{\mathrm{T}} = \left(\mathbb{A}\mathbb{B}^{\mathrm{T}}\right)^{\mathrm{T}}$.

2.2.4 Determinants

Theorem 2.2.1

$$\det\left(\mathbb{A}^{-1}\right) = (\det\mathbb{A})^{-1} \tag{2.32}$$

The determinant of a matrix $\mathbb{A}$ is defined by

$$\det\mathbb{A} \triangleq \sum_{j} (-1)^{t(j)} a_{1j_1} a_{2j_2} \ldots a_{nj_n} \tag{2.33}$$

where the number j below identifies the jth permutation of the $n!$ permutations of the numbers 1 through n

$$j = (j_1, j_2, \ldots j_k, \ldots, j_n) \quad j \text{ varies over all n!permutations of } 1, 2, \ldots, n \tag{2.34}$$

and $t(j)$ is the total number of inversions in the permutation j. For example, for $n = 3$, there are six permutations of numbers 1, 2, and 3. Then,

$$(j_1, j_2, j_3) = (1, 2, 3); (1, 3, 2); (2, 1, 3); (2, 3, 1); (3, 1, 2); (3, 2, 1) \tag{2.35}$$

The number j is used for the identification purpose, and the order in which a permutation is identified is not significant. For example, the permutation (1, 2, 3) may be identified by $j = 1$ as in 1 = (1, 2, 3). The same permutation may be identified by 6 = (1, 2, 3), and the permutation (3, 2, 1) may be identified by 1 = (3, 2, 1). $a_{ij_k}$is the element of the ith row and the j_k^{th} column, where j_k is the j_k^{th} number in the permutation j.

For illustration purposes, identify the fourth permutation in the above equation by $j = 4$ to write 4 = (2, 3, 1) and consider the corresponding term in (2.33). For $j = 4$, we have

$$(j_1, j_2, j_3) = (2, 3, 1)$$

Then, $t(4)$ is obtained by counting the number of times the numbers in the permutation are inversed as follows. 2 and 1 are inversed and 3 and 1 are inversed, and so $t(4) = 2$:

$$(-1)^{t(4)} a_{1j_1} a_{2j_2} a_{3j_3} = (-1)^2 a_{12} a_{23} a_{31} = a_{12} a_{23} a_{31}$$

Theorem 2.2.2 If $\mathbb{A}$ is a diagonal matrix with the diagonal elements a_{ii}, $i = 1, \ldots,$ n, its inverse matrix is a diagonal matrix with the diagonal elements $\frac{1}{a_{ii}}, i = 1, 2, \ldots, n$ as follows:

$$\begin{aligned} M_{11} &= \det\begin{bmatrix} C_{22} & C_{23} \\ C_{32} & C_{33} \end{bmatrix} = C_{22}C_{33} - C_{23}C_{32} \\ M_{12} &= \det\begin{bmatrix} C_{21} & C_{23} \\ C_{31} & C_{33} \end{bmatrix} = C_{21}C_{33} - C_{23}C_{31} \\ M_{13} &= \det\begin{bmatrix} C_{21} & C_{22} \\ C_{31} & C_{32} \end{bmatrix} = C_{21}C_{32} - C_{22}C_{31} \\ M_{21} &= \det\begin{bmatrix} C_{12} & C_{13} \\ C_{32} & C_{33} \end{bmatrix} = C_{12}C_{33} - C_{13}C_{32} \\ M_{22} &= \det\begin{bmatrix} C_{11} & C_{13} \\ C_{31} & C_{33} \end{bmatrix} = C_{11}C_{33} - C_{13}C_{31} \\ M_{23} &= \det\begin{bmatrix} C_{11} & C_{12} \\ C_{31} & C_{32} \end{bmatrix} = C_{11}C_{32} - C_{12}C_{31} \\ M_{31} &= \det\begin{bmatrix} C_{12} & C_{13} \\ C_{22} & C_{23} \end{bmatrix} = C_{12}C_{23} - C_{13}C_{22} \\ M_{32} &= \det\begin{bmatrix} C_{11} & C_{13} \\ C_{21} & C_{23} \end{bmatrix} = C_{11}C_{23} - C_{13}C_{21} \\ M_{33} &= \det\begin{bmatrix} C_{11} & C_{12} \\ C_{21} & C_{22} \end{bmatrix} = C_{11}C_{22} - C_{12}C_{21} \end{aligned} \tag{2.47}$$

$$A_{ij} = (-1)^{i+j} M_{ij} \tag{2.48}$$

Substituting the above into the following equation, we obtain the adjoint matrix in which the minors are given by (2.47)

$$\begin{aligned} \operatorname{adj}\, \mathbb{C}_X \triangleq \left([A_{ij}]_{i,j=1}^{3}\right)^{\mathrm{T}} &= \begin{bmatrix} A_{11} & A_{12} & A_{13} \\ A_{21} & A_{22} & A_{23} \\ A_{31} & A_{32} & A_{33} \end{bmatrix}^{\mathrm{T}} = \begin{bmatrix} M_{11} & -M_{12} & M_{13} \\ -M_{21} & M_{22} & -M_{23} \\ M_{31} & -M_{32} & M_{33} \end{bmatrix} \\ &= \begin{bmatrix} M_{11} & -M_{21} & M_{31} \\ -M_{12} & M_{22} & -M_{32} \\ M_{13} & -M_{23} & M_{33} \end{bmatrix}^{\mathrm{T}} \end{aligned} \tag{2.49}$$

By substituting (2.46) and (2.49) into (8.44), we obtain

$$\mathbb{C}_X^{-1} = \frac{1}{\det \mathbb{C}_X} \operatorname{adj}\, \mathbb{C}_X$$

Example 2.2.5

For $n = 5$, determine the total number of inversions for the permutation $j = (2, 4, 3, 5, 1)$. 2 before 1 is one inversion; 4 before 3 and 1 is two inversions; 3 before 1 is one inversion; 5 before 1 is one inversion. $t(j) = 5$.

Theorem 2.2.3 If $\mathbb{A}$ is symmetric, its inverse matrix $\mathbb{A}^{-1}$ is also symmetric.

Proof By the definition of an inverse matrix, we have

$$\mathbb{A}^{-1}\mathbb{A} = \mathbb{I} \tag{2.50}$$

By taking the transposition of both sides of the above, we have

$$\left(\mathbb{A}^{-1}\mathbb{A}\right)^{\mathrm{T}} = \mathbb{A}^{\mathrm{T}}\left(\mathbb{A}^{-1}\right)^{\mathrm{T}} = \mathbb{I}^{\mathrm{T}} = \mathbb{I} \tag{2.51}$$

Since $\mathbb{A}$ is symmetric, substituting $\mathbb{A} = \mathbb{A}^{\mathrm{T}}$ into the above, we have

$$\mathbb{A}\left(\mathbb{A}^{-1}\right)^{\mathrm{T}} = \mathbb{I} \tag{2.52}$$

Multiplying both sides of the above from the right with $\mathbb{A}^{-1}$ yields

$$\mathbb{A}^{-1}\mathbb{A}(\mathbb{A}\mathbb{A}^{-1})^{\mathrm{T}} = \mathbb{A}^{-1}$$

which yields

$$\left(\mathbb{A}^{-1}\right)^{\mathrm{T}} = \mathbb{A}^{-1} \tag{2.53}$$

By the definition of the symmetric matrix given by (2.18), $\mathbb{A}^{-1}$ is symmetric.

Q.E.D.

2.2.6 *Matrix Diagonalization*

A matrix $\mathbb{A}$ can be diagonalized by obtaining a diagonalizing matrix $\mathbb{P}$ and performing the following matrix operation:

$$\mathbb{P}^{-1}\mathbb{A}\,\mathbb{P} = \mathbb{D}, \quad \det\ \mathbb{P} \neq 0$$

where $\mathbb{D}$ is a diagonal matrix with the eigenvalues of $\mathbb{A}$ as the diagonal elements as follows:

$$\mathbb{D} = \begin{bmatrix} \lambda_1 & \cdots & 0 \\ \vdots & \ddots & \vdots \\ 0 & \cdots & \lambda_n \end{bmatrix}, \lambda_i = \text{eigenvalue}, , i = 1, \ldots, n$$

To find the diagonalizing matrix $\mathbb{P}$, find n eigenvalues of $\mathbb{A}$ by solving the following equation:

$$\det\left(\mathbb{A} - \lambda\mathbb{I}\right) = 0$$

Then, obtain the eigenvectors corresponding to the eigenvalues from the following n linearly independent equations:

$$\mathbb{A}\boldsymbol{b}_i = \lambda_i \boldsymbol{b}_i, i = 1, \ldots, n$$

where the components of the eigenvectors are denoted as follows:

$$\boldsymbol{b}_i = \begin{bmatrix} b_{1i} \\ b_{2i} \\ \cdot \\ b_{ki} \\ \cdot \\ b_{ni} \end{bmatrix}, i = 1, \ldots, n$$

If $\mathbb{A}$ is symmetric, $\mathbb{P}$ is an orthogonal matrix, that is, $\mathbb{P}^{-1} = \mathbb{P}^{\mathrm{T}}$. In addition, $\mathbb{P}$ is orthonormal. The eigenvectors are orthogonal to one another and their norms are unity. Therefore, the inner products, $\langle .,. \rangle$, of the eigenvectors are given by

$$\langle \boldsymbol{b}_i, \boldsymbol{b}_j \rangle = \sum_{k=1}^{n} b_{ki} b_{kj} = \delta_{ij}$$

Example 2.2.6
Diagonalize

$$\mathbb{A} = \begin{bmatrix} 7 & -3 \\ 10 & -4 \end{bmatrix}$$

Solution
To find the eigenvalues, solve

$$\det(\mathbb{A} - \lambda \mathbb{I}) = \det\left\{ \begin{bmatrix} 7-\lambda & -3 \\ 10 & -4-\lambda \end{bmatrix} \right\} = (\lambda - 7)(\lambda + 4) + 30 = \lambda^2 - 3\lambda + 2$$
$$= (\lambda - 2)(\lambda - 1) = 0$$

Solving the above, we have the eigenvalues $\lambda_1 = 2, \quad \lambda_2 = 1$. Find the eigenvectors by solving the following equations corresponding to the two eigenvalues:

$$\mathbb{A}\boldsymbol{b}_1 = \lambda_1 \boldsymbol{b}_1 \qquad \mathbb{A}\boldsymbol{b}_2 = \lambda_2 \boldsymbol{b}_2$$

For $\lambda_1 = 2$

$$\begin{bmatrix} 7 & -3 \\ 10 & -4 \end{bmatrix} \begin{bmatrix} b_{11} \\ b_{21} \end{bmatrix} = 2 \begin{bmatrix} b_{11} \\ b_{21} \end{bmatrix}$$
$$7b_{11} - 3b_{21} = 2b_{11} \quad b_{11} = \frac{3}{5} b_{21}$$
$$10b_{11} - 4b_{21} = 2b_{21} \quad b_{11} = \frac{3}{5} b_{21}$$

The first eigenvector is given by the following, where α_1 is an arbitrary constant:

$$\boldsymbol{b}_1 = \begin{bmatrix} b_{11} \\ b_{21} \end{bmatrix} = \alpha_1 \begin{bmatrix} \frac{3}{5} \\ 1 \end{bmatrix} \tag{2.54}$$

For $\lambda_2 = 1$, solve

$$\begin{bmatrix} 7 & -3 \\ 10 & -4 \end{bmatrix} \begin{bmatrix} b_{12} \\ b_{22} \end{bmatrix} = 1 \begin{bmatrix} b_{12} \\ b_{22} \end{bmatrix}$$
$$7b_{12} - 3b_{22} = b_{12} \quad b_{12} = \frac{1}{2} b_{22}$$
$$10b_{12} - 4b_{22} = b_{22} \quad b_{12} = \frac{1}{2} b_{22}$$

The corresponding eigenvector is

$$\boldsymbol{b}_2 = \alpha_2 \begin{bmatrix} 1 \\ 2 \end{bmatrix} \tag{2.55}$$

where α_2 is an arbitrary constant. Setting $\alpha_1 = \alpha_2 = 1$, we obtain the following diagonalizing matrix:

$$\mathbb{P} = \begin{bmatrix} \frac{3}{5} & 1 \\ 1 & 2 \end{bmatrix} \tag{2.56}$$

Next, find the inverse matrix of $\mathbb{P}$ as follows:

$$\det \mathbb{P} = \frac{3}{5} \times 2 - 1 = \frac{1}{5}$$
$$\mathbb{P}^{-1} = \frac{1}{\frac{1}{5}} \begin{bmatrix} 2 & -1 \\ -1 & \frac{3}{5} \end{bmatrix} = \begin{bmatrix} 10 & -5 \\ -5 & 3 \end{bmatrix}$$

to obtain the diagonal matrix

$$\mathbb{D} = \mathbb{P}^{-1}\mathbb{A}\mathbb{P} = \begin{bmatrix} 10 & -5 \\ -5 & 3 \end{bmatrix} \begin{bmatrix} 7 & -3 \\ 10 & -4 \end{bmatrix} \begin{bmatrix} \frac{3}{5} & 1 \\ 1 & 2 \end{bmatrix} = \begin{bmatrix} 10 & -5 \\ -5 & 3 \end{bmatrix} \begin{bmatrix} \frac{6}{5} & 1 \\ 2 & 2 \end{bmatrix} = \begin{bmatrix} 2 & 0 \\ 0 & 1 \end{bmatrix} \tag{2.57}$$

We see that the diagonal elements of $\mathbb{D}$ are the two eigenvalues. We also see that $\mathbb{A}$ and $\mathbb{D}$ have the same determinant, that is, $\det \mathbb{A} = \det \mathbb{D} = 2$.

$$l_{ii} = \boldsymbol{x}^{\mathrm{T}}\boldsymbol{y} = \begin{bmatrix} k_{i1} & k_{i2} & . & k_{ii} & . & k_{im} \end{bmatrix} \begin{bmatrix} k'_{1i} \\ k'_{2i} \\ . \\ k'_{ii} \\ . \\ k'_{mi} \end{bmatrix} \tag{2.65}$$

The ith column vector $\boldsymbol{y}$ of the product of $\mathbb{K}'_{n\times m}$ is obtained by multiplying $\mathbb{M}_{n\times n}$ by the ith column of the transposed matrix of $\mathbb{K}_{m\times n}$ as follows:

$$\boldsymbol{y} = \mathbb{M}_{n\times n} \begin{bmatrix} k_{i1} \\ k_{i2} \\ . \\ k_{ii} \\ . \\ k_{im} \end{bmatrix} = \mathbb{M}_{n\times n}\boldsymbol{x} \tag{2.66}$$

Substituting (2.66) into (2.65), we obtain

$$l_{ii} = \boldsymbol{x}^{\mathrm{T}}\boldsymbol{y} = \boldsymbol{x}^{\mathrm{T}}\mathbb{M}_{n\times n}\boldsymbol{x}$$

Since $\mathbb{M}_{n\times n}$ is assumed to be nonnegative definite, the above quadratic form should be nonnegative

$$l_{ii} = \boldsymbol{x}^{\mathrm{T}}\mathbb{M}_{n\times n}\boldsymbol{x} \geq 0$$

Q.E.D.

2.2.9 Complex Conjugate of a Matrix

A complex conjugate of a matrix $\mathbb{A}$ is defined to be the matrix obtained by replacing all the elements of $\mathbb{A}$, a_{ij}'s, with their complex conjugates, a^*_{ij}'s, as follows:

$$\mathbb{A}^* = \left[a^*_{ij}\right], i = 1, \ldots, n; j = 1, \ldots, m \tag{2.67}$$

2.2.10 Matrix Identities for the Estimation Theory

In this section, we review several key matrix identities that are used in subsequent discussion of estimation theory, including Kalman filtering.

Given square matrices $\mathbb{A}$ and $\mathbb{B}$, the following matrix identities hold true:

$$\begin{aligned}
&\text{(a)}\quad (\mathbb{AB})^{-1} = \mathbb{B}^{-1}\mathbb{A}^{-1}\\
&\text{(b)}\quad \begin{bmatrix} \mathbb{A} & 0 \\ 0 & \mathbb{B} \end{bmatrix}^{-1} = \begin{bmatrix} \mathbb{A}^{-1} & 0 \\ 0 & \mathbb{B}^{-1} \end{bmatrix}\\
&\text{(c)}\quad (\mathbb{A}^{-1})^{\mathrm{T}} = (\mathbb{A}^{\mathrm{T}})^{-1}
\end{aligned} \tag{2.68}$$

Proof of (a) Let $\mathbb{M} = \mathbb{B}^{-1}\mathbb{A}^{-1}$. To show that $\mathbb{M} = (\mathbb{AB})^{-1}$, we need to show that, if $(\mathbb{AB})$ is multiplied by $\mathbb{M}$, the product should produce the identity matrix $\mathbb{I}$:

$$\begin{aligned}
\mathbb{M}(\mathbb{AB}) &= \mathbb{B}^{-1}\mathbb{A}^{-1}\mathbb{AB} = \mathbb{B}^{-1}\mathbb{B} = \mathbb{I}\\
(\mathbb{AB})\mathbb{M} &= \mathbb{ABB}^{-1}\mathbb{A}^{-1} = \mathbb{AA}^{-1} = \mathbb{I}
\end{aligned}$$

Q.E.D.

Proof of (b) Since

$$\begin{bmatrix} \mathbb{A}^{-1} & 0 \\ 0 & \mathbb{B}^{-1} \end{bmatrix}\begin{bmatrix} \mathbb{A}^{-1} & 0 \\ 0 & \mathbb{B}^{-1} \end{bmatrix}^{-1} = \mathbb{I}$$

by right multiplying by the identity matrix given by the above expression before taking the inverse operation, we have the following expression:

$$\begin{aligned}
\begin{bmatrix} \mathbb{A} & 0 \\ 0 & \mathbb{B} \end{bmatrix}^{-1} &= \left\{\begin{bmatrix} \mathbb{A} & 0 \\ 0 & \mathbb{B} \end{bmatrix}\begin{bmatrix} \mathbb{A}^{-1} & 0 \\ 0 & \mathbb{B}^{-1} \end{bmatrix}\begin{bmatrix} \mathbb{A}^{-1} & 0 \\ 0 & \mathbb{B}^{-1} \end{bmatrix}^{-1}\right\}^{-1}\\
&= \left\{\begin{bmatrix} \mathbb{AA}^{-1} & 0 \\ 0 & \mathbb{BB}^{-1} \end{bmatrix}\begin{bmatrix} \mathbb{A}^{-1} & 0 \\ 0 & \mathbb{B}^{-1} \end{bmatrix}^{-1}\right\}^{-1} = \left\{\begin{bmatrix} \mathbb{I} & 0 \\ 0 & \mathbb{I} \end{bmatrix}\begin{bmatrix} \mathbb{A}^{-1} & 0 \\ 0 & \mathbb{B}^{-1} \end{bmatrix}^{-1}\right\}^{-1}\\
&= \left\{\mathbb{I}\begin{bmatrix} \mathbb{A}^{-1} & 0 \\ 0 & \mathbb{B}^{-1} \end{bmatrix}^{-1}\right\}^{-1} = \begin{bmatrix} \mathbb{A}^{-1} & 0 \\ 0 & \mathbb{B}^{-1} \end{bmatrix}
\end{aligned}$$

Q.E.D.

Proof of (c)

$$\mathbb{A}^{\mathrm{T}}(\mathbb{A}^{\mathrm{T}})^{-1} = \mathbb{I} \qquad (\mathbb{A}^{-1}\mathbb{A})^{\mathrm{T}} = (\mathbb{I})^{\mathrm{T}} = \mathbb{I}$$

Hence

$$\mathbb{A}^{\mathrm{T}}(\mathbb{A}^{\mathrm{T}})^{-1} = (\mathbb{A}^{-1}\mathbb{A})^{\mathrm{T}}$$

On the other hand, by (2.20d),

$$(\mathbb{A}^{-1}\mathbb{A})^{\mathrm{T}} = \mathbb{A}^{\mathrm{T}}(\mathbb{A}^{-1})^{\mathrm{T}}$$

Hence

$$\mathbb{A}^{T}\left(\mathbb{A}^{T}\right)^{-1} = \mathbb{A}^{T}\left(\mathbb{A}^{-1}\right)^{T}$$

or

$$\left(\mathbb{A}^{T}\right)^{-1} = \left(\mathbb{A}^{-1}\right)^{T}$$

Q.E.D.

Identity 1

$$(\mathbb{A} + \mathbb{B}\mathbb{C}\mathbb{D})^{-1} = \mathbb{A}^{-1} - \mathbb{A}^{-1}\mathbb{B}(\mathbb{C}^{-1} + \mathbb{D}\mathbb{A}^{-1}\mathbb{B})^{-1}\mathbb{D}\mathbb{A}^{-1} \tag{2.69}$$

Proof A useful technique in manipulating matrices is to form the identity matrix $\mathbb{I}$ by multiplying a matrix appearing in the given expression by its inverse and use the identity matrix to left or right multiply the expression with the product. In this case, we will form $\mathbb{I} = \mathbb{A}\mathbb{A}^{-1}$ and left multiply the given expression with this product as follows:

$$(\mathbb{A} + \mathbb{B}\mathbb{C}\mathbb{D})^{-1} = \left\{(\mathbb{A}\mathbb{A}^{-1})(\mathbb{A} + \mathbb{B}\mathbb{C}\mathbb{D})\right\}^{-1} = \left\{\mathbb{A}(\mathbb{I} + \mathbb{A}^{-1}\mathbb{B}\mathbb{C}\mathbb{D})\right\}^{-1}$$

Using (2.68a), we rewrite the above as

$$(\mathbb{A} + \mathbb{B}\mathbb{C}\mathbb{D})^{-1} = (\mathbb{I} + \mathbb{A}^{-1}\mathbb{B}\mathbb{C}\mathbb{D})^{-1}\mathbb{A}^{-1} \tag{2.70}$$

Apply $(\mathbb{I} + \mathbb{K})^{-1} = \mathbb{I} - (\mathbb{I} + \mathbb{K})^{-1}\mathbb{K}$ to the right-hand side of the above equation with the following correspondence:

$$\mathbb{K} = \mathbb{A}^{-1}\mathbb{B}\mathbb{C}\mathbb{D}$$

to obtain the following equation:

$$(\mathbb{I} + \mathbb{A}^{-1}\mathbb{B}\mathbb{C}\mathbb{D})^{-1} = \mathbb{I} - (\mathbb{I} + \mathbb{A}^{-1}\mathbb{B}\mathbb{C}\mathbb{D})^{-1}\mathbb{A}^{-1}\mathbb{B}\mathbb{C}\mathbb{D}$$

Substituting the above in (2.70), we obtain the following equation:

$$\begin{aligned}(\mathbb{A} + \mathbb{B}\mathbb{C}\mathbb{D})^{-1} &= \left\{\mathbb{I} - \left(\mathbb{I} + \mathbb{A}^{-1}\mathbb{B}\mathbb{C}\mathbb{D}\right)^{-1}\mathbb{A}^{-1}\mathbb{B}\mathbb{C}\mathbb{D}\right\}\ \mathbb{A}^{-1} \\ &= \mathbb{A}^{-1} - \left(\mathbb{I} + \mathbb{A}^{-1}\mathbb{B}\mathbb{C}\mathbb{D}\right)^{-1}\mathbb{A}^{-1}\mathbb{B}\mathbb{C}\mathbb{D}\mathbb{A}^{-1} \\ &= \mathbb{A}^{-1} - \ \left\{\left(\mathbb{I} + \mathbb{A}^{-1}\ \mathbb{B}\mathbb{C}\mathbb{D}\right)^{-1}\mathbb{A}^{-1}\right\}\ \mathbb{B}\mathbb{C}\mathbb{D}\mathbb{A}^{-1}\end{aligned} \tag{2.71}$$

Apply the following correspondences to the terms inside the braces

$$\mathbb{M} \rightarrow \mathbb{A}^{-1} \qquad \mathbb{N} \rightarrow \mathbb{B}\mathbb{C}\mathbb{D}$$

to obtain

$$(\mathbb{I}+\mathbb{A}^{-1}\mathbb{BCD})^{-1}\mathbb{A}^{-1}=\mathbb{A}^{-1}(\mathbb{I}+\mathbb{BCDA}^{-1})^{-1}$$

Substituting the last expression into (2.71), we obtain the following:

$$(\mathbb{A}+\mathbb{BCD})^{-1}=\mathbb{A}^{-1}-\{\mathbb{A}^{-1}(\mathbb{I}+\mathbb{BCDA}^{-1})^{-1}\}\mathbb{BCDA}^{-1}$$

Now, shift the braces to the right by one letter to rewrite the above as follows:

$$(\mathbb{A}+\mathbb{BCD})^{-1}=\mathbb{A}^{-1}-\mathbb{A}^{-1}\{(\mathbb{I}+\mathbb{BCDA}^{-1})^{-1}\mathbb{B}\}\mathbb{CDA}^{-1} \tag{2.72}$$

Make the substitutions $\mathbb{M}\to\mathbb{B}$ and $\mathbb{N}\to\mathbb{CDA}^{-1}$in (8.105) and apply the resultant matrix identity to the terms inside the braces of the above equation to obtain the following equation:

$$(\mathbb{I}+\mathbb{BCDA}^{-1})^{-1}\mathbb{B}=\mathbb{B}(\mathbb{I}+\mathbb{CDA}^{-1}\mathbb{B})^{-1}$$

Substituting the above expression into (2.72), we obtain the following equation:

$$(\mathbb{A}+\mathbb{BCD})^{-1}=\mathbb{A}^{-1}-\mathbb{A}^{-1}\ \mathbb{B}(\mathbb{I}+\mathbb{CDA}^{-1}\mathbb{B})^{-1}\mathbb{CDA}^{-1} \tag{2.73}$$

We can manipulate the above equation to change it into the form of (2.69), which will be needed for the Kalman filter in a later section. This can be done by left multiplying the terms inside the parentheses of the inverse operation by the identity matrix $\mathbb{CC}^{-1}=\mathbb{I}$ before taking the inverse operation as follows:

$$\begin{aligned}(\mathbb{A}+\mathbb{BCD})^{-1}&=\mathbb{A}^{-1}-\mathbb{A}^{-1}\mathbb{B}\ \{\ (\mathbb{CC}^{-1})(\mathbb{I}+\mathbb{CDA}^{-1}\mathbb{B})\ \}^{-1}\ \mathbb{CDA}^{-1}\\&=\mathbb{A}^{-1}-\mathbb{A}^{-1}\mathbb{B}\ \{\ \mathbb{C}(\mathbb{C}^{-1}+\mathbb{C}^{-1}\mathbb{CDA}^{-1}\mathbb{B})\ \}^{-1}\mathbb{CDA}^{-1}\end{aligned} \tag{2.74}$$

Applying (2.68a) to the inverse operation of the terms inside the braces, we obtain the following:

$$\{\mathbb{C}(\mathbb{C}^{-1}+\mathbb{C}^{-1}\mathbb{CDA}^{-1}\mathbb{B})\}^{-1}=(\mathbb{C}^{-1}+\mathbb{IDA}^{-1}\mathbb{B})^{-1}\mathbb{C}^{-1}$$

Substituting the above equation into (2.74), we obtain the following equation:

$$\begin{aligned}(\mathbb{A}+\mathbb{BCD})^{-1}&=\mathbb{A}^{-1}-\mathbb{A}^{-1}\mathbb{B}\ (\mathbb{C}^{-1}+\mathbb{DA}^{-1}\mathbb{B})^{-1}\mathbb{C}^{-1}\mathbb{CDA}^{-1}\\&=\mathbb{A}^{-1}-\mathbb{A}^{-1}\mathbb{B}(\mathbb{C}^{-1}+\mathbb{DA}^{-1}\mathbb{B})^{-1}\mathbb{DA}^{-1}\end{aligned}$$

Q.E.D.

Identity 2

$$(\mathbb{A}+\mathbb{BCD})^{-1}\mathbb{BC}=\mathbb{A}^{-1}\mathbb{B}(\mathbb{C}^{-1}+\mathbb{DA}^{-1}\mathbb{B})^{-1} \tag{2.75}$$

Proof

$$\begin{aligned}(\mathbb{A}+\mathbb{BCD})^{-1}\mathbb{BC}&=\left\{(\mathbb{AA}^{-1})(\mathbb{A}+\mathbb{BCD})\right\}^{-1}\mathbb{BC}=\left\{\mathbb{A}(\mathbb{A}^{-1}\mathbb{A}+\mathbb{A}^{-1}\mathbb{BCD})\right\}^{-1}\mathbb{BC}\\&=\left\{\mathbb{A}(\mathbb{I}+\mathbb{A}^{-1}\mathbb{BCD})\right\}^{-1}\mathbb{BC}=\left\{(\mathbb{I}+\mathbb{A}^{-1}\mathbb{BCD})^{-1}\mathbb{A}^{-1}\right\}\mathbb{BC}\end{aligned} \tag{2.76}$$

To proceed with the proof, we use the following matrix identity, the proof of which is given later in this book:

$$(\mathbb{I}+\mathbb{MN})^{-1}\mathbb{M}=\mathbb{M}(\mathbb{I}+\mathbb{NM})^{-1}$$

Make the substitutions $\mathbb{M}\rightarrow\mathbb{A}^{-1}$andN $\rightarrow\mathbb{BCD}$ in the above identity and apply the resultant matrix identity to the terms inside the braces on the right-hand side of (2.76) to obtain the following equation:

$$(\mathbb{I}+\mathbb{A}^{-1}\mathbb{BCD})^{-1}\mathbb{A}^{-1}=\mathbb{A}^{-1}(\mathbb{I}+\mathbb{BCDA}^{-1})^{-1}$$

Substituting the above into (2.76), we obtain the following, where the brace boundary is shifted to the right by one letter:

$$(\mathbb{A}+\mathbb{BCD})^{-1}\mathbb{BC}=\mathbb{A}^{-1}\{(\mathbb{I}+\mathbb{BCDA}^{-1})^{-1}\mathbb{B}\}\mathbb{C}$$

Applying the above procedure again to the terms inside the braces, we obtain the following equation:

$$(\mathbb{A}+\mathbb{BCD})^{-1}\mathbb{BC}=\mathbb{A}^{-1}\mathbb{B}(\mathbb{I}+\mathbb{CDA}^{-1}\mathbb{B})^{-1}\mathbb{C}$$

Left multiply the term in the parentheses in the right-hand side of the above equation before taking the inversion by the identity matrix $\mathbb{CC}^{-1}=\mathbb{I}$ as follows:

$$\begin{aligned}(\mathbb{A}+\mathbb{BCD})^{-1}\mathbb{BC}&=\mathbb{A}^{-1}\mathbb{B}\left\{(\mathbb{CC}^{-1})(\mathbb{I}+\mathbb{CDA}^{-1}\mathbb{B})\right\}^{-1}\mathbb{C}\\&=\mathbb{A}^{-1}\mathbb{B}\left\{\mathbb{C}(\mathbb{C}^{-1}+\mathbb{C}^{-1}\mathbb{CDA}^{-1}\mathbb{B})\right\}^{-1}\mathbb{C}=\mathbb{A}^{-1}\mathbb{B}(\mathbb{C}^{-1}+\mathbb{DA}^{-1}\mathbb{B})^{-1}\mathbb{C}^{-1}\mathbb{C}\\&=\mathbb{A}^{-1}\mathbb{B}(\mathbb{C}^{-1}+\mathbb{DA}^{-1}\mathbb{B})^{-1}\end{aligned}$$

Q.E.D.

Theorem 2.2.6 If an arbitrary matrix $\mathbb{A}$ is multiplied by its transposed matrix $\mathbb{A}^{\mathrm{T}}$, the result is a symmetric matrix. Let $\mathbb{M}=\mathbb{AA}^{\mathrm{T}}$. Then, $\mathbb{M}$ is a symmetric matrix.

Proof By using (2.20a) and (d), we have

$$\mathbb{M}^{\mathrm{T}} = \left(\mathbb{A}\mathbb{A}^{\mathrm{T}}\right)^{\mathrm{T}} = \left(\mathbb{A}^{\mathrm{T}}\right)^{\mathrm{T}}\mathbb{A}^{\mathrm{T}} = \mathbb{A}\mathbb{A}^{\mathrm{T}} = \mathbb{M} \tag{2.77}$$

Therefore, by the definition of the symmetric matrix, $\mathbb{M}$ is symmetric.

Q.E.D.

2.3 Set Theory Basics

This section defines set, subset, class, partition, Cartesian product, and set equality. This section then defines the union, intersection, and complement operations performed on sets.

2.3.1 *Definition of Set*

Any collection of distinct objects into a whole entity is called a set. This collection is identified by a pair of braces enclosing the objects as in {object 1, object 2, ...}. The objects included in a set are called the elements of the set. A set may be defined by enumerating all the elements of the set, e.g., {apple, banana, mango}. The order of the elements in a set is not a factor defining the set, and two sets {mango, apple, banana} and {apple, banana, mango} are the same. {apple, banana, mango, apple} is not a proper set because "apple" is repeated unless the two apples are distinguished so that one of them is identified by a different name.

Instead of enumerating the elements, a set may also be defined by specifying the conditions that an object must satisfy to be included in the set, that is, the set membership conditions. In everyday language, one might use the word "set" loosely as in "a set of tall students in the class." In set theory, however, such a use of the term is not accepted. How could a student be qualified to be tall to be included in this set? In set theory, it must be unambiguous as to whether an object can be included in a set or not. For example, "a set of students taller than 6 feet" is a proper set.

The following examples of sets will be used in subsequent discussions on probability:

{all possible outcomes of throwing a die} = {1, 2, 3, 4, 5, 6}
{all possible outcomes of tossing a coin} = {heads, tails}
{all positive integers} = $\{i, i > 0, i = \text{integer}\}$
{all real numbers between 3 and 7} = $\{x, 3 < x < 7\}$

A set S with elements $s_1, s_2, \ldots, s_n$ is denoted by $S = \{s_1, s_2, \ldots, s_n\}$. The membership of an element of a set is denoted by $\in$ as in $s_2 \in S$, which reads "s_2 is a member of S." An arbitrary member of S is represented by an element with an arbitrary indexing subscript i as in s_i or, sometimes, without any indexing subscript as in s. So, to say "an arbitrary element of S," $s_i \in S$ or $s \in S$ is used. The nonmembership to a set is denoted by $\notin$ as in $s_2 \notin S$, which reads "s_2 is not a member of S."

A space is a set of all elements under consideration and is denoted by Ω. For example, the first two examples given above are the spaces for the die-throwing and coin-tossing experiments.

A null set or an empty set is a set that contains no elements and is denoted by $\{\varnothing\}$.

2.3.2 Subset

A set E is called a subset of S if and only if every element of E is also an element of S. The symbol $\subset$ is used to show the subset relationship as follows:

$$E \subset S$$

The set relationship $E \subset S$ is read "E is a subset of S." $E \subset S$ is also read "E is included in S" or "E is contained in S." In the set relationship $E \subset S$, S is called a superset. $E \subset S$ is read "S is a superset of E," "S includes E," or "S contains E."

Suppose that two sets E and S are defined by enumerating the elements of the sets such as two large databases of telephone numbers. Procedurally speaking as in a computer algorithm, to show that E is a subset of S, one may take each element of E and compare it with the elements of S one by one. If there is a match for all elements of E, E is a subset of S.

Suppose that two sets E and S are defined by specifying the conditions of the sets. A typical method of proving analytically that E is a subset of S is to prove that an arbitrary member of E is also a member of S. In other words, if the premise that $(s \in E\ )$ implies $(s \in S)$ is proven, then $(E \subset S)$ is proven, and we can write the following inference:

$$(s \in E \Rightarrow s \in S) \quad \Rightarrow \quad E \subset S \tag{2.78}$$

In the expression in the parentheses in the above equation, $s \in E$ reads "s is an arbitrarily selected member of E." The expression $(s \in E \Rightarrow s \in S)$ reads: "if s is an arbitrarily selected member of E, then s is also a member of S." If this statement is proved, any member of E is also a member of S, and, therefore, the right-hand side expression $E \subset S$ is proven.

The proof that $s \in E$ implies $s \in S$, that is, the left-side inference in parentheses, goes as follows. Since s is a member of E, s must satisfy the membership condition of E. To prove that s is also a member of S, it suffices to prove that s also satisfies the membership condition of S. Since this can be proven for an arbitrary member of s, by proving that $s \in E$ implies $s \in S$, that is, $(s \in E \Rightarrow s \in S)$, in the above expression, the right-side relation $E \subset S$ is proven.

Example 2.3.1
Suppose that sets S and E are defined as follows:

$$S = \{\text{all positive integers}\}$$
$$E = \{\text{all integers greater than 0 and less than 10}\}$$

Prove that

$$E \subset S$$

Proof Let $s \in E$. Then, we know that

$$0 < s < 10 \text{ and } s \text{ is an integer}$$

Therefore, s is a positive integer. Therefore, $s \in S$. Since s is an arbitrary member of E, the above conclusion can be made for any member of E. Therefore, all members of E are also members of S, and, by the definition of a subset, E is a subset of S.

Q.E.D.

Now, change the set E slightly in the above example as follows:

$$E = \{\text{all integers greater than or equal to 0 and less than 10}\}$$

Then, E is not a subset of S because the above inference is not true if $s = 0$. Therefore, the above statement cannot be made for an arbitrarily selected s.

The converse of (2.78) is also true; that is, if E is a subset of S, then every member of E is also a member of S:

$$E \subset S \Rightarrow (s \in E \Rightarrow s \in S) \tag{2.79}$$

If s is an arbitrary member of S, it does not necessarily imply that s is also a member of a subset of S:

$$s \in S \nRightarrow s \in E \tag{2.80}$$

By the definition of a subset, an arbitrary set A is a subset of itself A, and a space is a subset of itself, and the null set is a subset of any arbitrary set including Ω:

$$A \subset A \qquad \Omega \subset \Omega \qquad \{\varnothing\} \subset A \qquad \{\varnothing\} \subset \Omega \tag{2.81}$$

Example 2.3.2

Consider the following set:

$A = \{\text{Sue, Tom, Pete}\}$

There are eight possible subsets of A as follows:

$E_1 = \{\text{Sue}\}$ $E_2 = \{\text{Tom}\}$ $E_3 = \{\text{Pete}\}$ $E_4 = \{\text{Sue, Tom}\}$
$E_5 = \{\text{Sue, Pete}\}$ $E_6 = \{\text{Tom, Pete}\}$ $E_7 = \{\text{Sue, Tom, Pete}\} = A$ $E_8 = \{\varnothing\}$

Example 2.3.3

Let

$X = \{\text{all classmates in Eatontown}\} = \{\text{Sue, Tom, Pete, John, Mary, Joan}\}$
$M = \{\text{male classmates in Eatontown}\} = \{\text{Tom, Pete, John}\}$

To determine whether M is a subset of X or not, take each element of M and check whether it is in X or not. In this example, M is a subset of X. Hence, for example, "Tom is a member of M" implies "Tom is a member of X." On the other hand, "Sue is a member of X" does not necessarily imply "Sue is a member of M."

If a set is a subset of a set A, the membership condition for the subset is stricter than or equal to that for A. For example:

$B = \{\text{all male classmates in Eatontown}\} = \{\text{Tom, Pete, John}\}$

The membership condition for A is to live in Eatontown. The membership condition for B is to live in Eatontown and, in addition, must be a male.

All Possible Subsets

For a space with n elements $\Omega = \{s_1, s_2, \ldots, s_i, \ldots, s_n\}$, how many subsets of Ω would be possible? This general question often arises in probability analysis.

To solve this problem, suppose that a particular subset of Ω is created by selecting particular elements from Ω for inclusion in the subset:

$$\begin{array}{l} \Omega = \{s_1, s_2, \ldots, s_i, \ldots, s_n\} \\ \qquad\quad\ \downarrow \qquad\ \downarrow \\ E = \quad \{ s_2, s_i\} \end{array} \tag{2.82}$$

Consider each of the n elements of Ω one at a time and decide on the binary choice for the element regarding whether to include it in a subset or not. A particular series of n decisions would create a particular subset. By considering the n elements of Ω as an n-bit number and setting the n bits to 1 or 0—1 for inclusion and 0 for no inclusion—we have 2^n binary numbers, each of which corresponds to a particular subset, yielding 2^n possible subsets. To illustrate this process, consider

$$\Omega = \{s_1, s_2, s_3\}$$

Setting the three bits representing the three elements of Ω to 1 or 0 yields eight binary numbers ($2^3 = 8$), each of which corresponds to a specific subset as follows:

111 $A = \{s_1, s_2, s_3\} = \Omega$
110 $B = \{s_1, s_2\}$
101 $C = \{s_1, s_3\}$
100 $D = \{s_1\}$
011 $E = \{s_2, s_3\}$
010 $F = \{s_2\}$
001 $G = \{s_3\}$
000 $H = \{\varnothing\}$

The above method of creating subsets is useful as a computer algorithm for automatically generating all possible subsets of a given set.

All Possible Subsets Consisting of *k* Elements

For a space with n elements $\Omega = \{s_1, s_2, \ldots, s_i, \ldots, s_n\}$, how many subsets of Ω consisting of k elements would be possible? The answer to the question is simply the k-combinations of the n elements given by

$$\binom{n}{k} = \frac{n!}{k!(n-k)!} \tag{2.83}$$

2.3.3 *Set Operations*

Union

The set operation "union" is a binary operation performed on two sets, A and B, to produce a third set, C, which is a set consisting of the elements belonging to either A or B or both as illustrated in Fig. 2.1b, which is written as follows:

$$C = A \cup B$$

The word "union" is also used as the resultant set itself produced by performing a union on A and B, that is, $A \cup B$ represents the set:

$$A \cup B = \{\text{set of elements belonging to } A \text{ or} B \text{ or both}\} \tag{2.84}$$

Consider the union of three sets as $A_1 \cup A_2 \cup A_3$. Since the union is a binary operation, this is a shorthand notation of two binary operations as in

$$C = A_1 \cup A_2 \qquad A_1 \cup A_2 \cup A_3 = C \cup A_3 \tag{2.85}$$

or

$$C = A_2 \cup A_3 \qquad A_1 \cup A_2 \cup A_3 = A_1 \cup C \tag{2.86}$$

Note that the notation $A \cup B$ represents the union operation between A and B and at the same time represents the resultant set produced by the union operation.

If $s \in A \cup B$, then s is a member of at least A or B, that is, it belongs to A or B or both:

$$s \in A \cup B \quad \Rightarrow \quad s \in A \ \text{or} \ s \in B \quad \text{or} \quad s \in A \ \text{and} \ s \in B \tag{2.87}$$

Given a set C, if $s \in C$ implies that s is a member of at least A or B, then $C = A \cup B$.

$$(s \in C \quad \Rightarrow \quad s \in A \ \text{or} \ s \in B \quad \text{or} \quad s \in A \ \text{and} \ s \in B) \qquad \Rightarrow \qquad C = A \cup B \tag{2.88}$$

Consider the process of forming $A \cup B$. $A \cup B$ may be created by including all the elements of both sets, A and B, and eliminating the duplicate elements between A and B. For example, given $A = \{$apple, pear, banana$\}$ and $B = \{$banana, orange$\}$, to create $A \cup B$, first, merge the two sets as $\{$apple, pear, banana, banana, orange$\}$ and then eliminate one of the duplicate element "banana" to yield $A \cup B = \{$apple, pear, banana, orange$\}$. Keeping this process in mind, we see the following set identity:

$$A \cup A = A \qquad A \cup \{\varnothing\} = A \qquad A \cup \Omega = \Omega, \text{if } A \subset \Omega \tag{2.89}$$

Example 2.3.4
Given $\Omega = \{a, b, c, d, e, f, g, h, i, j, k\}$, consider two subsets:

$$A = \{a, b, c, d\} \qquad B = \{e, f, g\}$$

Then

$$A \cup B = \{a, b, c, d, e, f, g\} \qquad A \cup A = \{a, b, c, d\}$$
$$A \cup \{\varnothing\} = \{a, b, c, d\} \qquad A \cup \Omega = \{a, b, c, d, e, f, g, h, i, j, k\}$$

Intersection
The set operation "intersection" is a binary operation performed on two sets, A and B, to produce a third set, C, which is a set consisting of the elements belonging to both A and B, which is written as follows:

$$C = A \cap B = \{s, s \text{ is } a \text{ member of both } A \text{ and } B\} \tag{2.90}$$

The intersection of A and B is a set of members satisfying the conditions of both A and B. See Fig. 2.1c

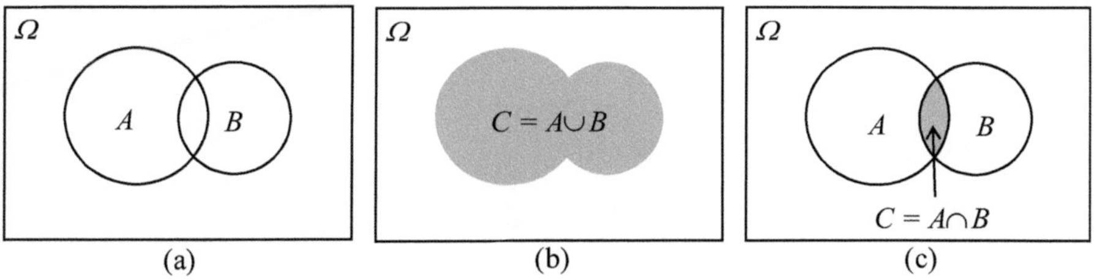

Fig. 2.1 (**a**) Space Ω and its subsets *A* and *B*, (**b**) Union of *A* and *B*, (**c**) Intersection of *A* and *B*

Example 2.3.5

$$S_1 = \{\text{students with an } A \text{ in physics}\}$$
$$S_2 = \{\text{students with an } A \text{ in math}\}$$
$$C = S_1 \cap S_2 = \{\text{students with an } A \text{ in both physics and math}\}$$

If $s \in A \cap B$, then s is a member of both A and B

$$s \in A \cap B \;\Rightarrow\; s \in A \text{ and } s \in B \tag{2.91}$$

Given a set C, if $s \in C$ implies that s is a member of both A and B, then $C = A \cap B$.

$$(s \in C \;\Rightarrow\; s \in A \text{ and } s \in B) \;\Rightarrow\; C = A \cap B \tag{2.92}$$

$A \cap B$ may be created as follows. Take each element of A and compare it with each element of B to see if there is a match between the two. If there is a match, include the element in $A \cap B$; otherwise, do not include it in $A \cap B$. Keeping this process in mind, we see the following set identity:

$$A \cap A = A \qquad A \cap \{\varnothing\} = \{\varnothing\} \qquad \Omega \cap A = A, \text{if } A \subset \Omega \tag{2.93}$$

Example 2.3.6
Given $\Omega = \{a, b, c, d, e, f, g, h, i, j, k\}$, consider three subsets:

$$A = \{a, b, c, d\} \qquad B = \{e, f, g\} \qquad C = \{b, c, d, e, f\}$$

Then

$$A \cap B = \{\varnothing\} \qquad A \cap C = \{b, c, d\} \qquad A \cap A = \{a, b, c, d\}$$
$$A \cap \{\varnothing\} = \{\varnothing\} \qquad A \cap \Omega = \{a, b, c, d\}$$

Mutually Exclusive Sets
A and B are mutually exclusive if their intersection is null (i.e, no common elements):

$$A \cap B = \{\varnothing\}$$

Sets $S_1, S_2, \ldots, S_n$ are mutually exclusive if, for all combinations of i and j for $i \neq j$, S_i and S_j are mutually exclusive:

$$S_i \cap S_j = \{\varnothing\}, \quad \text{for all } i, j \quad \text{for } i \neq j$$

Complement

The complement is a unary operation, which is performed on a single set. The complement of A is defined with respect to Ω. The complement of A, denoted by $\bar{A}$, is a set of all elements of Ω that are not in A.

By the definition of the complement, A and $\bar{A}$ do not have common elements, and, therefore, they are mutually exclusive:

$$A \cap \bar{A} = \{\varnothing\}$$

Referring to Fig. 2.2, we see that the union of A and $\bar{A}$ would completely fill Ω, that is,

$$A \cup \bar{A} = \Omega$$

Since Ω is a set of all the elements under consideration, there are no elements other than the elements of Ω. Therefore, the complement of Ω has no elements, that is,

$$\overline{\Omega} = \{\varnothing\}$$

By similar reasoning, the complement of the null set is Ω, that is,

$$\overline{\{\varnothing\}} = \Omega$$

Finally, we see that the complement of "the complement of a set A" is the original set A:

$$\bar{\bar{A}} = A$$

Let

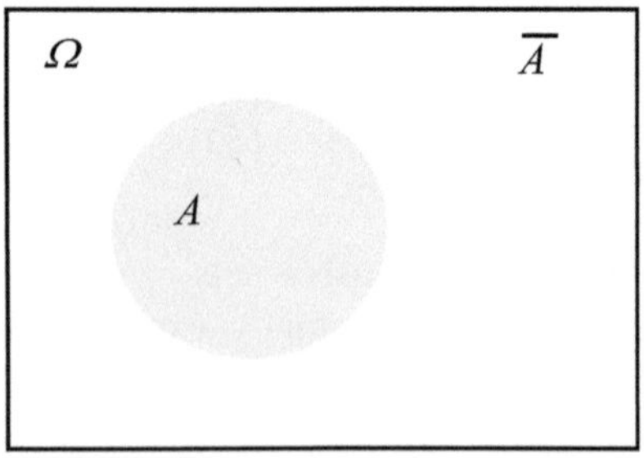

Fig. 2.2 Space Ω, subset A and its complement $\bar{A}$

$$\Omega = \{a, b, c, d, e, f, g, h, i, j, k\} \qquad A = \{a, b, c, d\}$$

Then

$$\begin{aligned} \bar{A} &= \{e, f, g, h, i, j, k\} \qquad A \cap \bar{A} = \{\varnothing\} \\ A \cup \bar{A} &= \{a, b, c, d, e, f, g, h, i, j, k\} \qquad \overline{\Omega} = \{\varnothing\} \\ \overline{\{\varnothing\}} &= \{a, b, c, d, e, f, g, h, i, j, k\} \end{aligned}$$

2.3.4 Set Equality

Two sets X and Y are said to be "equal" if and only if X is a subset of Y and Y is a subset of X:

$$\begin{aligned} &(X \subset Y \quad \text{and} \quad Y \subset X) \Rightarrow X = Y \\ &X = Y \Rightarrow (X \subset Y \quad \text{and} \quad Y \subset X) \end{aligned}$$

The set equality $X = Y$ is proven by proving that the left-hand side of the equality sign is a subset of the right-hand side, and the right-hand side is a subset of the left-hand side. This process of proving set equality is demonstrated by the following two examples.

Example 2.3.7
Mary received two separate lists of names, L_1 and L_2, as follows:

$L_1 = \{$Sue, Tom, Pete, Jane, Mat, Mike, Joan$\}$
$L_2 = \{$Tom, Pete, Mike, Jane, Sue, Mat, Joan, Joe$\}$

Mary wants to determine whether the two lists are the same $L_1 = L_2$. To do that, Mary needs to see if L_1 is a subset of L_2 and L_2 is a subset of L_1, that is, Mary needs to show that $L_1 \subset L_2$ and $L_2 \subset L_1$. To show $L_1 \subset L_2$, for each name on L_1, check whether the name is on L_2. To show $L_2 \subset L_1$, for each name on L_2, check whether the name is on L_1. Mary finds that $L_1 \subset L_2$ but $L_2 \not\subset L_1$, so $L_1 \neq L_2$.

Example 2.3.8

$$A \cap (B \cup C) = (A \cap B) \cup (B \cup C)$$

To show the above set equality, prove
$A \cap (B \cup C) \quad \subset \quad (A \cap B) \cup (A \cap C)$ and
$(A \cap B) \cup (A \cap C) \quad \subset \quad A \cap (B \cup C)$

Class
A set X is called a class if its elements are sets as in $X = \{A, B, C\}$, where A, B, and C are sets. To give an example, given $\Omega = \{1, 2, 3, 4, 5, 6\}$, consider the following

subsets of Ω: {1, 2}; {1, 2, 3}; and {3, 4, 5, 6}. Let a set Z = [{1, 2}, {1, 2, 3}, {3, 4, 5, 6}]. Then, Z is a class.

Partition

Consider the following analogy to facilitate the discussion of the concept of partition. Suppose that you are partitioning a large hall into three rooms by setting up walls. The three rooms will occupy the entire space of the hall, that is, no space is left unused. Therefore, when the walls are removed, the original space of the hall is recovered. Since the hall is compartmentalized by the walls, there is no overlapping space common to any two rooms, that is, no two rooms can overlap each other. Finally, there are more than one ways of partitioning the hall into three rooms (Fig. 2.3).

A partition Z of the space Ω is a class of mutually exclusive subsets A_i of Ω, where the union of all subsets A_i yields the original space Ω:

$$Z = \{A_1, A_2, \ldots, A_n\}$$

Z is a partition of Ω if and only if (Fig. 2.4)

$$\begin{aligned} &A_i \subset \Omega \text{ for all } \quad i = 1, \ldots, n \\ &A_i \cap A_j = \{\varnothing\} \text{ for all } \quad i,j, \quad i \neq j \\ &A_1 \cup A_2 \cup \ldots \cup A_n = \Omega \end{aligned} \tag{2.94}$$

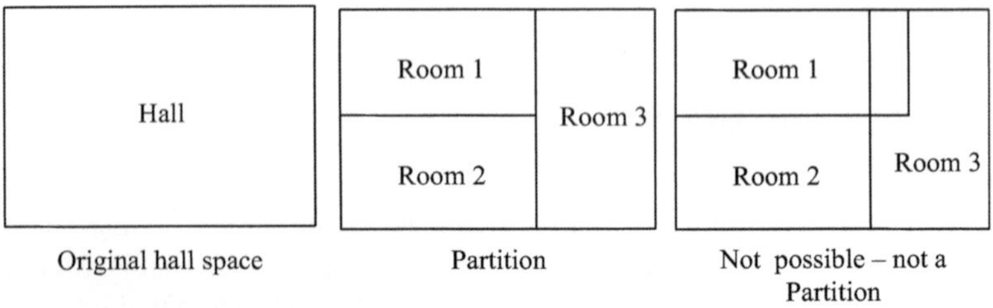

Fig. 2.3 Partitioning a hall

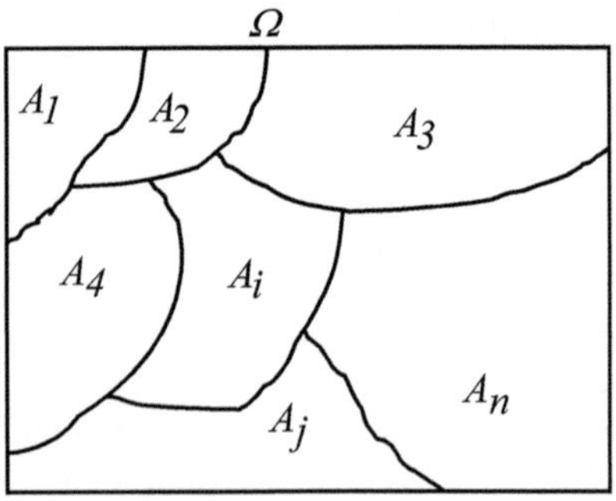

Fig. 2.4 Partition of Ω by n subsets

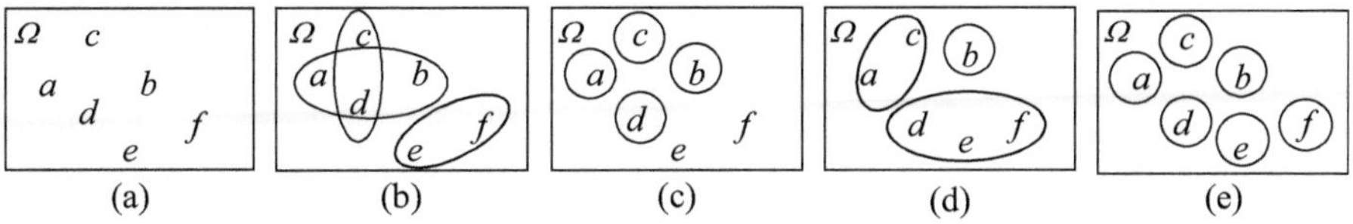

Fig. 2.5 **(a)** $\{a, b, c, d, e, f\}$, **(b)** $\{\{a, b, d\}, \{c, d\}, \{e, f\}\}$, **(c)** $\{\{a\}, \{b\}, \{c\}, \{d\}\}$, **(d)** $\{\{a, c\}, \{b\}, \{d, e, f\}\}$, **(e)** $\{\{a\}, \{b\}, \{c\}, \{d\}, \{e\}, \{f\}\}$

Example 2.3.9
Let the space be

$$\Omega = \{a, b, c, d, e, f\}$$

Consider the following classes:

$$C_1 = \{\{a, b, d\}, \{c, d\}, \{e, f\}\} \qquad C_2 = \{\{a\}, \{b\}, \{c\}, \{d\}\}$$
$$C_3 = \{\{a, c\}, \{b\}, \{d, e, f\}\} \qquad C_4 = \{\{a\}, \{b\}, \{c\}, \{d\}, \{e\}, \{f\}\}$$

Figure 2.5(a) shows the space Ω to be partitioned; (b) and (c) show the elements of C_1 and C_2; and (d) and (e) show the elements of C_3 and C_4. The figure illustrates that C_1 and C_2 are not partitions and that C_3 and C_4 are partitions.

2.3.5 *Cartesian Product*

The Concept
Consider the following construction of sets to facilitate the understanding of the concept of the Cartesian product. Consider two sets X and Y with m and n elements, respectively, as follows:

$$X = \{x_1, x_2, \ldots, x_i, \ldots, x_m\}$$
$$Y = \{y_1, y_2, \ldots, y_i, \ldots, y_n\}$$

Referring to the figure shown below, construct a table, or $m \times n$ matrix, having m rows and n columns as follows:

		Y					
		y_1	y_2	.	y_i	.	y_n
X	x_1	m_{11}	m_{12}		m_{1j}		m_{1n}
	x_2	m_{21}	m_{22}		m_{2j}		m_{2n}
	.						
	x_i	m_{i1}	m_{i2}		m_{ij}		m_{in}
	.						
	x_m	m_{m1}	m_{m2}		m_{mj}		m_{mn}

Label the row and column headings by the elements of X and Y, respectively. Denote the table entry corresponding to the ith row and the jth column by m_{ij}. Finally, construct the table entry m_{ij} as a pair (x_i, y_j) by taking the ith row heading, x_i, and the jth column heading, y_j—in that order. The pair (x_i, y_j) is an "ordered pair," where the order of the elements in the pair is a factor in distinguishing the pairs as distinct objects. Therefore, $(x_i, y_j) \neq (y_i, x_j)$.

Let Z be the collection of all the entries of the table as follows:

$$Z = \{(x_1, y_1), (x_2, y_2), \ldots, (x_i, y_i), \ldots, (x_m, y_m)\}$$

The set Z is referred to as the Cartesian product of X and Y and is denoted by $Z = X \times Y$.

Definition of the Cartesian Product

We formally define the Cartesian product. The Cartesian product of X and Y, denoted by $X \times Y$, is a set of all ordered pairs formed by taking one element from X and one element from Y—in that order—such as (x_i, y_j)

$$\begin{aligned} X \times Y &= \{(x_1, y_1), (x_2, y_2), \ldots, (x_i, y_i), \ldots, (x_m, y_m)\} \\ &= \{(x_i, y_i), x_i \in X, y_i \in Y \text{ for all } i, j\} \end{aligned}$$

The two sets X and Y can be sets of any objects. X and Y may be sets of real discrete numbers, in which case x_i and y_i are discrete on real lines. The two sets X and Y can also be sets of real numbers from continuums R_1 and R_2 as follows:

$$\begin{aligned} X &= \{x, x \in R_1\} \\ Y &= \{y, y \in R_2\} \\ X \times Y &= \{(x, y), x \in X, y \in Y\} \end{aligned}$$

Since the order in selecting the elements to form pairs is part of the definition of the Cartesian product,

$$X \times Y \neq Y \times X$$

Example 2.3.10

Referring to Fig. 2.6, let the two sets X and Y are defined as follows:

$$\begin{aligned} X &= \{\text{all real numbers on } x\text{-axis}\} \\ Y &= \{\text{all real numbers on } y\text{-axis}\} \end{aligned}$$

Then $X \times Y$ is the set of all (x, y) pairs on the plane.

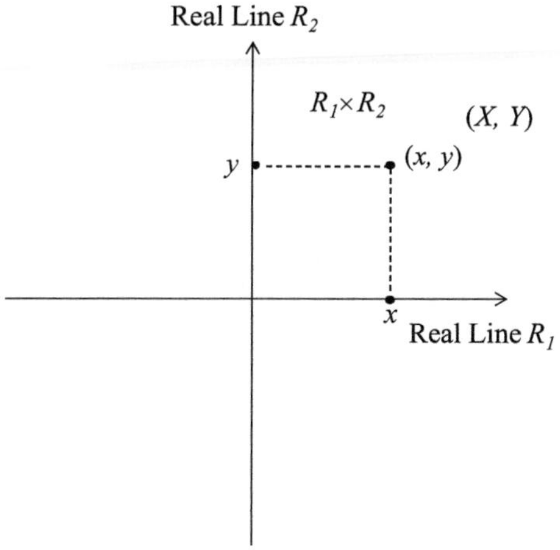

Fig. 2.6 Illustration of Cartesian product

Generalization of the Cartesian Product

The Cartesian product of n sets, $X_1, X_2, \ldots, X_n$, with $N_1, N_2, \ldots, N_n$ elements, respectively, is a set of all ordered n-tuples, $(x_{11}, x_{22}, \ldots, x_{ij}, \ldots, x_{nNn})$, where x_{ij} is the jth element of the ith set X_i:

$$X = X_1 \times X_2 \times \ldots \times X_n = \{\ldots, (x_{11}, x_{12}, \ , x_{1n}, x_{ij}, \ldots x_{nN_nN}), \ldots\} \quad (2.95)$$

Chapter 3
Probability Theory

This chapter defines the central concepts and terms used in probability theory including random experiment, space, event, elementary event, combined experiment, Cartesian product, etc. This chapter presents the axiomatic formulation of probability theory based on three axioms and shows how set operations are used in probability theory.

3.1 Random Experiments

A random experiment consists of executing a certain action or procedure under controlled conditions and taking an observation or a measurement on the outcome produced by the action. The experiment is called random because the outcome is unpredictable.

3.1.1 Space Ω

In probability theory, we are concerned about random experiments in which all possible outcomes are known in advance. The set of all possible outcomes of a random experiment is called the sample space, or simply the space, and is denoted by

$$\Omega = \{s_1, s_2, \ldots, s_i, \ldots, s_n\}$$

where s_i's are experimental outcomes. Sometimes, the space Ω is used to denote the experiment as in "experiment Ω_1," "experiment Ω_2," etc. An example of a random

K.I. Park, *Fundamentals of Probability and Stochastic Processes with Applications to Communications*, https://doi.org/10.1007/978-3-319-68075-0_3

experiment is a die-throwing experiment in which a die is thrown and the number of dots that the top side shows is observed. The space of this experiment is $\Omega = \{1, 2, 3, 4, 5, 6\}$, where the six numbers denote all possible numbers of dots that may show after a die is thrown.

Each execution of a random experiment produces one outcome. A single execution of a random experiment that produces an outcome is called a trial. For example, in a die-throwing experiment, a trial produces exactly one of the six possible outcomes in Ω.

3.1.2 Event

In the dictionary, the word *event* is defined as *outcome*. We have already encountered the word *outcome* while discussing a random experiment. Therefore, we need to have a clear understanding of the difference between the two words,*event* and *outcome*, before we proceed.

An event is the result of an experiment that is of interest or concern. To take an example, suppose that, in a die-throwing game, you would win \$10 if your die-throwing shows an outcome less than four. Here, the word "outcome" is a specific showing of the die face. The event of your interest is "winning \$10." Your die-throwing, a trial, would produce one outcome. If that outcome is either 1, 2, or 3, you would win the prize: the event of "winning \$10" would occur if the outcome of the trial is 1, 2, or 3. Among all possible outcomes of the die-throwing experiment, that is, $\Omega = \{1, 2, 3, 4, 5, 6\}$, there are three specific outcomes, 1, 2, and 3, that would make the event happen. These three numbers are members of a subset of Ω, $A = \{1, 2, 3\}$. The event "winning \$10" is represented by the subset A.

So, "event" is defined as follows: an event is a subset of the space Ω consisting of the elements that make the event happen. To form the subset defining an event, consider the elements of Ω and determine whether the elements would make the event happen or not. If yes, the elements are included in the subset.

The event consisting of all possible elements of an experiment, that is, Ω, is called a certain event and the event which has no element, that is, $\{\varnothing\}$, an impossible event.

An event consisting of a single element is called an elementary event, for example, in a die-throwing experiment, $\{1\}$, $\{2\}$, $\{3\}$, $\{4\}$, $\{5\}$, and $\{6\}$, and in a coin-tossing experiment, {heads} and {tails}. A key distinction to make here is that an element written by itself as "*1*," "2," etc. is an outcome, whereas a single outcome shown in braces as in $\{1\}$ is an elementary event.

3.1.3 Combined Experiments

Whether an event occurs or not is determined by the single outcome of a trial of an experiment. If an event under consideration involves the outcomes of multiple trials of a single experiment or a single or multiple trials of multiple experiments, a new experiment may be defined by combining the original experiments. This new experiment may be called a combined experiment in which a new space is defined as the set of all possible combinations of the outcomes of the individual trials of the original experiments. With this definition, the single outcome produced by a trial of this combined experiment is a unique sequence of the individual outcomes of the original experiments, and the event of the combined experiment is determined by this single outcome of a trial of the combined experiment.

For example, suppose that the event under consideration is determined by the sequence of the outcomes of n trials of a single experiment, e.g., throwing a die n times. A combined experiment may be defined by defining a single trial of the experiment as a sequence of n trials of the original experiment. The space of the combined experiment consists of all possible ordered sequences, that is, n-tuples, of the elements of the original space.

3.1.4 Probabilities and Statistics

In probability analysis, one begins by assigning probabilities to elementary events, or, by using the known probabilities of other events, if the event under consideration is based on the events for which the probabilities have already been determined. For example, in a die-throwing experiment, first, the probabilities of the elementary events of the six sides, that is, 1/6 for each side, are assigned. Without these initial assignments, one cannot proceed to address more complex problems associated with the outcomes of a die-throwing experiment. The only other option is to try the experiment many times and count the frequencies of the outcomes of interest. For example, in a die-throwing experiment, to determine the probability of an odd number, a die must be thrown many times, and the frequencies of odd numbers must be counted. Even then, a question remains as to how many times the die must be thrown before the probability can be determined.

This dilemma can be avoided by taking one's a priori judgment about the probabilities of the elementary events. The axiomatic approach to be discussed in the next section allows that a probability analysis can begin by one's a priori assignment of the probabilities of the elementary events.

Statistics deals with analyzing the frequencies of the outcomes. Therefore, statistics can provide one with the basis of making a priori judgments, for example, on the probabilities of elementary events.

3.2 Axiomatic Formulation of Probability Theory

The axiomatic formulation of probability theory was introduced by the Russian mathematician Kolmogoroff in 1933. In this approach, all possible outcomes of an experiment form a space Ω. Events are defined by the subsets of the space Ω.

Probabilities are determined for the events. Probabilities are "assigned" to the elementary events in Ω as the starting point of the probability analysis. The events and the probabilities must obey a set of axioms presented below.

Given two events A and B in Ω, the probabilities of A and B are denoted by $P(A)$ and $P(B)$. $P(A)$ and $P(B)$ are real numbers, referred to as probability measures, and must obey the following rules:

Axiom I

$$P(A) \geq 0 \tag{3.1}$$

Axiom II

$$P(\Omega) = 1 \tag{3.2}$$

Axiom III

$$\text{If } A \cap B = \{\varnothing\}, \text{then } P(A \cup B) = P(A) + P(B) \tag{3.3}$$

Axiom I states that the probability measure assigned to an event is nonnegative. Axiom II states that the probability measure assigned to a certain event is 1. Finally, Axiom III states that, if two events A and B are mutually exclusive with the probabilities $P(A)$ and $P(B)$, respectively, the probability that either A or B or both would occur, that is, the probability of the event $A \cup B$, is the sum of the two probabilities $P(A)$ and $P(B)$.

Example 3.2.1

To illustrate Axiom III, consider that A represents the event that Tom will attend a conference in Philadelphia tomorrow at 9 AM and B the event that Tom will travel to Boston tomorrow at 9 AM. Assume that $P(A) = 0.1$ and $P(B) = 0.2$. Clearly, A and B are mutually exclusive because Tom cannot be at two places at the same time. Then, the probability that Tom will either attend the conference in Philadelphia or travel to Boston is the sum of the two probabilities, 0.3.

While Axioms I and II give the rules for assigning probability measures, Axiom III gives the rule for deriving the probability measure for a complex event $A \cup B$ from the probabilities of A and B. In the axiomatic approach, these three axioms are all one needs to formulate a probability problem. The above three axioms together with the employment of set theory are sufficient for developing probability theory.

A generalization of Axiom III is given by the theorem below.

Theorem 3.2.1 For a finite number of mutually exclusive sets $A_i \cap A_j = \{\varnothing\}$ for all $i, j = 1, 2, \ldots, n, \; i \neq j$

$$P(A_1 \cup A_2 \cup \ldots \cup A_n) = P(A_1) + P(A_2) + \ldots + P(A_n) \tag{3.4}$$

Proof First, consider the case of three mutually exclusive sets, A, B and C, that is,

$$A \cap B = \{\varnothing\} \quad A \cap C = \{\varnothing\} \quad B \cap C = \{\varnothing\}$$

We have the following relationship:

$$(A \cup B) \cap C = (A \cap C) \cup (B \cap C) = \{\varnothing\} \cup \{\varnothing\} = \{\varnothing\}$$

That is, $(A \cup B)$ and C are two mutually exclusive sets. These two mutually exclusive sets satisfy Axiom III as follows:

$$P[(A \cup B) \cap C] = P(A \cup B) + P(C)$$

Applying Axiom III to the first term on the right-hand side, that is, $P[(A \cup B)] = P(A) + P(B)$, we obtain the following equation:

$$P[(A \cup B) \cap C] = P(A) + P(B) + P(C)$$

Continuing this process for n mutually exclusive sets $A_1, A_2, \ldots, A_n$, we prove the theorem by mathematical induction.

Q.E.D.

Example 3.2.2

In this example, we illustrate the significance of the three axioms in formulating a probability problem using a die-throwing experiment. Suppose that you would win \$10, if the number of dots shown after throwing the die is less than four, and would win a trip to Philadelphia if it is more than four dots. What is the probability that you would win \$10, a trip to Philadelphia or both? We will formulate and solve this problem using the three axioms.

For this problem, the space is $\Omega = \{1, 2, 3, 4, 5, 6\}$. In the axiomatic approach, the formulation of a probability problem starts with the assignments of the probability measures for the basic events, whether they are elementary events or the events for which a priori probabilistic information is known. For this problem, we will assign 1/6 for each of the six possible outcomes: $P(\{i\}) = 1/6, i = 1 \sim 6$. In the absence of any a priori information about the six elementary events, 1/6 is a reasonable assignment and satisfies Axiom I. If a priori information, e.g., past experimental data, is available about the die used in the experiment, different probabilities may be assigned. In any event, the key point here is that the formulation starts with the assignments of the probabilities.

There are two events of interest, "winning \$10" and "winning a trip to Philadelphia." An event is defined by a subset of Ω consisting of the outcomes that would make the event happen. What outcomes would make the event "winning \$10" happen? Outcomes *1*, *2*, or *3* would make you win \$10. So,

$$A = \text{event of``winning \$10''} = \{1, 2, 3\}$$

You would win a trip to Philadelphia if *5* or *6* shows. So,

$$B = \text{event of ``winning a trip to Philadelphia''} = \{5, 6\}$$

The event that you will win $10 or a trip to Philadelphia or both would then be represented by the union of A and B and $A \cup B$, and we need to determine $P(A \cup B)$.

We see that A and B are mutually exclusive, that is, $A \cap B = \{\varnothing\}$. Therefore, by Axiom III, we have

$$P(A \cup B) = P(A) + P(B)$$

Now, set A can be expressed as a union of three elementary events as follows:

$$A = \{1, 2, 3\} = \{1\} \cup \{2\} \cup \{3\}$$

where the elementary events are mutually exclusive, that is, $\{i\} \cap \{j\} = \{\varnothing\}$, for $i \neq j$.

So, by Theorem 3.2.1, we have

$$\begin{aligned} P(A) &= P[\{1, 2, 3\}] = P[\{1\} \cup \{2\} \cup \{3\}] \\ &= P[\{1\}] + P[\{2\}] + P[\{3\}] = \frac{3}{6} \end{aligned}$$

Similarly,

$$P(B) = P[\{5, 6\}] = P[\{5\} \cup \{6\}] = P[\{5\}] + P[\{6\}] = \frac{2}{6}$$

Therefore,

$$P(A \cup B) = P(A) + P(B) = \frac{3}{6} + \frac{2}{6} = \frac{5}{6}$$

The event represented by the union of two events A and B would occur if A or B or both would occur. For mutually exclusive A and B, the probability that "both" A and B would occur is zero. For the current problem, the probability of winning both would be zero because A and B are mutually exclusive.

In addition to Theorem 3.2.1, it is convenient to establish several key theorems up front that follow from the three probability axioms. These theorems are discussed below.

Theorem 3.2.2

$$P[\{\varnothing\}] = 0 \tag{3.5}$$

Equation (3.5) states that the probability of an impossible event is 0.

Two observations are made regarding this theorem. First, unlike the probability measure 1 of a certain event, which is "assigned" by Axiom II, the probability measure of 0 can be "derived" from the axioms. Second, one might wonder why this is not included as an axiom. Since this measure can be derived from the above axioms, it would be superfluous to include this as an axiom. As can be seen from the proof given below, this theorem could have been included as Axiom II, and the current Axiom II could be derived instead. In any event, it is not necessary to include both.

Proof For an arbitrary set A, the following equations hold true:

$$A = A \cup \{\varnothing\} \qquad A \cap \{\varnothing\} = \{\varnothing\}$$

Hence, from Axiom III, it follows that

$$P(A) = P[A \cup \{\varnothing\}] = P(A) + P[\{\varnothing\}]$$

Rearranging the terms of the above equation, we obtain the following equation:

$$P(\{\varnothing\}) = P(A) - P(A) = 0$$

Q.E.D.

Theorem 3.2.3 For any two events A and B, that is, $A \subset \Omega$ and $B \subset \Omega$,

$$P(A \cup B) = P(A) + P(B) - P(A \cap B) \leq P(A) + P(B) \tag{3.6}$$

Proof Referring to Fig. 3.1d, it can be shown that

$$A \cup B = A \cup \left(\bar{A} \cap B\right)$$

$$P(A \cup B) = P\left[A \cup \left(\bar{A} \cap B\right)\right]$$

Since A and $\bar{A} \cap B$ are mutually exclusive, that is, $A \cap (\bar{A} \cap B) = \{\varnothing\}$, using Axiom III, we can rewrite the above equation as follows:

$$P(A \cup B) = P(A) + P\left(\bar{A} \cap B\right) \tag{3.7}$$

On the other hand, referring to Fig. 3.1d, we can express B as the union of two intersections as follows:

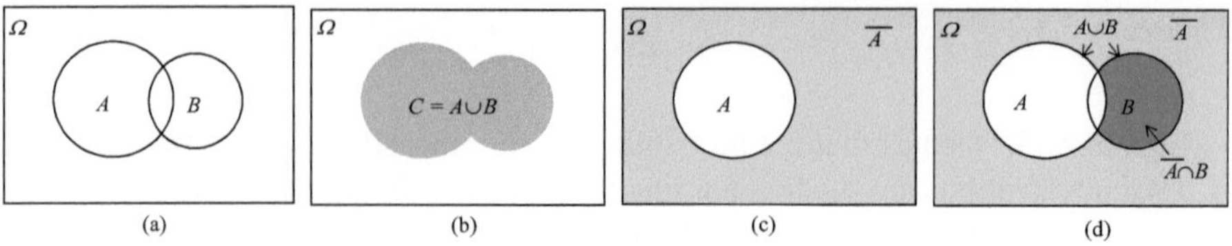

Fig. 3.1 (**a**) Subsets A and B, (**b**) $A \cup B$, (**c**) Subset A and its complement $\bar{A}$, (**d**) $\bar{A} \cap B$

$$B = (A \cap B)\ \left(\bar{A} \cap B\right)$$
$$P(B) = P\left[(A \cap B) \cup \left(\bar{A} \cap B\right)\right]$$

Since $(A \cap B)$ and $(\bar{A} \cap B)$ are mutually exclusive, that is, $(A \cap B) \cap (\bar{A} \cap B) = \{\varnothing\}$, using Axiom III, we can rewrite the above equation as follows:

$$P(B) = P[(A \cap B)\,] + P\left[\,\left(\overline{A} \cap B\right)\right]$$

Rearranging the terms of the above equation, we obtain the following equation:

$$P\left[\,\left(\bar{A} \cap B\right)\right] = P(B) - P[(A \cap B)\,] \tag{3.8}$$

Substituting (3.8) into (3.7), we obtain

$$P(A \cup B) = P(A\,) + P(B) - P[(A \cap B)\,]$$

Q.E.D.

Furthermore, by Axiom I, $P(A \cap B) \geq 0$, and, thus, the last equation yields the following relationship:

$$P(A\,) + P(B) - P[(A \cap B)\,] \leq P(A\,) + P(B)$$

Q.E.D.

Theorem 3.2.4

$$P(A) \leq 1 \tag{3.9}$$

Proof For an arbitrary subset A of Ω, that is, $A \subset \Omega$,

$$A \cup \bar{A} = \Omega$$

Hence, using Axiom III, we obtain the following equation:

$$P\left(A \cup \bar{A}\right) = P(\Omega) = 1$$

Since A and $\bar{A}$ are mutually exclusive, that is, $A \cap \bar{A} = \{\varnothing\}$, from Axiom III, we can rewrite the above equation as follows:

$$P(A \cup \bar{A}) = P(A) + P(\bar{A}) = 1$$

or

$$P(A) = 1 - P(\bar{A})$$

Q.E.D.

Furthermore, since, from Axiom I, $P(\bar{A}) \geq 0$, we have

$$P(A) \leq 1$$

Q.E.D.

This theorem shows that, if the three axioms are followed, the probability measure derived for any arbitrary event cannot be greater than 1. Once again, including this statement as an axiom would be superfluous.

Theorem 3.2.5 If $B \subset A$,

$$P(B) \leq P(A) \tag{3.10}$$

Proof If B is a subset of A, referring to Fig. 3.2c, we can express A as the union of two mutually exclusive sets as follows:

$$A = B \cup (\bar{B} \cap A)$$

Hence, we have

$$P(A) = P[B \cup (\bar{B} \cap A)]$$

As can be seen in Fig. 3.2c, since B and $(\bar{B} \cap A)$ are mutually exclusive, that is, $B \cap (\bar{B} \cap A) = \{\varnothing\}$, we can rewrite the above equation as follows:

$$P(A) = P(B) + P(\bar{B} \cap A)$$

or

$$P(B) = P(A) - P(\bar{B} \cap A)$$

Since, by Axiom I, $P(\bar{B} \cap A) \geq 0$, the above equation yields the following relationship:

$$P(B) \leq P(A)$$

Q.E.D.

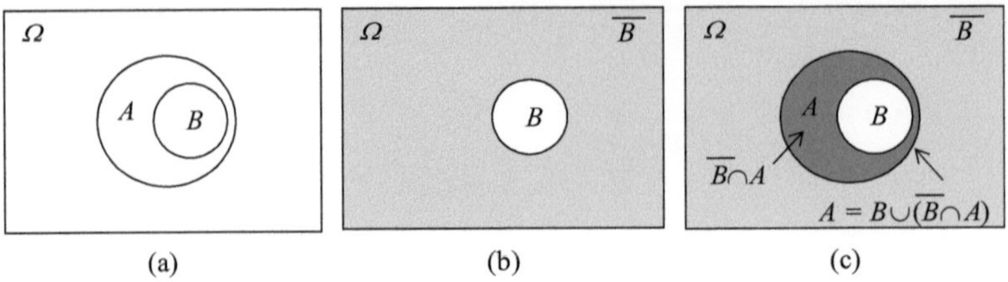

Fig. 3.2 (**a**) $B \subset A$, (**b**) Subset B and its complement $\bar{B}$, (**c**) $A = B \cup (\bar{B} \cap A)$

In sum, the following steps are taken to formulate a probability problem based on the axiomatic approach:

Define the experiment and the probability space Ω.
Assign the probabilities of the elementary events.
Define the event.
Determine the probability of the event.

The following example is a simple probability problem that can be solved without elaborate formulation. However, we will deliberately go through the above steps to illustrate the axiomatic approach to probability formulation.

Example 3.2.3
In a die-throwing game, a number less than 5 wins. Find the probability of winning the game.

Solution
Determine the space:

$$\Omega = \{1, 2, 3, 4, 5, 6\}$$

Assign the probabilities to the elementary events:

$$P[\{1\}] = P[\{2\}] = P[\{3\}] = P[\{4\}] = P[\{5\}] = P[\{6\}] = \frac{1}{6}$$

Define the event by selecting the elements of Ω that would make the event happen:

$$A = \{1, 2, 3, 4\}$$

Find $P(A)$.

The elementary events $\{1\}$, $\{2\}$, $\{3\}$, and $\{4\}$ are mutually exclusive. Using Axiom III, we obtain the following result:

$$P(A) = P[\{1,2,3,4\}] = P[\{1\} \cup \{2\} \cup \{3\} \cup \{4\}]$$
$$= P[\{1\}] + P[\{2\}] + P[\{3\}] + P[\{4\}] = \frac{4}{6}$$

3.3 Conditional Probability

Consider two events A and B defined in the space Ω and the event given by the intersection $A \cap B$. The probability measures of these events in Ω are denoted by $P(A)$, $P(B)$, and $P(A \cap B)$, respectively. Now, let us consider the ratio of $P(A \cap B)$ to $P(A)$ and that to $P(B)$ as follows assuming that $P(A)$ and $P(B)$ are not zero:

$$\frac{P(A \cap B)}{P(B)} \quad \text{and} \quad \frac{P(A \cap B)}{P(A)} \quad \text{where } P(A) \neq 0, P(B) \neq 0 \tag{3.11}$$

In the first ratio, consider that, given B, that is, for a fixed B, A is varied, that is, the ratio is a function of A. Similarly, in the second ratio, the ratio is a function of B for a fixed A. For the time being, let us denote these two quantities by $R[A$ given $B]$ and $R[B$ given $A]$, respectively. We now show that these two quantities are also probability measures in Ω satisfying Axioms I, II, and III. We show this using the first ratio, $R[A$ given $B]$, as A varies with B fixed.

First, the ratio $R[A$ given $B]$ satisfies Axiom I given by (3.1) as follows:

By Axiom I, $P(A \cap B) \geq 0$ and $P(B) \geq 0$. Therefore, if $P(B) \neq 0$,

$$\frac{P(A \cap B)}{P(B)} \geq 0 \tag{3.12}$$

Q.E.D.

Next, the ratio $R[A$ given $B]$ satisfies Axiom II given by (3.2) as follows:

Consider the case $A = \Omega$ so that the numerator of the ratio becomes $P(\Omega \cap B)$. We know that, since $B \subset \Omega$, $P(\Omega \cap B) = P(B)$. Hence,

$$\frac{P(\Omega \cap B)}{P(B)} = \frac{P(B)}{P(B)} = 1 \tag{3.13}$$

Q.E.D.

Finally, the ratio $R[A$ given $B]$ satisfies Axiom III given by (3.3) as follows:

Consider that A is equal to the union of two mutually exclusive sets C and D in Ω as follows:

$$A = C \cup D, \text{where } C \cup D = \{\varnothing\}$$

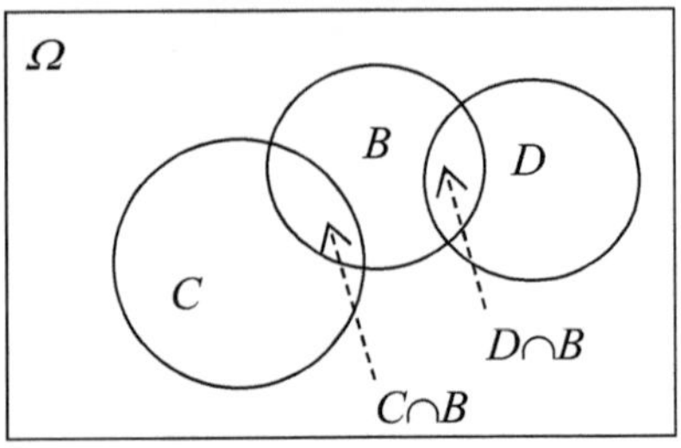

Fig. 3.3 $(C \cap B) \cup (D \cap B) = \{\varnothing\}$

and consider the following expression:

$$\frac{P[(C \cup D) \cap B]}{P(B)} \tag{3.14}$$

We have the following set identity:

$$(C \cup D) \cap B = (C \cap B) \cup (D \cap B)$$

Referring to Fig. 3.3, we see that, since C and D are mutually exclusive, $(C \cap B)$ and $(D \cap B)$ are mutually exclusive, that is, $(C \cap B) \cup (D \cap B) = \{\varnothing\}$.

Therefore, by Axiom III, we obtain the following equation:

$$\begin{aligned} P[(C \cup D) \cap B] &= P[(C \cap B) \cup (D \cap B)] \\ &= P[(C \cap B)] + P[(D \cap B)] \end{aligned}$$

Hence,

$$\begin{aligned} \frac{P[(C \cup D) \cap B]}{P(B)} &= \frac{P[(C \cap B)] + P[(D \cap B)]}{P(B)} \\ &= \frac{P[(C \cap B)]}{P(B)} + \frac{P[(D \cap B)]}{P(B)} \end{aligned} \tag{3.15}$$

Q.E.D.

3.3.1 *Definition of the Conditional Probability*

We have shown that the ratio of the probability of the intersection of two arbitrary events, A and B, in Ω, that is, $(A \cup B)$, to the probability of either A or B, is also a probability. In fact, these ratios, $R[B$ given $A]$ or $R[A$ given $B]$, are given a special name, "conditional probability," and are denoted by the following notations:

$$P(A|B) \triangleq \frac{P(A \cap B)}{P(B)} \quad \text{where } P(B) \neq 0 \tag{3.16}$$

$$P(B|A) \triangleq \frac{P(A \cap B)}{P(A)} \quad \text{where } P(A) \neq 0 \tag{3.17}$$

The first conditional probability as defined above can be interpreted as the probability of event A given that event B has occurred, and, similarly, the second conditional probability, as the probability of event B given that event A has occurred.

To understand this interpretation and how the name "conditional probability" is given to the quantity, consider the following: the probability of event A in the sample space Ω, $P(A)$, is the probability measure given to A with respect to the total measure 1 of Ω, that is, $P(A)/P(\Omega) = P(A)/1$. The probability measure that event A would occur due to the outcomes from the overlapping portion between A and B would be $P(A \cap B)$. Now, consider the probability measure of the event A if the space for A is restricted to B. This last probability measure would be the probability measure $P(A \cap B)$ relative to the probability measure of B, $P(B)$, that is, the ratio defined as the conditional probability $P(A/B)$ (Fig. 3.4).

3.3.2 Total Probability Theorem

Theorem 3.3.1 Let $B_1, B_2, \ldots, B_i, \ldots, B_n$ be n subsets of the space Ω that form a partition of Ω. Consider an arbitrary event A in Ω with a nonzero probability, $P(A) > 0$. Then, $P(A)$ can be expressed in terms of the conditional probabilities given B_i, $i = 1, 2, \ldots, n$ as follows:

$$\begin{aligned} P(A) = {} & P(A|B_1)P(B_1) + P(A|B_2)P(B_2) + \cdots \\ & + P(A|B_i)P(B_i) + \cdots + P(A|B_n)P(B_n) \end{aligned} \tag{3.18}$$

The last equation is referred to as the total probability theorem.

Proof Referring to Fig. 3.5, we can express event A as the union of n intersections as follows:

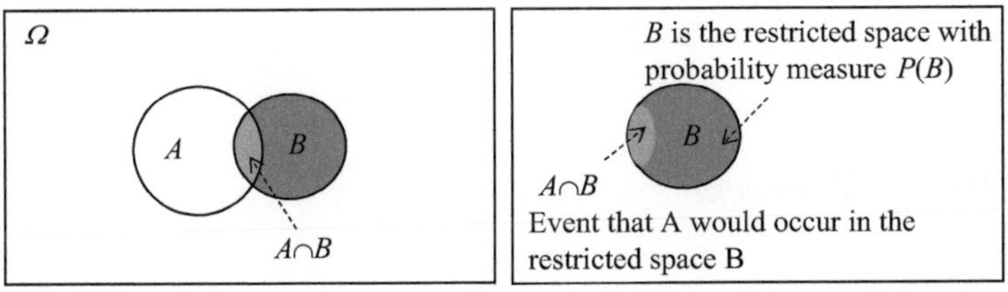

Fig. 3.4 $(A \cap B)$

Fig. 3.5 $A=(A\cap B_1)\cup(A\cap B_2)\cup\ldots\cup(A\cap B_i)\cup\ldots\cup(A\cap B_n)$

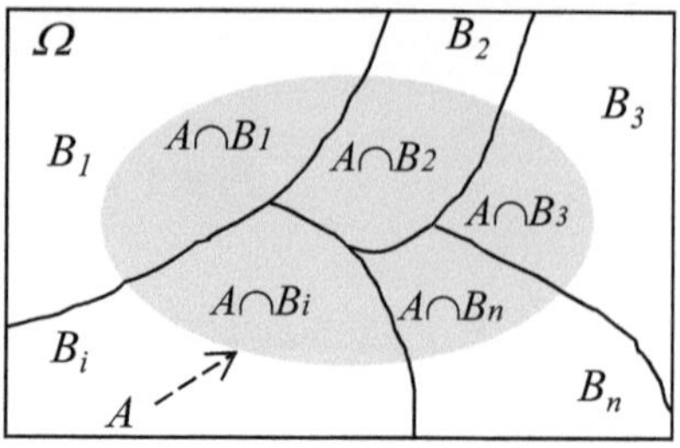

$$A=(A\cap B_1)\cup(A\cap B_2)\cup\ldots\cup(A\cap B_i)\cup\ldots\cup(A\cap B_n)$$

These n intersections are mutually exclusive, that is, $(A\cap B_i)\cap(A\cap B_j)=\{\varnothing\}$, $i\neq$ j. By Axiom III, we obtain the following equation:

$$\begin{aligned}P(A)&=P[(A\cap B_1)\cup(A\cap B_2)\cup\cdots\cup(A\cap B_i)\cup\cdots\cup(A\cap B_n)]\\&=P[(A\cap B_1)]+P[(A\cap B_2)]+\cdots+P[(A\cap B_i)]\\&\quad+\cdots\ +P[(A\cap B_n)]\end{aligned}$$

The right-hand side of the above equation can be expressed in the form of the total probability theorem as follows. First, by the definition of the conditional probability, we have the following equation

$$P(A|B_i)\triangleq\frac{P(A\cap B_i)}{P(B_i)}\quad\text{where } P(B_i)\neq 0$$

Next, rearranging the terms of the above equation, we obtain the following expression:

$$P(A\cap B_i)=P(A|B_i)P(\mathrm{B}_i)\quad\text{for } i=1,\ \ldots,n$$

Finally, substituting the above expression in the right-hand side of the equation for $P(A)$, we obtain the following equation

$$\begin{aligned}P(A)=&\,P(A|B_1)P(B_1)+P(A|B_2)P(B_2)\\&+\cdots P(A|B_i)P(B_i)+\cdots P(A|B_n)P(B_n)\end{aligned}$$

Q.E.D.

3.3.3 *Bayes' Theorem*

Theorem 3.3.2 Let $B_1, B_2, \ldots, B_i, \ldots, B_n$ are n subsets of the space Ω that form a partition of Ω. Consider an arbitrary event A with a nonzero probability, $P(A)>0$. The conditional probability of B_i given A is given by the following equation:

$$P(B_i|A) = \frac{P(B_i)P(A|B_i)}{P(B_1)P(A|B_1) + P(B_2)P(A|B_2) + \cdots P(B_n)P(A|B_n)} \tag{3.19}$$

This theorem is referred to as the Bayes' theorem and is used to determine the probability that a given event A implies the subset B_i of the partition. For example, given that a product is found to be defective, denoted by event A, the theorem can be used to calculate the probability that the defective product is from supplier B_i, when the defect data for each supplier, $P(A|B_i)$, is available.

Proof The left-hand side of the above equation is the conditional probability of B_i given A, which is given by the following equation by definition:

$$P(B_i|A) = \frac{P(B_i \cap A)}{P(A)} \tag{3.20}$$

On the other hand, the conditional probability of A given B_i is given by

$$P(A|B_i) = \frac{P(B_i \cap A)}{P(B_i)} \tag{3.21}$$

Rearranging the terms of the last equation, we obtain the following equation:

$$P\ (B_i \cap A) = P(B_i)P(A|B_i) \tag{3.22}$$

Substituting (3.22) into (3.20), we have

$$P(B_i|A) = \frac{P(B_i)P(A|B_i)}{P(A)} \tag{3.23}$$

substituting (3.18) into (3.23) yields (3.19).

Q.E.D.

Example 3.3.1

A reliability problem. A component is randomly selected from a batch of 10,000 pieces supplied by five different factories. The following table shows the factory data of failure statistics of the component and the number of pieces supplied by the factories. Suppose that the randomly selected component has just failed. What is the probability that the failed component is from Factory A?

Factory	#Supplied	Probability of failure	Factory	#Supplied	Probability of failure
A	1000	P(fail|A) = 1.3×10^{-6}	D	2000	P(fail|D) = 1.4×10^{-6}
B	3000	P(fail|B) = 1.2×10^{-6}	E	1000	P(fail|E) = 1.5×10^{-6}
C	3000	P(fail|C) = 1.1×10^{-6}			

From the number of components supplied by each factory given above, we have

$P(A) = 1000/10{,}000 = 0.1 \quad P(B) = 3000/10{,}000 = 0.3 \quad P(C) = 3000/10{,}000 = 0.3$
$P(D) = 2000/10{,}000 = 0.2 \quad P(E) = 1000/10{,}000 = 0.1$

Using the Bayes' theorem and substituting the above probabilities and the failure statistics of each factory given by the above table, we obtain the following solution:

$$P(A|\text{fail}) = \frac{P(A)P(\text{fail}|A)}{P(\text{fail}|A)P(A) + (\text{fail}|B)P(B) + P(\text{fail}|C)P(C) + (\text{fail}|D)P(D) + (\text{fail}|E)P(E)}$$

$$= \frac{1.3\times10^{-6}\times0.1}{1.3\times10^{-6}\times0.1 + 1.2\times10^{-6}\times0.3 + 1.1\times10^{-6}\times0.3 + 1.4\times10^{-6}\times0.2 + 1.5\times10^{-6}\times0.1}$$

$$= 0.104$$

Example 3.3.2

A communications signal detection problem. A total of 4000 characters have been received from four different sources as follows. The probabilities of character "a" from the four sources are given. Out of the total 4000 characters received, a randomly selected character is found to be "a." What is the probability that this character came from Source A?

Source	#Characters sent	Probability of "a"	Source	#Characters sent	Probability of "a"
A	500	P(a|A) = 0.1	C	2000	P(a|C) = 0.3
B	1000	P(a|B) = 0.2	D	500	P(a|D) = 0.4

Based on the number of characters sent by each source, we have

$$P(A) = \frac{500}{4000} = \frac{1}{8} \quad P(B) = \frac{1000}{4000} = \frac{2}{8} \quad P(C) = \frac{2000}{4000} = \frac{4}{8} \quad P(D) = \frac{500}{4000} = \frac{1}{8}$$

By the Bayes' theorem, we obtain the following solution:

$$P(A|a) = \frac{0.1\times\frac{1}{8}}{0.1\times\frac{1}{8} + 0.2\times\frac{2}{8} + 0.3\times\frac{4}{8} + 0.4\times\frac{1}{8}} = 0.476$$

3.3.4 *Independence of Events*

In our everyday language, we say that two events A and B are independent if the occurrence of A has no effect on the occurrence of B and vice versa. This lexical definition of the word *independence* can be expressed in terms of the conditional probability as follows:

$$P(A|B) = P(A) \tag{3.24}$$
$$P(B|A) = P(B) \tag{3.25}$$

By the definition of the conditional probability, these two equations read as follows: the probability of A's occurrence, $P(A)$, stays unchanged regardless of B's occurrence, that is, $P(A|B)$, and the probability of B's occurrence, $P(B)$, stays unchanged regardless of A's occurrence. By combining the definition of the conditional probability given by (3.16) and (3.17) and the above statements of independence, we obtain the following equation:

$$P(A|B) = \frac{P(A \cap B)}{P(B)} = P(A) \tag{3.26}$$

$$P(B|A) = \frac{P(A \cap B)}{P(A)} = P(B) \tag{3.27}$$

From these, we have the following relationship for two independent events A and B:

$$P(A \cap B) = P(A)P(B) \tag{3.28}$$

Definition of Independence
Two events A and B are said to be independent if

$$P(A \cap B) = P(A)P(B) \tag{3.29}$$

This definition of independence is consistent with the definition of the conditional probability.

3.4 Cartesian Product

This section defines a special type of set called the Cartesian product. To illustrate the Cartesian product, consider the following example:

Example 3.4.1
You are to throw a coin and then throw a die. If the coin shows the heads side and the die shows less than four dots, you win a prize. What is the probability of winning the prize? This problem can be formulated using the axiomatic probability approach as follows: first, formulate the game as a combined experiment of two separate experiments of coin-throwing and die-throwing experiments with spaces Ω_1 and Ω_2, respectively, as follows:

		Ω_2					
		1	2	3	4	5	6
Ω_1	heads	(heads, 1)	(heads, 2)	(heads, 3)	(heads, 4)	(heads, 5)	(heads, 6)
	tails	(tails, 1)	(tails, 2)	(tails, 3)	(tails, 4)	(tails, 5)	(tails, 6)

Fig. 3.6 Cartesian product $\Omega_1 \times \Omega_2$

$$\Omega_1 = \{\text{heads}, \text{tails}\} \qquad \Omega_1 = \{1, 2, 3, 4, 5, 6\}$$

where the numbers in Ω_2 represent the number of dots on the die.

The space Ω of all possible outcomes of the game is the Cartesian product of Ω_1 and Ω_2 as follows (Fig. 3.6):

$$\begin{aligned}\Omega &= \Omega_1 \times \Omega_2 \\ &= \{(\text{heads}, 1), (\text{heads}, 2), (\text{heads}, 3), (\text{heads}, 4), (\text{heads}, 5), (\text{heads}, 6), (\text{tails}, 1), \\ &\quad (\text{tails}, 2), (\text{tails}, 3), (\text{tails}, 4), (\text{tails}, 5), (\text{tails}, 6)\}\end{aligned}$$

The event A "winning the prize" consists of the following elements of Ω:

$$A = \{(\text{heads}, 1), (\text{heads}, 2), (\text{heads}, 3)\}$$

We will return to this example to calculate the probability of the event A after discussing more on the combined experiment.

Figure 3.7 illustrates that, if E and F are subsets of X and Y, respectively, their Cartesian product, $E \times F$, is a subset of $X \times Y$.

$$E \subset X \text{ and } F \subset Y \quad \Rightarrow \quad E \times F \subset X \times Y$$

Consider two spaces, Ω_1 and Ω_2

$$\Omega_1 = \{x_1, x_2, \ldots, x_i, \ldots, x_m\} \qquad \Omega_2 = \{y_1, y_2, \ldots, y_i, \ldots, y_n\}$$

and the events represented by the following two Cartesian products:

$$E \times \Omega_2, \text{where } E \subset \Omega_1 \qquad \Omega_1 \times F, \text{where } F \subset \Omega_2$$

The event $E \times \Omega_2$ would occur if an element of E and an element of Ω_2 occur to form a pair, which is a member of the Cartesian product $E \times \Omega_2$. Since Ω_2 is a space, any element of Ω_2 paired with an element of E would make the event $E \times \Omega_2$ happen. By this reasoning, we establish

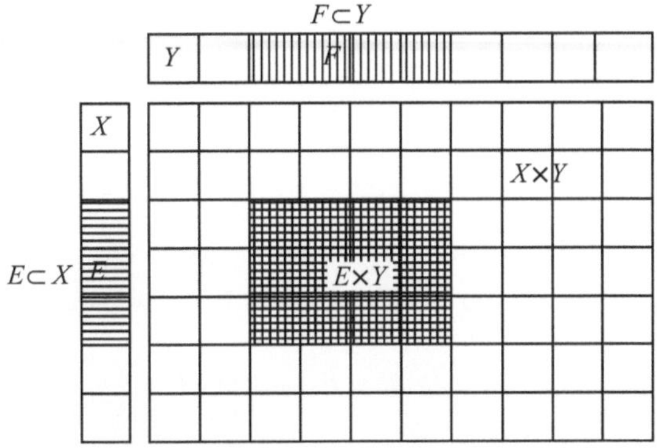

Fig. 3.7 $E \times F \subset X \times Y$

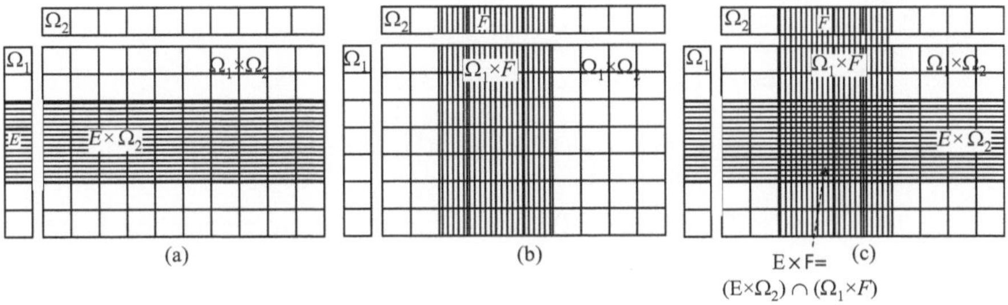

Fig. 3.8 (**a**) $E \times \Omega_2$, (**b**) $\Omega_1 \times F$, (**c**) $E \times F = (E \times \Omega_2) \cap (\Omega_1 \times F)$

$$P(E \times \Omega_2) = P(E) \tag{3.30}$$

$$P(\Omega_1 \times F) = P(F) \tag{3.31}$$

where $P(E)$ and $P(F)$ are probabilities of the events E and F defined in Ω_1 and Ω_1, respectively.

Figure 3.8 illustrates that the Cartesian product $E \times F$ can be expressed as the intersection of the two Cartesian products $E \times \Omega_2$ and $\Omega_1 \times F$ as follows:

$$E \times F = (E \times \Omega_2) \cap (\Omega_1 \times F) \tag{3.32}$$

Assume that the two individual experiments with spaces Ω_1 and Ω_2, respectively, are independent, that is, an outcome from Ω_1 has no effect on the outcome from Ω_2. Under this condition, the two events $E \times \Omega_2$ and $\Omega_1 \times F$ are independent.

From (3.32), we obtain the following equation:

$$\begin{aligned} P(E \times F) &= P[(E \times \Omega_2) \cap (\Omega_1 \times F)] \\ &= P(E \times \Omega_2) P(\Omega_1 \times F) \end{aligned} \tag{3.33}$$

Applying (3.30) and (3.31) to the above equation, we obtain the following equation:

$$P(E \times F) = P(E)P(F) \tag{3.34}$$

Consider the case where E and F are elementary events in Ω_1 and Ω_2, respectively, as follows:

$$E = \{x_i\} \qquad\qquad F = \{y_j\}$$

Substituting the above into (3.34), we obtain the following equation:

$$P(E \times F) = P\left[\{x_i\} \times \{y_j\}\right] = P[\{x_i\}]P\left[\{y_j\}\right] \tag{3.35}$$

To illustrate the concepts of (3.24) and (3.35), return to Example 3.4.1 and find the probability of the event A, where

$$\begin{aligned} A &= \{(\text{heads}, 1), (\text{heads}, 2), (\text{heads}, 3)\} \\ P(A) &= P[\{(\text{heads}, 1), (\text{heads}, 2), (\text{heads}, 3)\}] \\ &= P[\{(\text{heads}, 1)\}] + P[\{(\text{heads}, 2)\}] + P[\{(\text{heads}, 3)\}] \end{aligned} \tag{3.36}$$

Note that $\{(\text{heads}, 1)\}$, $\{(\text{heads}, 2)\}$, and $\{(\text{heads}, 3)\}$ are elementary events in the combined experiment space Ω.

Using (3.35), we can express the probabilities of the elementary events of the set A as the products of the probabilities of the elementary events of Ω_1 and Ω_2, respectively, as follows:

$$P[\{(\text{heads}, 1)\}] = P[\{\text{heads}\} \times \{1\}] = P[\{\text{heads}\}]P[\{1\}] \tag{3.37}$$

$$P[\{(\text{heads}, 2)\}] = P[\{\text{heads}\} \times \{2\}] = P[\{\text{heads}\}]P[\{2\}] \tag{3.38}$$

$$P[\{(\text{heads}, 3)\}] = P[\{\text{heads}\} \times \{3\}] = P[\{\text{heads}\}]P[\{3\}] \tag{3.39}$$

where $\{\text{heads}\}$ is an elementary event in Ω_1 and $\{1\}$, $\{2\}$, and $\{3\}$ are elementary events in Ω_2.

Assume the following probabilities for these elementary events in the two separate spaces:

$$P[\{\text{heads}\}] = \frac{1}{2}$$

$$P[\{1\}] = P[\{2\}] = P[\{3\}] = \frac{1}{6}$$

Substituting these probabilities into (3.37) through (3.39) and the resultants in (3.36), we have

$$P(A) = \frac{1}{2}\left(\frac{1}{6} + \frac{1}{6} + \frac{1}{6}\right) = \frac{1}{4}$$

To illustrate (3.30), in Example 3.4.1, change the event as follows: You win the prize if coin-throwing shows heads regardless of the outcome of die-throwing. This event, denoted by B, is

$$\begin{aligned} B &= \{(\text{heads}, 1), (\text{heads}, 2), (\text{heads}, 3), (\text{heads}, 4), (\text{heads}, 5), (\text{heads}, 6)\} \\ &= \{\text{heads}\} \times \Omega_2 \end{aligned}$$

Since event B occurs if coin-throwing shows heads regardless of the outcome of die-throwing, $P(B)$ should be equal to $P[\{\text{heads}\}]$. This is confirmed by applying (3.30) to the above as follows:

$$P(B) = P[\{\text{heads}\} \times \Omega_2] = P[\{\text{heads}\}]P[\Omega_2] = P[\{\text{heads}\}] = \frac{1}{2}$$

Similarly, to illustrate (3.31), suppose that you win a prize if die-throwing shows six regardless of the outcome of coin-throwing. This event, denoted by C, is

$$C = \{(\text{heads}, 6), (\text{tails}, 6)\} = \Omega_2 \times \{6\}$$

Since event C occurs if die-throwing shows six regardless of the outcome of coin-throwing, $P(C)$ should be equal to $P[\{6\}]$. This is confirmed by applying (3.34) to the above as follows:

$$P(C) = P[\Omega_2 \times \{6\}] = P[\Omega_2]P[\{6\}] = \frac{1}{6}$$

Chapter 4
Random Variables

This chapter defines random variables and discusses the various methods used to analyze the statistical properties of random variables. It defines the cumulative distribution function and the probability density function of random variables. It discusses the random variables treated individually as well as those treated jointly and the random variables defined as functions of other random variables. Both real and complex random variables are treated.

4.1 Definition of a Random Variable

Before defining a random variable, the concept of a function is briefly reviewed first. The function f is a relation that maps an object to another object. Figure 4.1 illustrates two cases of the function. In Fig. 4.1a, the function f maps an object s in general, e.g., an outcome of an experiment, to a real number X. In Fig. 4.1b, the function f maps a real number x to another real number y. s and x are called an independent variable and X and y, a dependent variable. The set of all objects Ω or the set of numbers on the real line over which x varies is called the domain of s or x, and the set to which s or x is mapped, the range of s or x.

We apply the concept of function defined above to the case of random experiments. The space of a random experiment Ω is the domain D. The independent variable is the outcome of the experiment, which is an object in general or a real number.

The experimental outcome, denoted by s, is the independent variable, and the real number which the experimental outcome is mapped to, denoted by X, is the dependent variable.

$$X = f(s) \tag{4.1}$$

K.I. Park, *Fundamentals of Probability and Stochastic Processes with Applications to Communications*, https://doi.org/10.1007/978-3-319-68075-0_4

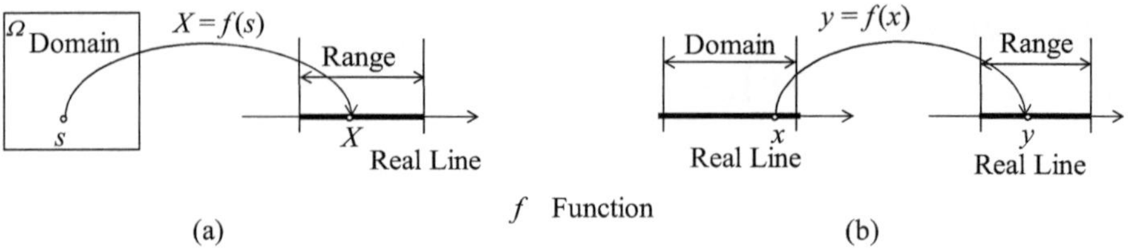

Fig. 4.1 (**a**) Mapping of s to X by $f(s)$, (**b**) mapping of x to y by $f(x)$

Ω Ball colors	Real Line $X = f(s)$
red	1
white	1
black	2
purple	2
orange	3
blue	3
.	.
.	.
.	.

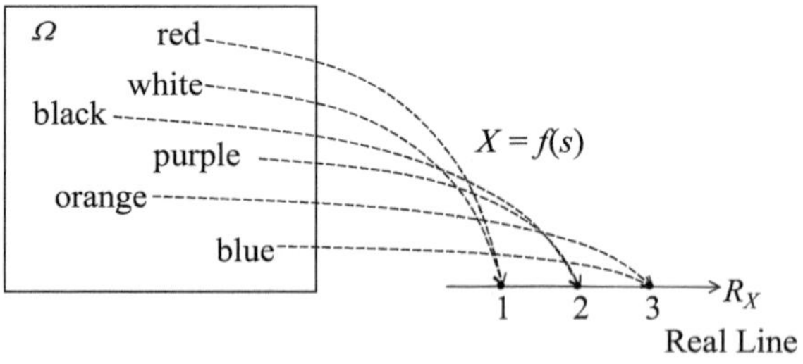

Fig. 4.2 Mapping of six different color balls, s, to three real numbers by f(s)

The experimental outcome s is random, and, therefore, the dependent variable X is random. The dependent variable X in (4.1) is defined as the random variable and will be abbreviated as RV. The function f may map multiple outcomes to a same value, that is, f may be "many to one." However, f does not map s to multiple values, that is, f is not "one to many." In this book, all RVs are denoted by capital letters such as X, Y, Z, etc.

Three examples of RV are considered below. The first example is a case in which the experimental outcomes are not real numbers, and the second and the third examples are cases in which the experimental outcomes are real numbers.

Example 4.1.1

A box contains balls of six different colors. A ball is drawn at random from the box. The following payoffs are defined. $1 is paid off if the color of the ball drawn is red or white; $2, if black or purple; and $3, if orange or blue. Since the outcome of ball-drawing is random, the amount of payoff is random. The random outcomes are mapped to real numbers, which are the payoff dollar amounts. Figure 4.2 illustrates the mapping by the function f from the space Ω to a real line.

Example 4.1.2

Suppose that you stand at a toll booth and count the number of cars passing through the toll booth in each 1-minute interval from 9:00 A.M. to 9:10 A.M. Each 1-minute count would be a random number and would be an integer.

Any integer would be a possible outcome. So, all possible outcomes are non-negative positive integers. Consider the function *f* which maps the outcome to the same number as follows. In other words, the experimental space Ω and the range *R* are the same sets. Since the 1-minute counts can be any non-negative integer, the space Ω is as follows:

$$\Omega = \{\text{all possible counts}\} = \{i, i = \text{integer} \geq 0\}$$

Example 4.1.3

Consider taking temperature measurements, that is, random outcomes *s*, at regular intervals and mapping each outcome, that is, each reading, to a value equal to the reading itself. Since the temperature readings can be any real numbers, the experimental space Ω is as follows:

$$\Omega = \{\text{all possible temperature readings}\} = \{x, x = \text{real number} \geq 0\}$$

Figure 4.3 illustrates the functional mapping from Ω to R_X for the last two examples, in which the outcome *s* is mapped to itself by $X = f(s) = s$.

For both examples, the experimental space Ω and the range to which Ω is mapped to by the function *f* are the same, that is, $\Omega = R_X$. The specific value the RV *X* takes by the mapping *f* is denoted by a lowercase letter *x*. Once *s* is fixed, *x* is fixed by *f*(*s*) and, therefore, is no longer random.

Ω Temp Reading, *s*	Real Line $X = f(s)$
60.3	60.3
63.9	63.9
59.7	59.7
.	.
.	.

(a)

Ω 1-min count, *s*	Real Line $X = f(s)$
3	3
7	7
0	0
2	2
6	6
.	.
.	.

(b)

Fig. 4.3 (**a**) Mapping of random temperature readings to the same numbers, (**b**) Mapping of random counts to the same numbers

For the last two examples, R_X is a set of real numbers. In Fig. 4.3a:

$$R_X = \{x, -\infty \le x \le +\infty\}; \text{real line from} - \infty \text{ to} + \infty$$

In Fig. 4.3b:

$$R_X = \{x, x \text{ is an integer} \ge 0\}$$

Complex RV. The complex RV X is defined as follows:

$$X = X_r + jX_i \qquad X_r, X_i \text{ are real RVs} \tag{4.2}$$

The complex conjugate of X, denoted by X^*, is given by:

$$X^* = X_r - jX_i \tag{4.3}$$

4.2 Random Variables Treated Singly

This section discusses the RVs treated singly and defines the cumulative distribution function (CDF) and the probability density function (*pdf*) of the RV treated singly.

4.2.1 Cumulative Distribution Function

Definition of the CDF

Let us apply the axiomatic probability formulation to the RV X. Consider a random experiment in which a number is selected at random from the real line $R = \{x, -\infty \le x \le +\infty\}$. Each trial produces a single outcome, which is a real number. This experimental outcome is denoted by RV X. The space for this experiment is $R = \{x, -\infty \le x \le +\infty\}$. An event is defined by "a subset of the space consisting of the elements that make the event happen." One specific type of subset of R is the left-open interval $\{\lambda, \text{all } \lambda \text{ such that } \lambda \le x\}$. This subset defines the event that the RV X is less than or equal to x. This event is denoted by $\{X \le x\}$.

$$\{X \le x\} = \{\lambda, \text{all } \lambda \text{ such that } \lambda \le x\}$$

$\{X \le x\}$ is a new notation that needs to be defined. $\{X \le x\}$ has a dual meaning. $\{X \le x\}$ is used to denote the event as well as the set defining the event. First, $\{X \le x\}$ is used to represent the set on the right-hand side of the above expression and is read "$\{X \le x\}$ *is a set of all numbers less than or equal to x*." Second, the same notation $\{X \le x\}$ is used to denote the event that RV X is less than or equal to

x and is read "$\{X \leq x\}$ *is the event that RV X is less than or equal to x.*" For example, for $x_1 < x_2$, we say that $\{X \leq x_1\}$ is a subset of $\{X \leq x_2\}$. In this instance, what we mean is that the set defining the event $\{X \leq x_1\}$ is a subset of the set defining the event $\{X \leq x_2\}$, if $x_1 < x_2$, that is, the set $\{\lambda$, all λ such that $\lambda < x_1\}$ is a subset of the set $\{\lambda$, all λ such that $\lambda < x_2\}$, if $x_1 < x_2$.

The cumulative distribution function, abbreviated as the CDF, of the RV X, which is denoted by $F_X(x)$, is defined as the probability of the event $\{X \leq x\}$ as follows:

$$F_X(x) \triangleq P[\{X \leq x\}] \tag{4.4}$$

The right boundary point x of the left-open interval, which is shown as the argument of the CDF, is arbitrary and can be treated as a variable. The subscript X is used to denote that the CDF is for RV X. Since $F_X(x)$ is defined as a probability, it must satisfy the probability axioms given by (3.1) through (3.3).

Per Axiom I:

$$F_X(x) \geq 0 \tag{4.5}$$

Per Axiom II:

$$F_X(+\infty) \leq 1 \tag{4.6}$$

The two sets $\{X \leq x\}$ and $\{X > x\}$ are mutually exclusive, that is, $\{X \leq x\} \cap \{X > x\} = \{\varnothing\}$, and the union of the two sets is the entire real line from $-\infty$ to $+\infty$, that is, the space Ω. Therefore, by Axiom III, we obtain the following equation:

$$P(\Omega) = P[\{X \leq x\} \cup \{X > x\}] = P[\{X \leq x\}] + P[\{X > x\}] = 1 \tag{4.7}$$

or

$$P[\{X > x\}] = 1 - P[\{X \leq x\}] \tag{4.8}$$

Rewriting the right-hand side of (4.8) using the definition of the CDF, we obtain the following equation:

$$P[\{X > x\}] = 1 - F_X(x) \tag{4.9}$$

Consider two specific values of X, x_1 and x_2, such that $x_1 < x_2$. Referring to Fig. 4.4a, we see that the set $\{X \leq x_1\}$ is a subset of $\{X \leq x_2\}$, that is:

$$\{X \leq x_1\} \subset \{X \leq x_2\}$$

By (3.10), we have the following equation:

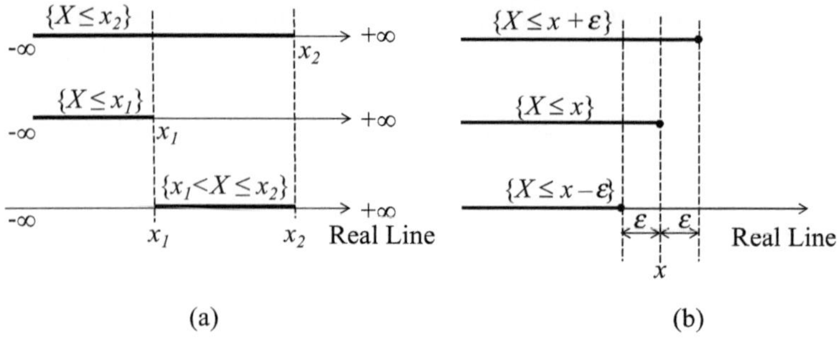

Fig. 4.4 (**a**) $\{X \le x_1\} \subset \{X \le x_2\}$, (**b**) $\{X \le x\} \subset \{X \le (x + \varepsilon)\}$

$$P[\{X \le x_1\}] \le P[\{X \le x_2\}]$$

Rewriting the above equation using the definition of the CDF, we obtain the following equation:

$$F_X(x_1) \le F_X(x_2) \ \text{ if } x_1 < x_2 \tag{4.10}$$

This last property of the CDF is important in characterizing the graph of the CDF. It shows that, as x is increased, the CDF $F_X(x)$ cannot decrease, although it can stay the same; that is, $F_X(x)$ is a nondecreasing function of x.

Continuity of a Function

Before going into the next property of the CDF, we first need to discuss the concept of continuity of a function. A function $y = f(x)$ is defined to be "continuous" at a point $x = c$ if the following conditions are met. First, the function must have a finite value at $x = c$, $f(c)$. Second, the function must have a finite limiting value as x approaches c, that is, $\lim_{x \to c} f(x)$ must exist. Finally, this limiting value and the finite value of the function at $x = c$ of the first condition must coincide, that is:

$$\lim_{x \to c} f(x) = f(c)$$

A function may be continuous from the right or from the left, depending on the direction from which the limiting condition is met. A function may be continuous from the right but not from the left and vice versa. A function is continuous if and only if it is continuous both from the right and from the left. This concept is illustrated by the figure below.

In all three examples of Fig. 4.5, a finite value of $f(c)$ exists. In Fig. 4.5 (a), $\lim_{\varepsilon \to 0} f(c + \varepsilon) = f(c)$, and the function is continuous from the right; similarly, in (b), the function is continuous from the left; and in (c), the function is continuous both from the right and from the left, and, therefore, is continuous.

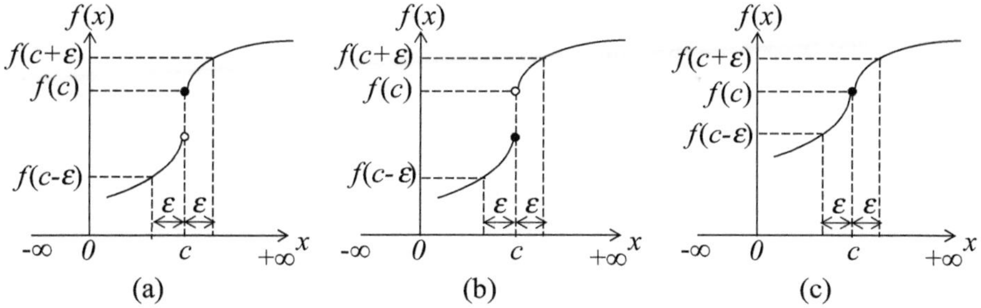

Fig. 4.5 (**a**) $f(x)$ continuous from the right with discontinuity at $x = c$, (**b**) $f(x)$ continuous from the left with discontinuity at $x = c$, (**c**) $f(x)$ continuous with no discontinuity at $x = c$

Right-Continuity Property of the CDF

Regardless of an RV, its CDF is continuous from the right, that is:

$$\lim_{\varepsilon\to 0} F_X(x+\varepsilon) = F_X(x)$$

However, a CDF is not necessarily continuous from the left.

$$\lim_{\varepsilon\to 0} F_X(x-\varepsilon) \neq F_X(x)$$

If a CDF is continuous both from the right and from the left, then it is continuous.

A rigorous proof of the right-continuity property of the CDF involves the convergence of sets using the measure theory, which is not addressed here. However, a heuristic approach is discussed to provide an insight into this property.

By the definition of the CDF, we have the following equations:

$$\begin{gathered} F_X(x+\varepsilon) = P[\{X \le (x+\varepsilon)\}] \\ F_X(x) = P[\{X \le x\}] \\ F_X(x-\varepsilon) = P[\{X \le (x-\varepsilon)\}] \end{gathered}$$

Consider the three sets defining the events, $\{X \le (x+\varepsilon)\}$, $\{X \le x\}$, and $\{X \le (x-\varepsilon)\}$, and their set relations depicted in Fig. 4.4b. With respect to the set $\{X \le x\}$, $\{X \le (x+\varepsilon)\}$ is a superset, whereas $\{X \le (x-\varepsilon)\}$ is a subset. Recall that the CDF of RV X is defined as the probability of the event that the RV X is less than or equal to x, that is, $F_X(x) \triangleq P[\{X \le x\}]$. According to this definition, the set defining the event for the CDF, $\{X \le x\}$, includes the exact point $X = x$. In other words, the set $\{X \le x\}$ is a "left-open, right-closed" subset of R, which includes the "exact" point x as the limit closing the subset on the right. Because of this, no matter how closely the superset $\{X \le (x+\varepsilon)\}$ closes in to its subset $\{X \le x\}$ from the right as $\varepsilon \to 0$, the limiting set $\lim_{\varepsilon\to 0}\{X \le (x+\varepsilon)\}$ still contains the exact point $X = x$. Therefore, we can heuristically understand that $\lim_{\varepsilon\to 0} F_X(x+\varepsilon) = F_X(x)$ and the right-continuity of $F_X(x)$.

Now, let us look at the set relation between the two sets, $\{X \le (x-\varepsilon)\}$ and $\{X \le x\}$. $\{X \le (x-\varepsilon)\}$ is a subset of $\{X \le x\}$. As $\varepsilon \to 0$, the subset $\{X \le (x-\varepsilon)\}$

approaches its superset $\{X \le x\}$. Unlike the first case where the superset approaches its subset $\{X \le x\}$ from the right, in this case, the subset $\{X \le (x-\varepsilon)\}$ approaches from the left its superset $\{X \le x\}$, which includes the exact point $X = x$ as the right limiting point. Because of this, no matter how closely $\{X \le (x-\varepsilon)\}$ approaches $\{X \le x\}$, it still would not include the exact point $X = x$. Since the definition of CDF $F_X(x)$ demands that the point $X = x$ be included in the set defining the event, the limiting function $\lim_{\varepsilon \to 0} F_X(x-\varepsilon)$ does not necessarily equal $F_X(x)$, that is, $\lim_{\varepsilon \to 0} F_X(x-\varepsilon) \neq F_X(x)$.

In summary, a CDF is right-continuous. However, a CDF is not necessarily left-continuous. If a CDF is left-continuous, it is continuous. The converse is also true, that is, if a CDF is continuous, it is left-continuous.

Probability of Exact Value *x*

We examine closely the probability that RV X takes an exact value x, $P[\{X = x\}]$.

Since $\{X \le (x-\varepsilon)\} \subset \{X \le x\}$, $\{X \le x\}$ can be expressed as the union of two mutually exclusive sets as follows:

$$\{X \le x\} = \{X \le (x-\varepsilon)\} \cup \{(x-\varepsilon) < X \le x\}$$

where

$$\{X \le (x-\varepsilon)\} \cap \{(x-\varepsilon) < X \le x\} = \{\varnothing\}$$

Taking the probabilities of the left- and right-side events, we have:

$$\begin{aligned} P[\{X \le x\}] &= P[\{X \le (x-\varepsilon)\} \cup (x-\varepsilon) < X \le x\}] \\ &= P[\{X \le (x-\varepsilon)\}] + P[\{(x-\varepsilon) < X \le x\}] \end{aligned}$$

Using the definition of the CDF, we can rewrite the above equation as follows:

$$F_X(x) = F_X(x-\varepsilon) + P[\{(x-\varepsilon) < X \le x\}] \tag{4.11a}$$

Taking the limit $\varepsilon \to 0$ of the above equation, we have:

$$F_X(x) = \lim_{\varepsilon \to 0} F_X(x-\varepsilon) + \lim_{\varepsilon \to 0} P[\{(x-\varepsilon) < X \le x\}]$$

or, rearranging the terms:

$$\lim_{\varepsilon \to 0} F_X(x-\varepsilon) = F_X(x) - \lim_{\varepsilon \to 0} P[\{(x-\varepsilon) < X \le x\}] \tag{4.11b}$$

As $\varepsilon \to 0$, the set $\{(x-\varepsilon) < X \le x\}$ approaches the set $\{X = x\}$, $\lim_{\varepsilon \to 0} \{(x-\varepsilon) < X \le x\} = \{X = x\}$, and (4.11b) is rewritten as follows:

$$\lim_{\varepsilon \to 0} P[\{(x-\varepsilon) < X \le x\}] = P[\{X = x\}]$$

Substituting the above expression into (4.11b), we obtain the following equation:

$$\lim_{\varepsilon \to 0} F_X(x - \varepsilon) = F_X(x) - P[\{X = x\}] \tag{4.11c}$$

or, rearranging the terms:

$$P[\{X = x\}] = F_X(x) - \lim_{\varepsilon \to 0} F_X(x - \varepsilon) \tag{4.12}$$

The last equation gives the probability that RV X takes the exact value x, $P[\{X=x\}]$, and gives the following important properties of a CDF.

For a Continuous CDF
As discussed earlier, if a CDF is continuous at x, it is left-continuous, that is, $\lim_{\varepsilon \to 0} F_X(x - \varepsilon) = F_X(x)$. Substituting this equation into (4.12), we obtain:

$$P[\{X = x\}] = 0$$

This result shows that, if an RV X takes a value over a continuum, the probability that X would be equal to an exact value x is zero. Only the probability that X would fall in an interval can be nonzero. The converse is also true as follows. If $P[\{X=x\}]=0$, the CDF is continuous at x.

This result agrees with the common sense notion of a continuous RV. Suppose that you are waiting for a bus that would arrive in a certain time interval, say, 9 AM to 10 AM. The probability that it will arrive at an exact instant is zero however small the error may be, e.g., in nanosecond.

For a Discontinuous CDF
If a CDF is discontinuous at x, then, the CDF is not continuous from the left, and we have the inequality:

$$\lim_{\varepsilon \to 0} F_X(x - \varepsilon) \neq F_X(x)$$

Comparing the above inequality with (4.12), we see that the value of $P[\{X=x\}]$ is equal to the magnitude of the discontinuity or "jump" at x on the CDF as illustrated in Fig. 4.6.

Continuous RV
The RV X which has a continuous CDF is referred to as the continuous RV. Continuous $F_X(x)$ is illustrated in Fig. 4.7 by using two cases of continuum on R over which the RV X takes on a value. The first case is a continuum without intermittent spaces; the second case is a continuum with intermittent spaces. In the first case, $F_X(x)$ monotonically increases. In the second case, $F_X(x)$ monotonically increases except over intermittent spaces over which $F_X(x)$ stays constant. In both cases, however, as x is varied, $F_X(x)$ is continuous and has no discontinuity.

As discussed earlier, for a continuous RV X, the probability that X will take a specific value x is zero, that is:

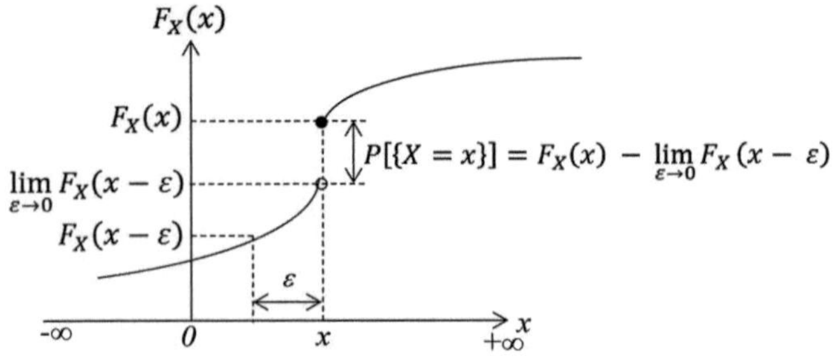

Fig. 4.6 Discontinuity at x with the magnitude of the jump equal to $P[\{X = x\}]$

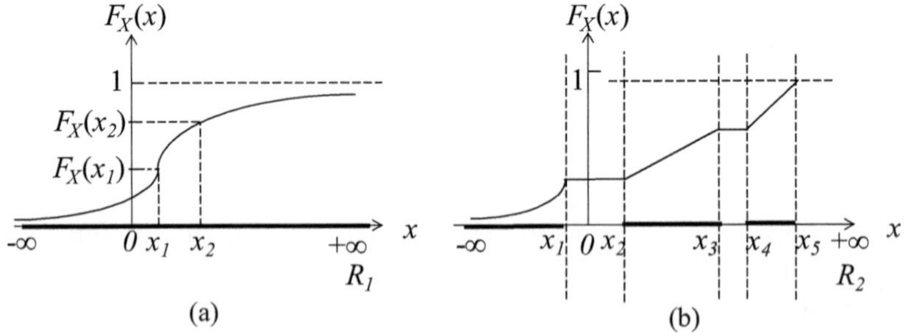

Fig. 4.7 (**a**) CDF over a continuum without intermittent spaces, (**b**) CDF over a continuum with intermittent spaces

$$P[\{X = 0\}] = 0 \tag{4.13}$$

Discrete RV

In this case, R is a set of discrete points on the real line. In general, R contains an infinite number of discrete points. For illustration purposes, consider R containing five discrete points as follows:

$$R = \{x_1, x_2, \ldots, x_5\} = \Omega$$

Figure 4.8a shows these five discrete points.

Let the probabilities that the RV X will take these values be given by the following equations:

$$P[\{X = x_i\}] = p_i, i = 1, \ldots, 5$$

where

$$p_1 + p_2 + p_3 + p_4 + p_{p5} = 1$$

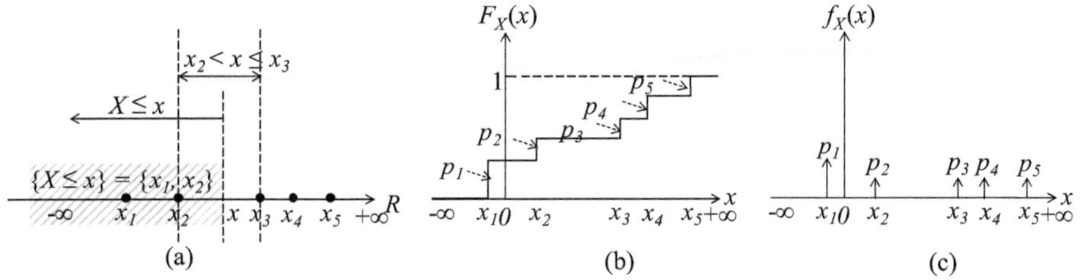

Fig. 4.8 (**a**) Five discrete points on the real line, (**b**) Discrete CDF, (**c**) Probabilities at the discrete points

In Fig. 4.8a, starting from an arbitrary point x below the point x_1, slide x towards $+\infty$ and consider the event $\{X \leq x\}$. We see that, if $x \leq x_1$, the subset $\{X \leq x\}$ in R contains no point, that is, $\{X \leq x\} = \{\varnothing\}$. If x is in the interval $x_1 < x \leq x_2$, the subset $\{X \leq x\}$ contains one point, x_1, $\{X \leq x\} = \{x_1\}$. If x is in the interval $x_2 < x \leq x_3$, the subset $\{X \leq x\}$ contains two points x_1, and x_2, $\{X \leq x\} = \{x_1, x_2\}$. Continue this process until x passes x_5. Then, if $x_5 < x$, $\{X \leq x\}$ contains all five points, that is, $\{X \leq x\} = \Omega$. The result of this process is summarized below.

For $-\infty < x \leq x_1, \{X \leq x\} = \{\varnothing\}$

$$P[\{X \leq x\}] = P[\{\varnothing\}] = 0$$

For $x_1 < x \leq x_2, \{X \leq x\} = \{x_1\}$

$$P[\{X \leq x\} = P[\{x_1\}] = p_1$$

For $x_2 < x \leq x_3, \{X \leq x\} = \{x_1, x_2\}$

$$P[\{X \leq x\}] = P[\{x_1, x_2\}] = P[\{x_1\}] + P[\{x_2\}] = p_1 + p_2$$

For $x_3 < x \leq x_4, \{X \leq x\} = \{x_1, x_2, x_3\}$

$$P[\{X \leq x\}] = P[\{x_1, x_2, x_3\}] = P[\{x_1\}] + P[\{x_2\}] + P[\{x_3\}] = p_1 + p_2 + p_3$$

For $x_4 < x \leq x_5, \{X \leq x\} = \{x_1, x_2, x_3, x_4\}$

$$P[\{X \leq x\}] = P[\{x_1, x_2, x_3, x_4\}] = P[\{x_1\}] + P[\{x_2\}] + P[\{x_3\}] + P[\{x_4\}]$$
$$= p_1 + p_1 + p_3 + p_4$$

For $x_5 < x \leq +\infty, \quad \{X \leq x\} = \{x_1, x_2, x_3, x_4, x_5\} = \Omega$

$$P[\{X \leq x\}] = P[\{x_1, x_2, x_3, x_4, x_5\}] = P[\{x_1\} + P[\{x_2\} + P[\{x_3\} + P[\{x_4\} + P[\{x_5\}]$$
$$= p_1 + p_1 + p_3 + p_4 + p_5 = P(\Omega) = 1$$

Figure 4.8b plots the above cumulative probabilities. The figure illustrates that, in the case of R of discrete points, as x is increased, $F_X(x)$ jumps at a discrete point by an amount equal to the probability measure of that point, stays unchanged until x reaches the next discrete point where it jumps by the amount equal to the probability measure of the point, and so on. $F_X(x)$ for R of discrete points is a step function. The RV which has a step function $F_X(x)$ is referred to as the discrete RV. Figure 4.8c shows the probabilities at the discrete points corresponding to Fig. 4.8b.

Consider the event that the RV X will take on a value less than or equal to a specific value x. In Example 4.1.3, for example, this event would occur if the random temperature reading turns out to be less than or equal to x.

"event that temperature reading is less than or equal to a specific value x"
= {set of all real numbers less than or equal to x} $= \{s, s \leq x\}$

To simplify the notation, the last set is denoted by $\{X \leq x\}$. The probability of this event is referred to as the CDF of RV X. Now, treat the value x as a variable. Then, $F_X(x)$ is a function of variable x.

For a given RV X, $F_X(x)$ provides the complete characterization of the probabilistic behavior of X. More about this will be discussed later with applications in stochastic processes.

Probability of the Event $\{x_1 \leq X \leq x_2\}$

Referring to Fig. 4.8a, we see that:

$$\{X \leq x_2\} = \{X \leq x_1\} \cup \{x_1 < X \leq x_2\}$$

and

$$\{X \leq x_1\} \cap \{x_1 < X \leq x_2\} = \{\emptyset\}$$

By Axiom III, we have:

$$\begin{aligned} P[\{X \leq x_2\}] &= P[\{X \leq x_1\} \cup \{x_1 < X \leq x_2\}] \\ &= P[\{X \leq x_1\}] + P[\{x_1 < X \leq x_2\}] \end{aligned}$$

Rewriting the above equation using the definition of the CDF, we obtain the following equation:

$$F_X(x_2) = F_X(x_1) + P[\{x_1 < X \leq x_2\}]$$

or, rearranging the terms:

$$P[\{x_1 < X \leq x_2\}] = F_X(x_2) - F_X(x_1) \tag{4.14}$$

4.2.2 *The Probability Density Function* (pdf)

Referring to Fig. 4.9a, consider a small interval of magnitude Δx from an arbitrary x to $x + \Delta x$ on the real line R and the probability that an RV X would fall in this small interval:

$$P[\{x < X \leq (x + \Delta x)\}]$$

The above probability would vary depending on the magnitude of the interval, that is, Δx: the bigger the interval is, the bigger the probability would be. Normalizing the

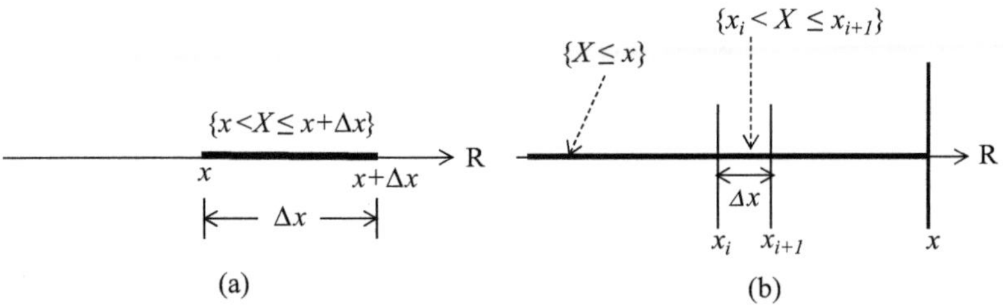

Fig. 4.9 (**a**) A small interval of magnitude Δx, (**b**) Division of $\{X \leq x\}$ into intervals of magnitude Δx

above probability by dividing it by the magnitude of the interval Δx, we have the "density" of the probability over the interval Δx as follows:

$$\frac{1}{\Delta x} P[\{x < X \leq (x + \Delta x)\}]$$

The probability density function (*pdf*) of the RV X, denoted by $f_X(x)$, is defined as the limiting value of the above density as Δx approaches 0, that is:

$$f_X(x) \triangleq \lim_{\Delta x \to 0} \frac{1}{\Delta x} P[\{x < X \leq (x + \Delta x)\}] \tag{4.15}$$

Given the definition of the *pdf* by (4.15), we have the following relationship:

$$P[\{x < X \leq (x + \Delta x)\}] \approx f_X(x)\Delta x \tag{4.16}$$

Note that the *pdf* is defined by (4.15) independently of the definition of the CDF given by (4.4). The next theorem shows the relationship between the two independent definitions.

Theorem 4.2.1 Given the CDF of an RV X as defined by (4.4), the *pdf* as defined by (4.15) can be obtained by taking the derivative of the CDF as follows:

$$f_X(x) = \frac{\mathrm{d}}{\mathrm{d}x} F_x(x) \tag{4.17}$$

Proof Substitute the following into (4.14)

$$x_1 = x \qquad x_2 = x + \Delta x$$

to write

$$P[\{x < X \leq x + \Delta x\}] = F_X(x + \Delta x) - F_X(x) \tag{4.18}$$

Substituting (4.18) into the defining equation of the *pdf* (4.15) yields:

$$f_X(x) = \lim_{\Delta x \to 0} \frac{F_X(x + \Delta x) - F_X(x)}{\Delta x} \tag{4.19}$$

The right-hand side of the last equation is the definition of the derivative of $F_X(x)$, as defined in differential calculus.

Q.E.D.

Theorem 4.2.2 This theorem is the converse of Theorem 4.2.1. Given the *pdf* $f_X(x)$, the CDF $F_X(x)$, defined by (4.4), can be obtained by taking the integral of the *pdf* as follows:

$$F_X(x) = \int_{-\infty}^{x} f_X(\lambda)d\lambda \tag{4.20}$$

Proof The proof can be given simply by taking the reverse process of Theorem 4.2.1 as follows. From (4.17), we have:

$$\mathrm{d}F_X(x) = f_X(x)\mathrm{d}x$$

Taking the integral of both sides of the above equation, we obtain the following equation:

$$F_X(x) = \int_{-\infty}^{x} f_X(\lambda)d\lambda$$

Q.E.D.

Note that, the above proof is based on the relationship given by (4.17) that has been derived from the definition of the CDF. In the following discussion, (4.20) is derived directly from the definition of the *pdf* given by (4.15).

The subset $\{X \leq x\}$ of the real line R is divided into small intervals of magnitude Δx as shown in Fig. 4.9b so that $\{X \leq x\}$ is expressed as the union of these small mutually exclusive intervals as follows:

$$\{X \leq x\} = \{x_1 < X \leq (x_1 + \Delta x)\} \cup \ldots \cup \{x_i < X \leq (x_i + \Delta x)\} \cup \ldots$$

where

$$\{x_i < X \leq (x_i + \Delta x)\} \cap \ \{x_j < X \leq (x_j + \Delta x)\} = \{\varnothing\} \text{ for } i \neq j$$

Since the small subintervals are mutually exclusive, by Axiom III given (3.3), we have the following equation:

$$\begin{aligned} F_X(x) &= P[\{X \leq x\}] \\ &= P[\{x_1 < X \leq (x_1 + \Delta x)\} \cup \ldots \cup \{x_i < X \leq (x_i + \Delta x)\} \cup \ldots] \\ &= P[\{x_1 < X \leq (x_1 + \Delta x)\} + \cdots + P[\{x_i < X \leq (x_i + \Delta x)\}] + \cdots \\ &= \textstyle\sum_i P[\{x_i < X \leq (x_i + \Delta x)\}] \end{aligned} \tag{4.21}$$

Using (4.16), we can approximate the probability that the RV X would fall in the ith interval in (4.21) as follows:

$$P[\{x_i < X \leq (x_i + \Delta x)\}] \approx f_X(x_i)\Delta x \tag{4.22}$$

Substituting (4.22) into (4.21) yields:

$$F_X(x) \approx \sum_i f_X(x_i)\Delta x \tag{4.23}$$

In the limit as $\Delta x \rightarrow 0$, the summation of the above equation becomes the integral as follows:

$$F_X(x) = \lim_{\Delta x \rightarrow 0} \sum_i f_X(x_i)\Delta x = \int_{-\infty}^{x} f_X(\lambda)\Delta\lambda$$

Q.E.D.

Properties of the *pdf*

Referring to Fig. 4.10, we see that the integral of the *pdf* from x_1 to x_2 is equal to the area under the *pdf* curve, $f_X(x)$, from x_1 to x_2. This area is equal to the probability that the RV X will fall in the interval $\{x_1 < X \leq x_2\}$ as follows:

$$F(x_2) - F(x_1) = \int_{x_1}^{x_2} f(x)\mathrm{d}x = P\left[\{x_1 < X \leq x_2\}\right] \tag{4.24}$$

For the discrete RV X, the analytical expression of its *pdf* is given in terms of the probability measures at the discrete points and the Dirac delta function denoted by δ as follows:

$$\begin{array}{c} f_X(x) = \sum_i p_i \delta(x - x_i) \\ p_i = P[\{X = x_i\}] \quad i = 1, \ldots \end{array} \tag{4.25}$$

The Dirac delta function has the following property.

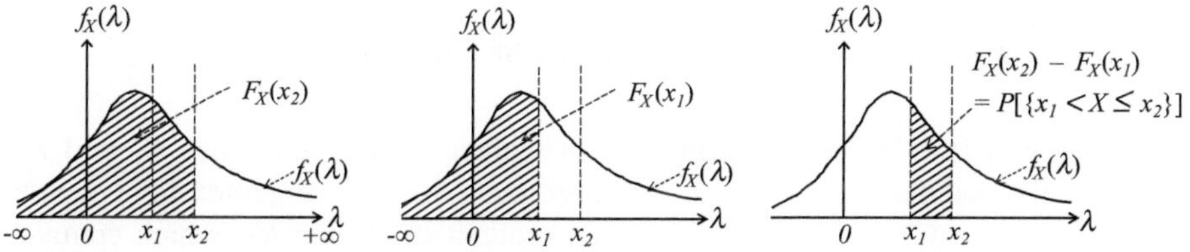

Fig. 4.10 The values of the integral shown as the shaded areas

$$\int_{-\infty}^{x} \delta(x - x_i)\mathrm{d}x = \begin{cases} 1 & \text{if} \quad x \geq x_i \\ 0 & \text{if} \quad x < x_i \end{cases} \tag{4.26}$$

$$\int_{-\infty}^{x} f(x)\delta(x - x_i)\mathrm{d}x = \begin{cases} f(x_i) & \text{if} \quad x \geq x_i \\ 0 & \text{if} \quad x < x_i \end{cases} \tag{4.27}$$

This property allows the *pdf* of the discrete RV to be integrated to generate the corresponding CDF as follows:

$$\begin{aligned} F_X(x) &= \int_{-\infty}^{x} f_X(\lambda)\mathrm{d}\lambda = \int_{-\infty}^{x} \sum_i p_i \delta(\lambda - x_i)\mathrm{d}\lambda = \sum_i \int_{-\infty}^{x} p_i \delta(\lambda - x_i)\mathrm{d}\lambda \\ &= \sum_i p_i \int_{-\infty}^{x} \delta(\lambda - x_i)\mathrm{d}\lambda = \sum_i p_i \quad \text{for } x_i < x \leq x_{i+1} \end{aligned} \tag{4.28}$$

The last equation is an analytical expression of a step function because as x moves into the next interval, a new probability p_i is added to the sum accumulated up to the current interval.

The total area under the *pdf* curve is equal to 1.

$$F_X(+\infty) = \int_{-\infty}^{+\infty} f(x)\mathrm{d}x = 1 \tag{4.29}$$

Since $F_X(x)$ is a nondecreasing function of x, its derivative, which is the *pdf*, $f_X(x)$, is non-negative. Hence:

$$f_X(x) \geq 0 \tag{4.30}$$

4.3 Random Variables Treated Jointly

This section discusses two RVs treated jointly and defines their joint CDF and joint *pdf*.

4.3.1 The Joint CDF of Two Random Variables

Consider two RVs X and Y taking on the values on the real lines R_1 and R_2, respectively, as the spaces. Now consider the two RVs together in a single, combined space Ω given by the Cartesian product of R_1 and R_2, which contains all possible ordered pairs (x, y) of R_1 and R_2. Consider the subset of $\Omega = R_1 \times R_2$ which contains the ordered pairs (x, y) satisfying the conditions $X \le x$ and $Y \le y$. Denote this subset by $\{X \le x, Y \le y\}$. The probability of this event is defined as the joint CDF of X and Y as follows:

$$F_{XY}(x, y) \triangleq P[\{X \le x, Y \le y\}] \tag{4.31}$$

Property 1

$$P[\{x_1 < X \le x_2, Y \le y\}] = F_{XY}(x_2, y) - F_{XY}(x_1, y) \tag{4.32}$$

$$P[\{X \le x, y_1 < Y \le y_2\}] = F_{XY}(x, y_2) - F_{XY}(x, y_1) \tag{4.33}$$

Proof Referring to Fig. 4.11a, we see that the following equations hold true.

$$\{X \le x_2, Y \le y\} = \{x_1 < X \le x_2, Y \le y\} \cup \{X \le x_1, Y \le y\}$$
$$\{x_1 < X \le x_2, Y \le y\} \cap \{X \le x_1, Y \le y\} = \{\varnothing\}$$

Using Axiom III given by (3.3) and the above equations, we obtain the following equations:

$$\begin{aligned} P[\,\{X \le x_2, Y \le y\}] &= P[\,\{x_1 < X \le x_2, Y \le y\} \cup \{X \le x_1, Y \le y\}] \\ &= P[\,\{x_1 < X \le x_2, Y \le y\} + P[\{X \le x_1, Y \le y\}] \end{aligned} \tag{4.34}$$

Rearranging the terms of (4.34) yields the following equation:

$$P\big[\,\{x_1 < X \le x_2, Y \le y\}\big] = P[\,\{X \le x_2, Y \le y\}] - P\big[\{X \le x_1, Y \le y\}\big] \tag{4.35}$$

Using the definition of the joint CDF given by (4.31), we can rewrite (4.35) as follows:

$$P[\{x_1 < X \le x_2, Y \le y\}] = F_{XY}(x_2, y) - F_{XY}(x_1, y)$$

Q.E.D.

Similarly,

$$P[\{X \le x, y_1 < Y \le y_2\}] = F_{XY}(x, y_2) - F_{XY}(x, y_1) \tag{4.36}$$

Property 2

$$\begin{aligned} P[\{x_1 < X \le x_2, y_1 < Y \le y_2\}] = &F_{XY}(x_2, y_2) - F_{XY}(x_1, y_2) \\ &- F_{XY}(x_2, y_1) + F_{XY}(x_1, y_1) \end{aligned} \tag{4.37}$$

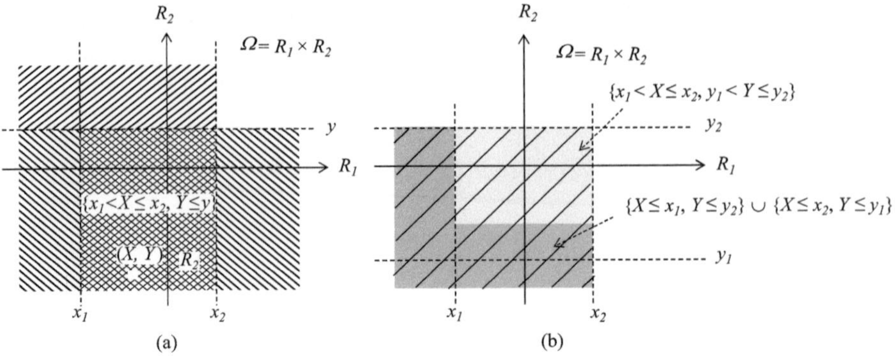

Fig. 4.11 (**a**) $\{X \le x_2, Y \le y\} = \{x_1 < X \le x_2, Y \le y\} \cup \{X \le x_1, Y \le y\}$, (**b**) $\{X \le x_2, Y \le y\} = \{x_1 < X \le x_2, Y \le y\} \cup \{X \le x_1, Y \le y\}$

Proof Referring to Fig. 4.11b, we can express the set $\{X \le x_2, Y \le y_2\}$, represented by the hatched area, as the union of two mutually exclusive sets $\{x_1 < X \le x_2, y_1 < Y \le y_2\}$, represented by the gray rectangular area, and $[\{X \le x_1, Y \le y_2\} \cup \{X \le x_2, Y \le y_1\}]$, represented by the darker L-shaped area as follows:

$$\begin{aligned}\{X \le x_2, Y \le y_2\} = \{&x_1 < X \le x_2, y_1 < Y \le y_2\} \cup \\ &[\{X \le x_2, Y \le y_1\} \cup \{X \le x_1, Y \le y_2\}]\end{aligned} \tag{4.38}$$

where

$$\{x_1 < X \le x_2, y_1 < Y \le y_2\} \cap [\{X \le x_2, Y \le y_1\} \cup \{X \le x_1, Y \le y_2\}] = \{\varnothing\}$$

By Axiom III given by (3.3), we obtain the probability of the event of (4.38) as follows:

$$\begin{aligned}P[\{X \le x_2, Y \le y_2\}] = &P[\{x_1 < X \le x_2, y_1 < Y \le y_2\}] \\ &+ P[\{X \le x_2, Y \le y_1\} \cup \{X \le x_1, Y \le y_2\}]\end{aligned} \tag{4.39}$$

By Theorem 3.2.3 given by (3.6), we can write the second term of the right-hand side of (4.39) as follows:

$$\begin{aligned}&P[\{X \le x_2, Y \le y_1\} \cup \{X \le x_1, Y \le y_2\}] \\ &= P[\{X \le x_2, Y \le y_1\}] + P[\{X \le x_1, Y \le y_2\}] \\ &\quad - P[\{X \le x_2, Y \le y_1\} \cap \{X \le x_1, Y \le y_2\}]\end{aligned} \tag{4.40}$$

Referring to Fig. 4.11b, we can write the intersection of the two sets in the last term of (4.40) as follows:

$$\{X \le x_1, Y \le y_2\} \cap \{X \le x_2, Y \le y_1\} = \{X \le x_1, Y \le y_1\} \tag{4.41}$$

with the corresponding probability as follows:

$$P[\{X \le x_1, Y \le y_2\} \cap \{X \le x_2, Y < y_1\}] = P[\{X \le x_1, Y \le y_1\}] \tag{4.42}$$

Substituting (4.42) into (4.40) and the resultant into (4.39), we can write (4.39) as follows:

$$\begin{aligned} P[\{X \le x_2, Y \le y_2\}] = {} & P[\{x_1 < X \le x_2, y_1 < Y \le y_2\}] + P[\{X \le x_1, Y \le y_2\}] \\ & + P[\{X \le x_2, Y \le y_1\}] - P[\{X \le x_1, Y \le y_1\}] \end{aligned} \tag{4.43}$$

Using the definition of the joint CDF given by (4.31), we rewrite (4.43) as follows:

$$\begin{aligned} F_{XY}(x_2, y_2) = {} & P[\{x_1 < X \le x_2, y_1 < Y \le y_2\}] \\ & + F_{XY}(x_1, y_2) + F_{XY}(x_2, y_1) - F_{XY}(x_1, y_1) \end{aligned} \tag{4.44}$$

Rearranging the terms of (4.44), we obtain the following equation:

$$\begin{aligned} P[\{x_1 < X \le x_2, y_1 < Y \le y_2\}] = {} & F_{XY}(x_2, y_2) - F_{XY}(x_1, y_2) \\ & - F_{XY}(x_2, y_1) + F_{XY}(x_1, y_1) \end{aligned}$$

Q.E.D.

Property 3
The event $\{X \le +\infty, Y \le +\infty\}$ is a certain event, and the following three events are impossible events:

$$\{X \le -\infty, Y \le y\} \qquad \{X \le x, Y \le -\infty\} \qquad \{X \le -\infty, Y \le -\infty\}$$

Therefore, we have the following properties of the joint CDF:

$$F_{XY}(+\infty, +\infty) = P[\{X \le +\infty, Y \le +\infty\}] = 1 \tag{4.45}$$

$$F_{XY}(-\infty, y) = P[\{X \le -\infty, Y \le y\}] = 0 \tag{4.46}$$

$$F_{XY}(x, -\infty) = P[\{X \le x, Y \le -\infty\}] = 0 \tag{4.47}$$

$$F_{XY}(-\infty, -\infty) = P[\{X \le -\infty, Y \le -\infty\}] = 0 \tag{4.48}$$

4.3.2 *Joint* pdf *of* X *and* Y

Referring to Fig. 4.12a, consider the probability that the pair of RVs (X, Y) would fall in a small rectangle of sides Δx and Δy. Divide this probability by the area of the rectangle, $\Delta x \Delta y$, to represent the normalized value, "density," as follows:

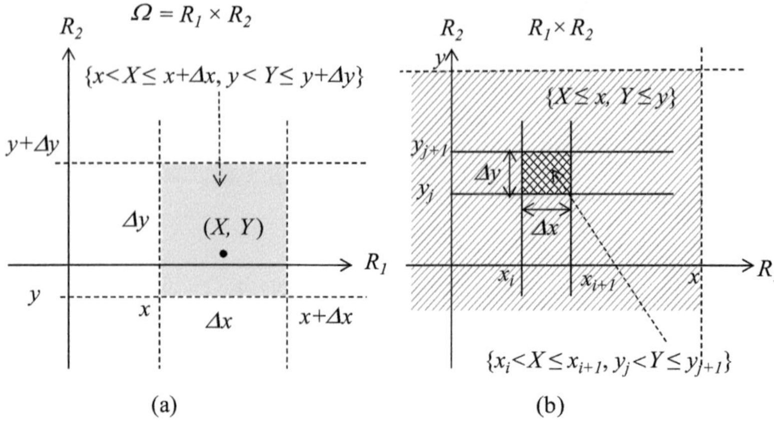

Fig. 4.12 (**a**) A small rectangle of sides Δx and Δy, (**b**) $\{x_i < X \leq x_i + \Delta x,\ y_i < Y \leq y_i + \Delta y\}$

$$\frac{P[\{x < X < x + \Delta x, y < Y < y + \Delta y\}]}{\Delta x \Delta y} \tag{4.49}$$

The joint *pdf* of X and Y is defined as the limiting value of (4.49) as $\Delta x \to 0$, $\Delta y \to 0$ as follows:

$$f_{XY}(x,y) \triangleq \lim_{\Delta x \to 0\ \Delta y \to 0} \frac{1}{\Delta x \Delta y} P[\{x < X \ \ (x + x), y < Y < (y + \Delta y)\}] \tag{4.50}$$

Given the definition of the joint *pdf* by (4.50), we have the following relationship:

$$P[\{x < X \leq (x + \Delta x), y < Y \leq (y + \Delta y)\}] \approx f_{XY}(x,y)\Delta x \Delta y \tag{4.51}$$

Note that the joint *pdf* is defined by (4.50) independently of the definition of the joint CDF given by (4.31). The next theorem shows the relationship between the two independent definitions.

Theorem 4.3.1 Given the joint CDF $F_{XY}(x, y)$, defined by (4.31), the joint *pdf* $f_{XY}(x, y)$, defined by (4.50), can be obtained by taking the second-order partial derivative of the CDF as follows:

$$f_{XY}(x,y) = \frac{\partial}{\partial y}\left(\frac{\partial F_{XY}(x,y)}{\partial x}\right) = \frac{\partial^2 F_{XY}(x,y)}{\partial x \partial y} \tag{4.52}$$

Proof Substitute the following into (4.37):

$$x_1 = x \qquad x_2 = x + \Delta x \qquad y_1 = y \qquad y_2 = y + \Delta y$$

to obtain the following equation:

$$\begin{aligned} &P[\{x < X < (x + \Delta x), y < Y < (y + \Delta y)\}] \\ &\quad = F_{XY}(x + \Delta x, y + \Delta y) - F_{XY}(x, y + \Delta y) - F_{XY}(x + \Delta x, y) + F_{XY}(x, y) \end{aligned} \tag{4.53}$$

Substituting (4.53) into the defining equation of the joint *pdf* (4.50), we obtain the following equation:

$$\begin{aligned} f_{XY}(x, y) \triangleq \lim_{\Delta x \to 0 \Delta y \to 0} \frac{1}{\Delta x \Delta y} [&F_{XY}(x + \Delta x, y + \Delta y) - F_{XY}(x, y + \Delta y) \\ &- F_{XY}(x + \Delta x, y) + F_{XY}(x, y)] \end{aligned} \tag{4.54}$$

On the other hand, in differential calculus, the partial derivative of $F_{XY}(x, y)$ with respect to x keeping y fixed is given by:

$$\frac{\partial F_{XY}(x, y)}{\partial x} = \lim_{\Delta x \to 0} \frac{F_{XY}(x + \Delta x, y) - F_{XY}(x, y)}{\Delta x} \tag{4.55}$$

Now, taking the partial derivative of (4.55) with respect to y keeping x fixed, we obtain the following equation:

$$\begin{aligned} \frac{\partial}{\partial y}\left(\frac{\partial F_{XY}(x, y)}{\partial x}\right) &= \frac{\partial^2 F_{XY}(x, y)}{\partial x \partial y} \\ &= \lim_{\Delta y \to 0} \frac{1}{\Delta y}\left[\lim_{\Delta x \to 0} \frac{1}{\Delta x}[\{F_{XY}(x + \Delta x, y + \Delta y) - F_{XY}(x, y + \Delta y)\}\right. \\ &\quad \left. - \{F_{XY}(x + \Delta x, y) - F_{XY}(x, y)\}]\right] \\ &= \lim_{\Delta x \to 0\ \Delta y \to 0} \frac{1}{\Delta x \Delta y}[F_{XY}(x + \Delta x, y + \Delta y) - F_{XY}(x, y + \Delta y) \\ &\quad - F_{XY}(x + \Delta x, y) + F_{XY}(x, y)] \end{aligned} \tag{4.56}$$

By comparing (4.56) with (4.54), we see that the left-hand sides of the two equations are equal as follows:

$$f_{XY}(x, y) \triangleq \frac{\partial^2 F_{XY}(x, y)}{\partial x \partial y}$$

Q.E.D.

Theorem 4.3.2 Given the joint *pdf* $f_{XY}(x,y)$, the joint CDF $F_{XY}(x,y)$, defined by (4.31), can be obtained by taking the double integral of the *pdf* as follows:

$$F_{XY}(x,y) = \int_{-\infty}^{y}\int_{-\infty}^{x} f_{XY}(\lambda,\delta)d\lambda\, \mathrm{d}\delta \tag{4.57}$$

This theorem is the converse of Theorem 4.3.1.

Proof The proof can be given simply by taking the reverse process of Theorem 4.3.1 given by (4.52), which has been derived from the definition of the CDF. In the following discussion, however, (4.57) is derived directly from the definition of the joint *pdf* given by (4.50).

Divide the subset $\{X \le x,\ Y \le y\}$ of $R_1 \times R_2$ into small mutually exclusive rectangular sets with sides Δx and Δy, $\{x_i < X \le x_i + \Delta x,\ y_i < Y \le y_i + \Delta y\}$, shown by the crosshatched area in Fig. 4.12b.

The joint CDF of RVs X and Y defined by (4.31) is given by the following double summation:

$$\begin{aligned} F_{XY}(x,y) &= P[\{X \le x, Y \le y\}] \\ &= P[\{x_1 < X < x_1 + \Delta x, y_1 < Y \le y_1 + \Delta y\} \cup \ldots \\ &\quad \cup \left\{x_i < X \le x_i + \Delta x, y_j < Y \le y_j + \Delta y\right\} \cup \ldots] \\ &= \sum\nolimits_j \sum\nolimits_i P\left[\left\{x_i < X \le x_i + \Delta x, y_j < Y \le y_j + \Delta y\right\}\right] \end{aligned} \tag{4.58}$$

Substituting (4.51) into (4.58), we obtain the following approximation of the joint *pdf*:

$$F_{XY}(x,y) \approx \sum_j \sum_i f_{XY}\left(x_i, y_j\right) \Delta x \Delta y \tag{4.59}$$

In the limit as $\Delta x \to 0$, $\Delta y \to 0$, the summation becomes an integral as follows:

$$F_{XY}(x,y) = \lim_{\Delta x \to 0,\, \Delta y \to 0} \sum\nolimits_j \sum\nolimits_i f_{XY}\left(x_i, y_j\right) \Delta x \Delta y = \int_{-\infty}^{y}\int_{-\infty}^{x} f_{XY}(\lambda,\delta)\mathrm{d}\lambda\ \mathrm{d}\delta$$

Q.E.D.

Theorem 4.3.3

$$f_X(x) = \int_{-\infty}^{+\infty} f_{XY}(x,y)\mathrm{d}y \tag{4.60a}$$

$$f_Y(y) = \int_{-\infty}^{+\infty} f_{XY}(x,y)\mathrm{d}x \tag{4.60b}$$

This theorem results from Theorem 4.3.2. $f_X(x)$ and $f_Y(y)$ are called the marginal *pdf*s of X and Y, respectively.

Proof In the Cartesian product $R_1 \times R_2$, we have the following relationship:

$$\{X \le x\} = \{X \le x, Y \le y\} \cup \{X \le x, Y > y\} = \{X \le x, Y \le +\infty\}$$

In (4.58), extend j to cover the entire real line R_2 of y from $-\infty$ to $+\infty$ so that the Δx-by-Δy cell becomes a strip of infinite length of width Δx as shown in Fig. 4.13a.

Then, in (4.58), the left-hand side becomes $F_X(x)$, and the summation over j in the right-hand side extends from $y = -\infty$ to $y = +\infty$. The index j extending the entire y axis is denoted by "j, $y = -\infty, +\infty$" in the following equation:

$$\begin{aligned} F_X(x) &= P[\{X \le x\}] \\ &= \sum_{j,y=\ -\infty,+\infty} \sum_i P\left[\left\{x_i < X \le (x_i + \Delta x), y_j < Y \le \left(y_j + \Delta y\right)\right\}\right] \end{aligned}$$

Using (4.50) for the probability in the double summation of the above equation and changing the order of the summation over i and j, we rewrite the above equation as follows:

$$\begin{aligned} F_X(x) &= \lim_{\Delta x \to 0,\, \Delta y \to 0} \sum_i \sum_{j,y\ =\ -\infty,+\infty} f_{XY}\left(x_i, y_j\right)\Delta y \Delta x \\ &= \lim_{\Delta x \to 0} \sum_i \left\{ \lim_{\Delta y \to 0} \sum_{j,y=-\infty,+\infty} f_{XY}\left(x_i, y_j\right)\Delta y \right\} \Delta x \\ &= \lim_{\Delta x \to 0} \sum_i \left\{ \int_{-\infty}^{+\infty} f_{XY}(x_i, y)\mathrm{d}y \right\} \Delta x \\ &= \int_{-\infty}^{x} \left\{ \int_{-\infty}^{+\infty} f_{XY}(\lambda, y)\mathrm{d}y \right\} \mathrm{d}\lambda \end{aligned} \tag{4.61}$$

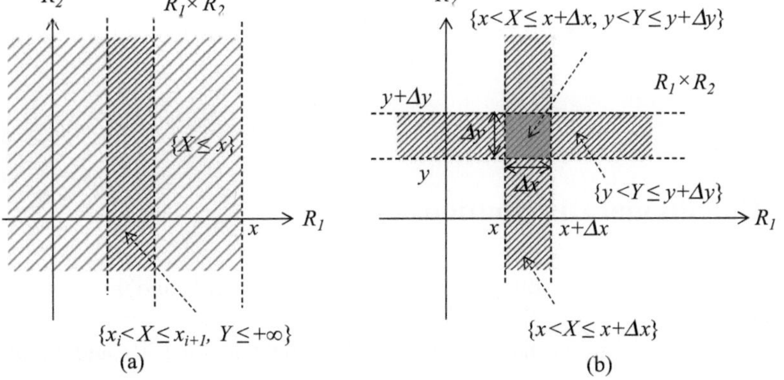

Fig. 4.13 (**a**) A strip of infinite length of width Δx, (**b**) $\{x < X \le x + \Delta x\} \cap \{y < Y \le y + \Delta y\} = \{x < X \le x + \Delta x, y < Y \le y + \Delta y\}$

Fig. 4.14 Joint *pdf* $f_{XY}(x, y)$

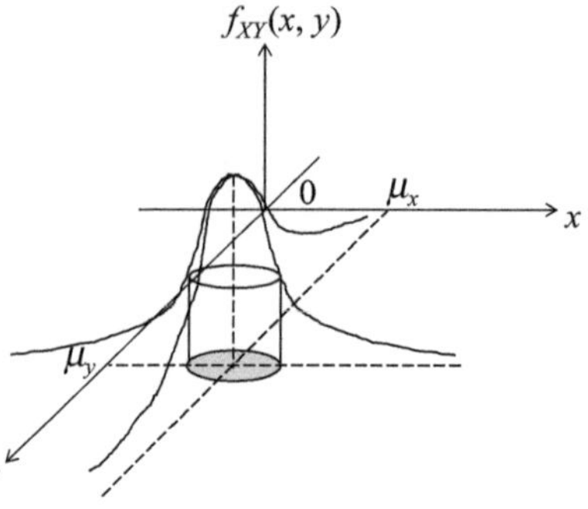

Since the left-hand sides of (4.61) and (4.20) are the same, by equating the right-hand sides of the two equations, we obtain the following equation:

$$\int_{-\infty}^{x} f_X(\lambda)d\lambda = \int_{-\infty}^{x} \left\{ \int_{-\infty}^{+\infty} f_{XY}(\lambda, y)\mathrm{d}y \right\} d\lambda$$

The above equation shows that the integrand of the left-hand side is equal to the integral in braces on the right-hand side as follows:

$$f_X(x) = \int_{-\infty}^{+\infty} f_{XY}(x, y)\mathrm{d}y$$

Q.E.D.

Equation of (4.60b) is proven similarly.

The plot of the joint *pdf* $f_{XY}(x, y)$ is a surface over the $x-y$ coordinate as illustrated in Fig. 4.14. One of the conditions that a joint *pdf* must satisfy is that the volume under the surface of $f_{XY}(x, y)$ must be equal to 1.

$$F_{XY}(+\infty, +\infty) = \int_{-\infty}^{+\infty} \int_{-\infty}^{+\infty} f_{XY}(x, y)\mathrm{d}x\mathrm{d}y = 1 \tag{4.62}$$

4.4 Conditional Distributions

This section defines the conditional CDF, the conditional *pdf*, and the independence of two RVs.

Consider RVs X and Y taking on the values on real lines R_1 and R_2 as their spaces, respectively, as follows:

$$\Omega_X = R_1 \qquad \Omega_Y = R_2 \tag{4.63}$$

Let A_X and B_Y be arbitrary events in R_1 and R_2, respectively, which are determined by the outcomes of X and Y, respectively, as follows:

$$A_X \in \Omega_X \qquad B_Y \in \Omega_Y \tag{4.64}$$

The conditional CDF of X given B_Y, denoted by $F_{X|Y}(x|B_Y)$, is defined as the conditional probability of $\{X \leq x\}$ given B_Y per the defining equation (3.16) as follows:

$$F_{X|Y}(x|B_Y) \triangleq P[\{X \leq x\}|B_Y] = \frac{P[\{X \leq x\} \cap B_Y]}{P(B_Y)} \quad \text{for } P(B_Y) \neq \{\varnothing\} \tag{4.65}$$

Similarly,

$$F_{Y|X}(y|A_X) \triangleq P[\{Y \leq y\}|A_X] = \frac{P[\{X \leq x\} \cap A_X]}{P(A_X)} \quad \text{for } P(A_X) \neq \{\varnothing\} \tag{4.66}$$

Theorem 4.3.4 Given two RVs X and Y, their individual CDFs, $F_X(x)$ and $F_Y(y)$, and their joint CDF, $F_{XY}(x, y)$, the conditional CDF of x given $\{Y \leq y\}$ is given by the following equation:

$$F_{X|Y}[x|\{Y \leq y\}] = \frac{F_{XY}(x, y)}{F_Y(y)} \tag{4.67}$$

Similarly,

$$F_{Y|X}[y|\{X \leq x\}] = \frac{F_{XY}(x, y)}{F_X(x)} \tag{4.68}$$

Proof Letting $B_Y = \{Y \leq y\}$ and substituting B_Y into (4.65), we obtain the following equation:

$$F_{X|Y}[x|\{Y \leq y\}] = \frac{P[\{X \leq x\} \cap \{Y \leq y\}]}{P[\{Y \leq y\}]} \tag{4.69}$$

In Fig. 4.15, the crosshatched area shows the following set relationship in the Cartesian product:

$$\{X \leq x\} \cap \{Y \leq y\} = \{X \leq x, Y \leq y\} \tag{4.70}$$

Substituting (4.70) into the set inside the probability of the numerator of (4.69), we obtain the following equation:

$$P[\{X \leq x\} \cap \{Y \leq y\}] = P[\{X \leq x, Y \leq y\}]$$

and rewrite (4.69) as follows:

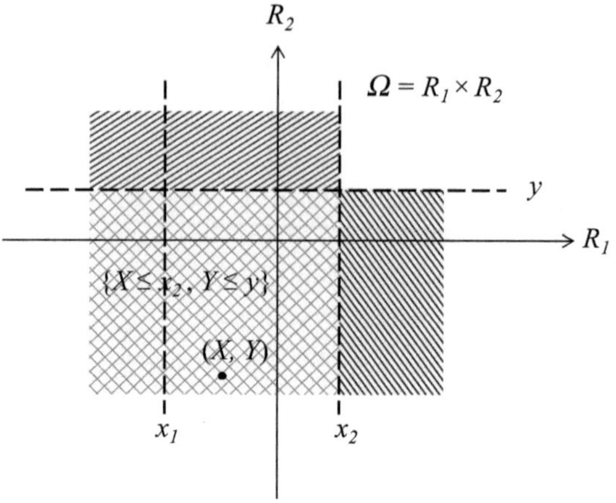

Fig. 4.15 $\{X \leq x\} \cap \{Y \leq y\} = \{X \leq x, Y \leq y\}$

$$F_{X|Y}(x|\{Y \leq y\}) = \frac{P[\{X \leq x, Y \leq y\}]}{P[\{Y \leq y\}]}$$

Using the definition of the CDF given by (4.4) and that of the joint CDF given by (4.31), we rewrite the above equation as follows:

$$F_{X|Y}(x|\{Y \leq y\}) = \frac{F_{XY}(x, y)}{F_Y(y)}$$

Q.E.D.

Equation (4.68) is proven similarly.

Theorem 4.3.5 Given two RVs X and Y, their individual CDFs, $F_X(x)$ and $F_Y(y)$, and their joint CDF, $F_{XY}(x, y)$, the conditional CDF of x given $\{y_1 < Y \leq y_2\}$ is given by the following equation:

$$F_{X|Y}(x|\{y_1 < Y \leq y_2\}) = \frac{F_{XY}(x, y_2) - F_{XY}(x, y_1)}{F_Y(y_2) - F_Y(y_1)} \tag{4.71}$$

$$F_{Y|X}(y|\{x_1 < X \leq x_2\}) = \frac{F_{XY}(x_2, y) - F_{XY}(x_1, y)}{F_X(x_2) - F_X(x_1)} \tag{4.72}$$

Proof

$$\begin{aligned} F_{X|Y}(x|\{y_1 < Y \leq y_2\}) &= P[\{X \leq x\}|\{y_1 < Y \leq y_2\}] \\ &= \frac{P[\{X \leq x, \ y_1 < Y \leq y_2\}]}{P[\{y_1 < Y \leq y_2\}]} \end{aligned} \tag{4.73}$$

Substituting (4.14) and (4.33) into (4.73), we obtain (4.71).

Q.E.D.

Equation (4.72) is proven similarly.

Conditional *pdf*

The conditional *pdf* of X with Y fixed at y, that is, $Y = y$, which is denoted by $f_{X|Y}(x|\, y)$, is defined by the following limiting value:

$$f_{X|Y}(x|y) = f_{X|Y}(x|\{Y = y\}) \triangleq \lim_{\Delta x \to 0,\, \Delta y \to 0} \frac{\left\{\frac{1}{\Delta x} P[\{x < X \leq x + \Delta x\}]\right\}}{P[\{y < Y \leq y + \Delta y\}]} \tag{4.74}$$

Theorem 4.3.6 The conditional *pdf* defined by (4.74) is obtained by the following equations:

$$f_{X|Y}(x|y) = \frac{f_{XY}(x, y)}{f_Y(y)} \tag{4.75a}$$

$$f_{Y|X}(y|x) = \frac{f_{XY}(x, y)}{f_X(x)} \tag{4.75b}$$

Proof From the definition given by (4.74), we have:

$$\begin{aligned} f_{X|Y}(x|y) &\triangleq \lim_{\Delta x \to 0,\, \Delta y \to 0} \frac{\left\{\frac{1}{\Delta x} P[\{x < X \leq x + \Delta x\}]\right\}}{P[\{y < Y \leq y + \Delta y\}]} \\ &= \lim_{\Delta x \to 0,\, \Delta y \to 0} \frac{\left\{\frac{1}{\Delta x} P[\{x < X \leq x + \Delta x\} \cap \{y < Y \leq y + \Delta y\}]\right\}}{P[\{y < Y \leq y + \Delta y\}]} \end{aligned} \tag{4.76}$$

Referring to Fig. 4.13b, we express the event in the numerator of (4.76) as follows:

$$\begin{aligned} &\{x < X \leq x + \Delta x\} \cap \{y < Y < y + \Delta y\} \\ &= \{x < X \leq x + \Delta x, y < Y \leq y + \Delta y\} \end{aligned} \tag{4.77}$$

Substituting (4.77) into (4.76), we have:

$$f_{X|Y}(x|y) = \lim_{\Delta x \to 0,\, \Delta y \to 0} \frac{\left\{\frac{1}{\Delta x} P[\{x < X \leq x + \Delta x, y < Y \leq y + \Delta y\}]\right\}}{P[\{y < Y \leq y + \Delta y\}]} \tag{4.78}$$

Substituting (4.53) and using (4.24) into (4.78), we obtain the following equation:

$$f_{X|Y}(x|y) = \lim_{\Delta x \to 0,\, \Delta y \to 0} \frac{\left\{\frac{1}{\Delta x}[F_{XY}(x+\Delta x, y+\Delta y) - F_{XY}(x+\Delta x, y) - F_{XY}(x, y+\Delta y) + F_{XY}(x,y)]\right\}}{F_Y(y+\Delta y) - F_Y(y)} \tag{4.79}$$

Divide the numerator and the denominator in (4.79) by Δy to obtain:

$$f_{X|Y}(x|y) = \lim_{\Delta x \to 0,\, \Delta y \to 0} \frac{\left\{\frac{1}{\Delta x \Delta y}[F_{XY}(x+\Delta x, y+\Delta y) - F_{XY}(x+\Delta x, y) - F_{XY}(x, y+\Delta y) + F_{XY}(x,y)]\right\}}{\left\{\frac{1}{\Delta y}[F_Y(y+\Delta y) - F_Y(y)]\right\}} \tag{4.80}$$

Equations (4.54) and (4.19) show that, as $\Delta x \to 0$ and $\Delta y \to 0$, (4.80) becomes (4.75a).

Q.E.D.

The second equation of (4.75b) is proven similarly.

4.4.1 *Independence of Two Random Variables*

The independence of events *A* and *B* has been defined by (3.29). The independence of RVs *X* and *Y* is defined by the events defined for *X* and *Y*. Let the spaces for *X* and *Y* be R_1 and R_2, respectively. Consider arbitrary subsets of R_1 and R_2 as follows:

$$A \in R_1 \qquad B \in R_2$$

Define two events in R_1 and R_2 as follows:

$$\{X \in A\} \qquad \{Y \in B\}$$

If these two events are independent, the RVs *X* and *Y* are said to be independent. If RVs *X* and *Y* are independent:

$$P[\{X \in A\} \cap \{Y \in B\}] = P[\{X \in A\}]\ P[\{Y \in B\}] \tag{4.81}$$

Theorem 4.3.7 If RVs *X* and *Y* are independent:

$$F_{XY}(x,y) = F_X(x) F_Y(y) \tag{4.82a}$$

$$f_{XY}(x,y) = f_X(x) f_Y(y) \tag{4.82b}$$

Proof In Fig. 4.15, we see that the following equation holds true.

$$\{X \le x, Y \le y\} = \{X \le x\} \cap \{Y \le y\} \tag{4.83}$$

By the definition of the joint CDF given by (4.31), we have the following equation:

$$F_{XY}(x, y) = P[\{X \le x, Y \le y\}]$$

Substituting (4.83) into the set inside the probability operator in the above equation, we obtain the following equation:

$$F_{XY}(x, y) = P[\{X \le x\} \cap \{Y \le y\}]$$

If RVs X and Y are independent, by (4.81), we can rewrite the above equation as follows:

$$F_{XY}(x, y) = P\big[\{X \le x\}\big]\, P\big[\,\{Y \le y\}\big] = F_X(x)F_Y(y)$$

This proves (4.82a).

Q.E.D.

Substituting (4.82a) into (4.52), we obtain the following equation:

$$\begin{aligned} f_{XY}(x, y) &= \frac{\partial^2 F_{XY}(x, y)}{\partial x \partial y} = \frac{\partial}{\partial y}\left\{\frac{\partial}{\partial x}F_X(x)F_Y(y)\right\} = \frac{\partial}{\partial y}\left\{F_Y(y)\frac{\partial}{\partial x}F_X(x)\right\} \\ &= f_X(x)f_Y(y) \end{aligned}$$

This proves (4.82b).

Q.E.D.

4.5 Functions of RVs

Given two RVs X and Y, define two new RVs W and Z by the functions $g(.)$ and $h(.)$ of X and Y as follows:

$$W = g(X, Y) \tag{4.84a}$$

$$Z = h(X, Y) \tag{4.84b}$$

The functions $g(.)$ and $h(.)$ transform X and Y as independent variables to W and Z as dependent variables. Denote the real lines of X, Y, W, and Z by R_X, R_Y, R_W, and R_Z. The spaces of (X, Y) and (W, Z), Ω_{XY} and Ω_{WZ}, are given by the Cartesian products of $R_X \times R_Y$ and $R_W \times R_Z$, respectively.

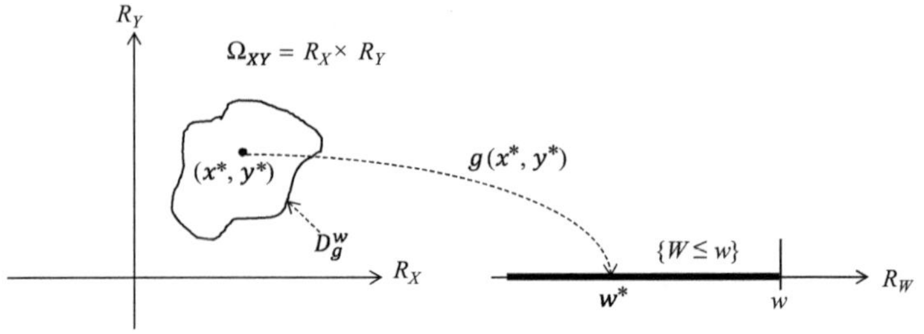

Fig. 4.16 Transformation of X and Y to W by the function $g(.,.)$

Figure 4.16 illustrates the transformation of X and Y to W by the function $g(.)$.

4.5.1 *CDFs of* W *and* Z

The distributions of W and Z are determined by the joint distribution of X and Y and the functions $g(.)$ and $h(.)$ as follows. $\{W \leq w\}$ denotes the set of real numbers less than or equal to w on R_W. $D_g{}^w$ denotes the set of ordered pairs (x, y) in the Cartesian product $R_X \times R_Y$ that are mapped by $g(.)$ to $\{W \leq w\}$ on R_W as follows:

$$\begin{aligned}\{W \leq w|w \in R_W\} &= \{g(X, Y) \leq w|w \in R_W\} \\ &= \left\{(x, y) \in D_g^w|\ (x, y) \in D_g^w \rightarrow g(x, y) \leq w\right\}\end{aligned} \tag{4.85}$$

By (4.4), the CDF of W is given by:

$$F_W\ (w) = P[\{W \leq w|w \in R_W\}] = \iint f_{XY}(x, y)\mathrm{d}x\mathrm{d}y \qquad \text{for } (x, y) \in D_g^w \tag{4.86}$$

Similarly,

$$F_Z\ (z) = P[\{Z \leq z|z \in R_Z\}] = \iint f_{XY}(x, y)\mathrm{d}x\mathrm{d}y \quad \text{for } (x, y) \in D_h^z \tag{4.87}$$

4.5.2 pdfs *of* W *and* Z

Figure 4.17 shows $\{w < W \leq w + \Delta w\}$, which is the set representing the event that W would fall in the small interval of length Δw between w and $w + \Delta w$. This set can also be represented as the intersection of $\{W \leq w\}$ and $\{W \leq w + \Delta w\}$. $D_g{}^{\Delta w}$ denotes the set of ordered pairs (x, y) in the Cartesian product $R_X \times R_Y$ that are

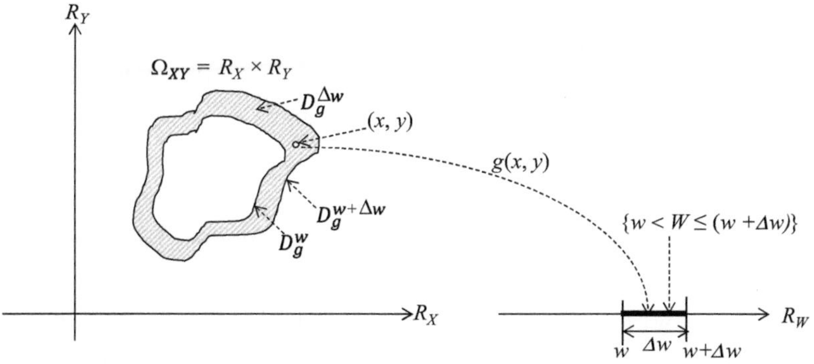

Fig. 4.17 Mapping of (x, y) to $\{w < W \leq w + \Delta w\}$ by gunction g(., .)

mapped by $g(.)$ to $\{w < W \leq w + \Delta w\}$, where the letter D denoting the set signifies that the set is in the domain of the function $g(., .)$.

Using (4.16), we can approximate the probability that RV W would fall in the small interval of length Δw as the product of the *pdf* of W and Δw as follows:

$$P[\{w < W \leq (w + \Delta w)\}] \approx f_W(w)\Delta w \tag{4.88}$$

As $\Delta w \to 0$, the left-hand side of (4.88) approaches the following integral:

$$\lim_{\Delta w \to 0} P[\{w \leq W \leq (w + \Delta w)\}] = \iint f_{XY}(x, y)\mathrm{d}x\mathrm{d}y \qquad \text{for } (x, y) \in D_g^{\Delta w} \tag{4.89}$$

Substituting (4.89) into (4.88) and taking $\Delta w \to 0$, we obtain the following equation:

$$\iint f_{XY}(x, y)\mathrm{d}x\mathrm{d}y = f_W(w)\mathrm{d}w \qquad \text{for } (x, y) \in D_g^{\Delta w} \tag{4.90}$$

Similarly,

$$\iint f_{XY}(x, y)\mathrm{d}x\mathrm{d}y = f_Z(z)\mathrm{d}z \qquad \text{for } (x, y) \in D_h^{\Delta z} \tag{4.91}$$

The CDF and the *pdf* of *W* Defined by a Function of One RV *X*

The RV W defined by a function of one RV X may be considered as a degenerate case of (4.84a) as follows:

$$W = g(X) \tag{4.92}$$

$\{W \leq w\}$ denotes the set of real numbers less than or equal to w on R_W. $D_g^{\,w}$ denotes the set of the real numbers on R_X that are mapped by $g(.)$ to $\{W \leq w\}$ on R_W.

$$\{W \le w | w \in R_W\} = \{g(X) \le w | w \in R_W\}$$
$$= \left\{x \in D_g^w | \, x \in D_g^w \rightarrow g(x) \le w\right\} \qquad (4.93)$$

By (4.4), the CDF of W is given by:

$$F_W(w) = P[\{W \le w | w \in R_W\}] = \int f_X(x) \mathrm{d}x \qquad \text{for } x \in D_g^w \qquad (4.94)$$

$\{w < W \le w + \Delta w\}$ is the set representing the event that W would fall in the small interval of length Δw between w and $w + \Delta w$, which can also be represented as the intersection of $\{W \le w\}$ and $\{W \le w + \Delta w\}$. ${D_g}^{\Delta w}$ denotes the set of x in R_X that are mapped by $g(.)$ to $\{w < W \le w + \Delta w\}$.

By (4.16), we have the following relation:

$$P[\{w < W \le (w + \Delta w)\}] \approx f_W(w) \Delta w \qquad (4.95)$$

As $\Delta w \rightarrow 0$, the left-hand side of (4.95) approaches the following integral:

$$\lim_{\Delta w \rightarrow 0} P[\{w \le W \le (w + \Delta w)\}] = \int f_X(x) \mathrm{d}x \qquad \text{for } x \in D_g^w \qquad (4.96)$$

4.5.3 *The Joint CDF of* W *and* Z

Figure 4.18 shows the set $\{W \le w, Z \le z\}$, which denotes the set of ordered pairs of random outcomes (W, Z) in $R_W \times R_Z$ with $W \le w$ and $Z \le z$. ${D_{gh}}^{wz}$ denotes the set of ordered pairs (x, y) in the Cartesian product $R_X \times R_Y$ that are mapped by $g(.,.)$ and $h(.,.)$ to $\{W \le w, Z \le z\}$ in $R_W \times R_Z$.

In Fig. 4.18, the shaded area on the right represents the following set:

$$\{W \le w, Z \le z | (w, z) \in R_W \times R_Z\} = \{g(X, Y) \le w, h(X, Y) \le z | (w, z) \in R_W \times R_Z\}$$
$$= \left\{(X, Y) \in D_{gh}^{wz} | \, (X, Y) \in D_{gh}^{wz} \rightarrow g(X, Y) \le w, h(X, Y) \le z\right\}$$
$$(4.97)$$

By (4.31), we have the following equation for the joint CDF of W and Z:

$$F_{WZ}(w, z) = P[\{W \le w, Z \le z\}] = \iint f_{XY}(x, y) \mathrm{d}x \mathrm{d}y \quad \text{where } (x, y) \in D_{gh}^{wz} \quad (4.98)$$

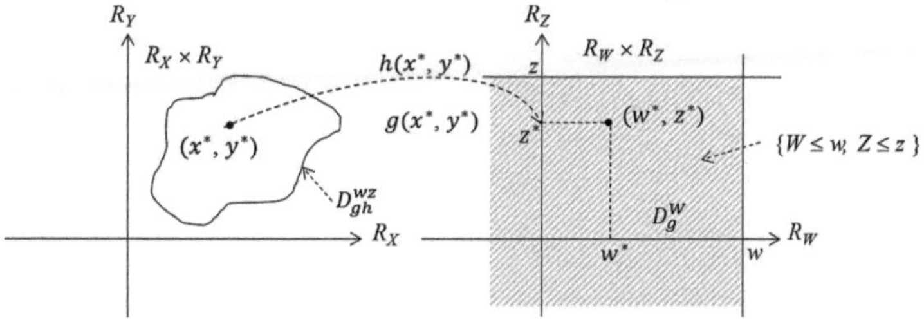

Fig. 4.18 Mapping of (x, y) to w by g(., .) and to z by h(., .)

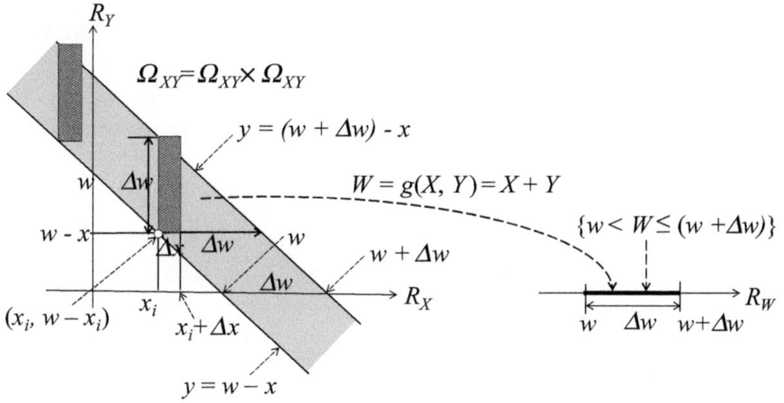

Fig. 4.19 Transformation of ordered pairs (X, Y) in $R_X \times R_Y$ to W in R_W by function $g(X, Y) = X + Y$

Theorem 4.5.1 If two RVs X and Y are independent, the *pdf* of the sum, $W = X + Y$, is given by the convolution of the *pdf*s of X and Y, $f_X(x)$ and $f_Y(y)$, as follows:

$$f_W(w) = f_X(x) * f_Y(x) = \int_{-\infty}^{+\infty} f_X(x) f_Y(w - x) \mathrm{d}x$$
$$= \int_{-\infty}^{+\infty} f_X(w - y) f_Y(y) \mathrm{d}y \tag{4.99}$$

This theorem is particularly important for linear systems analyses discussed in a later chapter. Often in physical systems, the random noise in a system appears as an independent additive noise to the signal, and the distribution of the signal plus the noise is analyzed using this theorem.

Proof Figure 4.19 illustrates the transformation of the ordered pairs (X, Y) in the Cartesian product $R_X \times R_Y$ to W in R_W by the function $W = g(X, Y) = X + Y$.

To determine the (X, Y) pairs that are transformed by $W = X + Y$ to a fixed point on R_W at $W = w$, solve $w = X + Y$ or $Y = w - X$. All the points on the line $y = w - x$

are transformed to the single point w. Similarly, all the points on the line $y = (w + \Delta w) - x$ are transformed to the point $(w + \Delta w)$. Therefore, all the points in the gray strip between these two lines are transformed to the points in the interval of length Δw from w to $w + \Delta w$. The probability that W would fall in this interval is equal to the probability that (X, Y) would fall in the gray strip.

Based on the geometry shown in Fig. 4.19, the horizontal and vertical widths of the gray strip are both equal to Δw. Divide this gray strip into small rectangles as shown by the hatched rectangles of width Δx and height Δw so that the area of the gray strip is approximated by the sum of the rectangles as x goes from $-\infty$ to $+\infty$. The probability that (X, Y) would fall in the rectangle with the base from x_i to $(x_i + \Delta x)$ and the height Δw is approximated as follows:

$$P\left[\{(X,Y)\in i\text{th rectangle}\right] \approx f_{XY}(x_i, y_i)\Delta x\Delta w = f_{XY}(x_i, w - x_i)\Delta x\Delta w$$

By summing the above expression over the index i to cover the entire gray strip, we obtain the following approximation to the probability that W would fall in the small interval:

$$\begin{aligned} P[\{w < W \le (w + \Delta w)\}] &= P[\{(X,Y)\in \text{grey strip}\}] \\ &\approx \sum_i f_{XY}(x_i, w - x_i)\Delta x\Delta w \end{aligned} \tag{4.100}$$

The above summation is an approximation of the probability because it assumes that the value of the joint *pdf* over the rectangle is fixed at the value at (x_i, y_i). Dividing both sides of (4.100) by Δw, we obtain the following relationship:

$$\frac{P[\{w < W \le (w + \Delta w)\}]}{\Delta w} \approx \sum_i f_{XY}(x_i, w - x_i)\Delta x \tag{4.101}$$

In (4.101), as $\Delta w \to 0$ and $\Delta x \to 0$, the left-hand side becomes the *pdf* of W, the right-hand side becomes the integral, and the approximation becomes the equality as follows:

$$\begin{aligned} f_W(w) &= \lim_{\Delta w \to 0,\, \Delta x \to 0} \frac{P[\{w < W \le (w + \Delta w)\}]}{\Delta w} \\ &= \int_{-\infty}^{+\infty} f_{XY}(x, w - x)\mathrm{d}x \end{aligned} \tag{4.102}$$

Finally, if X and Y are independent, their joint *pdf* is the product of the individual *pdf*s as follows:

$$f_{XY}(x, w - x) = f_X(x) f_Y(w - x) \tag{4.103}$$

Substituting (4.103) into (4.102), we obtain the following equation:

$$f_W(w) = \int_{-\infty}^{+\infty} f_X(x) f_Y(w - x)\mathrm{d}x$$

Q.E.D.

By creating the rectangles horizontally, the alternative form of the convolution can be proven.

Chapter 5
Characterization of Random Variables

An RV X is completely characterized by its CDF, $F_X(x)$ or, equivalently, the *pdf*, $f_X(x)$. The word "completely" means that the CDF $F_X(x)$ contains all the information necessary for calculating the probabilities of the events associated with the RV X and there is no additional information that can statistically augment the CDF. Similarly, two random variables taken together are completely characterized by their joint CDF, $F_{XY}(x, y)$, or joint *pdf*, $f_{XY}(x, y)$.

Certain parameters may be derived from the CDF or the *pdf* such as the mean, the variance, etc. If a CDF is available, these parameters can be derived, but the converse is not necessarily true; that is, given a mean and a variance of an RV X, the CDF of X cannot necessarily be constructed except in some important special cases to be discussed later.

5.1 Expected Value or Mean

The concept of the expected value of an RV is illustrated by an example first. Consider that, in a die-throwing experiment, a payoff is stipulated as \$10 times the number of dots produced by a trial. The space Ω of this experiment is

$$\Omega = \{1, 2, 3, 4, 5, 6\}$$

Let the probabilities of the elementary events be as follows:

$$P[\{i\}] = p_i = \frac{1}{6}, \quad i = \text{number of dots} \tag{5.1}$$

The payoff is an RV X defined by the following equation:

K.I. Park, *Fundamentals of Probability and Stochastic Processes with Applications to Communications*, https://doi.org/10.1007/978-3-319-68075-0_5

$$X = 10 \times i, \text{for } i = 1, 2, \ldots, 6 \quad (5.2)$$

Before throwing a die, how much payoff can a participant expect? As a measure of gauging the expectation, a quantity, called the "expected value," is defined as the weighted sum of the payoffs of individual outcomes, where $p_i s$ are used as the weights as follows:

$$\sum_{i=1}^{6} (10 \times i)p_i = 35$$

The expected value of this die-throwing experiment is \$35. This quantity is called the "expected value" or the mean.

Mean of Discrete RV
Let a discrete RV X have the following probabilities:

$$P[\{X = x_i\}] = p_i, \quad i = 1, \ldots$$

Denote the expectation operator by $E(.)$ The expected value of X is defined by the following equation:

$$E(X) \triangleq \sum_i x_i p_i \quad (5.3)$$

The above definition can be expressed in terms of the following integral:

$$E(X) \triangleq \int_{-\infty}^{+\infty} x f_X(x) \mathrm{d}x \quad (5.4)$$

The analytical expression of the *pdf* of the discrete RV X is given by (4.25). Substituting (4.25) into the above expression, we obtain the following equation:

$$E(X) = \int_{-\infty}^{+\infty} x \left\{ \sum_i p_i \delta(x - x_i) \right\} \mathrm{d}x = \sum_i x_i p_i \quad (5.5)$$

Mean of Continuous RV
For the case of a discrete RV X, the concept of its expected value could be more immediately grasped as demonstrated by the payoff example of a die-throwing experiment, which was then naturally extended to the general definition of the expected value. For the case of a continuous RV X, however, it is harder to grasp the significance of its expected value. In fact, we start with an approximation of the expected value of a continuous RV X by discretization and take its limiting case to obtain the expected value of the continuous RV X.

Fig. 5.1 Approximation of $P[x_i < X \leq x_{i+1}]$ by $f_X(x_i)\Delta x$

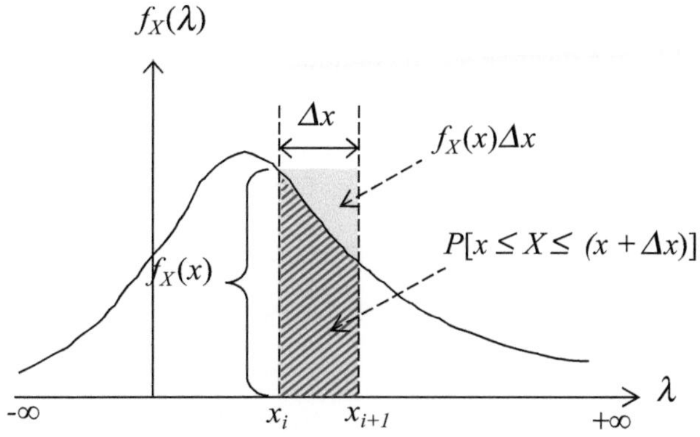

First, divide the real line R from $-\infty$ to $+\infty$ into small intervals of length Δx by placing discrete points x_i, x_{i+1}, ... Let the payoff value stay constant at x_i if the RV X falls in the interval of from x_i to x_{i+1}. Recalling the case of the discrete RV, the expected value of this case is given by the sum of x_is weighted by the probabilities that RV X would fall in the small intervals of length Δx. This is illustrated by Fig. 5.1.

Given the *pdf*, $f_X(x)$, the exact probability that the RV X would fall in the small interval of from x_i to x_{i+1} is the integral of $f_X(x)$ from x_i to x_{i+1}, which is equal to the area under the *pdf* curve shown by the hatched area in the figure. For a small interval of Δx, this exact area can be approximated by the rectangle with the height $f_X(x)$ and the base Δx, that is,

$$P[x_i < X \leq x_{i+1}] \approx f_X(x_i)\Delta x$$

Since RV X is assumed to be constant at x_i over this interval, the expected value of X over this interval can be approximated by the product of x_i and the approximate probability that X will fall in this interval by

$$x_i f_X(x_i)\Delta x$$

The expected value of the continuous RV X can be approximated by

$$E(X) \approx \sum_i x_i f_X(x_i)\Delta x \tag{5.6}$$

In the limit as $\Delta x \to 0$, (5.6) becomes the following integral:

$$\mu_X \triangleq E(X) = \int_{-\infty}^{+\infty} x f_X(x)\mathrm{d}x \tag{5.7}$$

Equation (5.7) defines the expected value or the mean, commonly denoted by μ, of both a discrete and a continuous RV.

Mean of Complex RV
For the complex RV X defined by (4.2), the mean is given by the means of the real and imaginary components as follows:

$$\mu_X = E(X) = E(X_r + jX_i) = E(X_r) + jE(X_i) = \mu_{X_r} + j\mu_{X_i} \tag{5.8}$$

where

$$\mu_{X_r} = \int_{-\infty}^{+\infty} x f_{X_r}(x) \mathrm{d}x \tag{5.9a}$$

$$\mu_{X_i} = \int_{-\infty}^{+\infty} x f_{X_i}(x) \mathrm{d}x \tag{5.9b}$$

where $f_{X_r}(x)$ and $f_{X_i}(x)$ are the *pdfs* of the real and imaginary components of X.

The mean of the complex conjugate of X, defined by (4.3), is the complex conjugate of the mean of X as given by the following equation:

$$\begin{aligned} \mu_X^* &= E(X^*) \\ &= E(X_r - jX_i) \\ &= E(X_r) - jE(X_i) \\ &= \mu_{X_r} - j\mu_{X_i} \end{aligned} \tag{5.10}$$

From the definition given by (5.7), the following equations can be derived. For a fixed value x' of an RV X, that is, $X = x'$, its expected value should be the fixed value itself

$$E(X) = x' \tag{5.11}$$

Proof

$$E(x') = \int_{-\infty}^{+\infty} x' f_X(x) \mathrm{d}x = x' \int_{-\infty}^{+\infty} f_X(x) \mathrm{d}x$$

Since, by (4.29), the integral in the above equation is equal to 1, we have the following result:

$$E(x') = x'$$

Q.E.D.

Similarly, the expected value of a constant is the constant itself, which can be proven by the same method as above.

$$E(\alpha) = \alpha, \quad \alpha = \text{constant} \tag{5.12}$$

Another rule that follows from the definition of the expectation operation is that, if an RV is modified by multiplying it by a constant, the expected value of the product is equal to the expected value of the original RV multiplied by the constant. To show this intuitive concept, take the same example of the coin-tossing payoff considered earlier. If the payoff is doubled on a different day, then the new expected payoff would be obtained by simply doubling the original payoff. In terms of the expectation operation, this rule is as follows:

$$E(\alpha X) = \alpha E(X), \quad \alpha = \text{constant} \tag{5.13}$$

Proof

$$E(\alpha X) = \int_{-\infty}^{+\infty} \alpha x f_X(x)\mathrm{d}x = \alpha \int_{-\infty}^{+\infty} x f_X(x)\mathrm{d}x = \alpha E(X)$$

Q.E.D.

It is useful for subsequent analyses to summarize several frequently used rules of the expectation operation by the following five theorems.

Theorem 5.1.1 Given $W = g(X)$ and the *pdf* of X, $f_X(x)$,

$$E[W] = \int_{-\infty}^{+\infty} g(x) f_X(x)\mathrm{d}x \tag{5.14}$$

By (5.4), the expected value of W is given by

$$E[W] = \int_{-\infty}^{+\infty} w f_W(w)\mathrm{d}w \tag{5.15}$$

Equation (5.14) allows the expected value of W to be obtained directly from $g(.)$ and $f_X(x)$ without having to obtain $f_W(w)$ as in (5.15).

Proof By (4.16), the probability of the event $\{w < W \leq w + \Delta w\}$ is approximated as follows:

$$P[w < W \leq (w + \Delta w)] \approx f_w(w)\Delta w \tag{5.16}$$

Multiply both sides of the above equation by w as follows:

$$wP[w < W \leq (w + \Delta w)] \approx w f_w(w)\Delta w \tag{5.17}$$

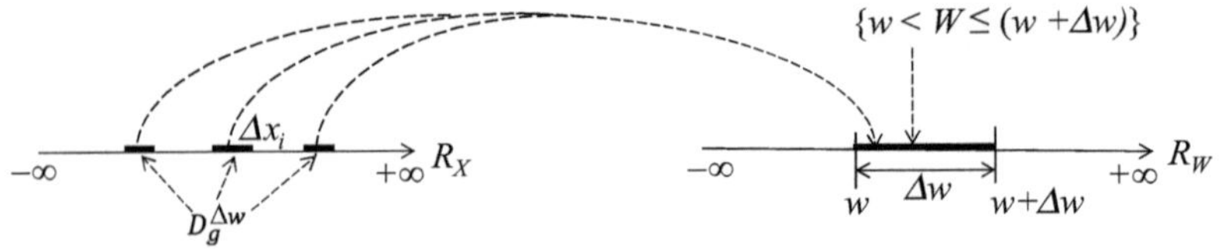

Fig. 5.2 Mapping of $Dg^{\Delta w}$on R_X to $\{w < W \leq w + \Delta w\}$ in R_W

In Fig. 5.2, $Dg^{\Delta w}$on R_X is the sum of the small intervals which are all mapped by $g(.)$ to the interval of length Δw from w to $(w + \Delta w)$ on R_W. Several intervals are shown for $Dg^{\Delta w}$ in the figure to illustrate that there could be multiple intervals on R_x that are mapped to the same single interval on R_W.

Referring to Fig. 5.2, we have the following relationship:

$$wf_w(w)\Delta w \approx \sum_{D_g^{\Delta w}} g(x) f_X(x_i)\Delta x_i \tag{5.18}$$

As Δw $\rightarrow$ 0, (5.18) becomes

$$wf_w(w)\mathrm{d}w = \int g(x) f_X(x_i)\mathrm{d}x_i \qquad \text{for } x_i \in D_g^{\Delta w} \tag{5.19}$$

Take the integral of both sides of (5.19). As w goes from $-\infty$ to $+\infty$, the right-hand side of (5.19) sums up all the infinitesimal intervals of dx_is, and the left-hand side becomes the expected value of W as follows:

$$E(W) = \int_{-\infty}^{+\infty} wf_w(w)\mathrm{d}w = \int_{-\infty}^{+\infty} g(x) f_X(x)\mathrm{d}x$$

Q.E.D.

Theorem 5.1.2 Given $W = g(X, Y)$ and the joint *pdf* of X and Y, $f_{xy}(x, y)$,

$$E[W] = \int_{-\infty}^{+\infty}\int_{-\infty}^{+\infty} g(x, y) f_{XY}(x, y)\mathrm{d}x\mathrm{d}y \tag{5.20}$$

Proof See the similar proof given for Theorem 5.1.1.

Theorem 5.1.3

$$E(X + Y) = E(X) + E(Y) \tag{5.21}$$

Proof Let

$$W = g(X, Y) = X + Y$$

By Theorem 5.1.2, we have the following equation:

$$
\begin{aligned}
E[W] &= \int_{-\infty}^{+\infty}\int_{-\infty}^{+\infty}(x+y)f_{XY}(x,y)\mathrm{d}x\mathrm{d}y \\
&= \int_{-\infty}^{+\infty}\int_{-\infty}^{+\infty}xf_{XY}(x,y)\mathrm{d}x\mathrm{d}y + \int_{-\infty}^{+\infty}\int_{-\infty}^{+\infty}yf_{XY}(x,y)\mathrm{d}x\mathrm{d}y \\
&= \int_{-\infty}^{+\infty}x\left\{\int_{-\infty}^{+\infty}f_{XY}(x,y)\mathrm{d}y\right\}\mathrm{d}x + \int_{-\infty}^{+\infty}y\left\{\int_{-\infty}^{+\infty}f_{XY}(x,y)\mathrm{d}x\right\}\mathrm{d}y \\
&= \int_{-\infty}^{+\infty}x\left\{\int_{-\infty}^{+\infty}f_{XY}(x,y)\mathrm{d}y\right\}\mathrm{d}x + \int_{-\infty}^{+\infty}y\left\{\int_{-\infty}^{+\infty}f_{XY}(x,y)\mathrm{d}x\right\}\mathrm{d}y
\end{aligned}
$$

Substituting (4.60b) into the above equation, we obtain the following equation:

$$
E(W) = \int_{-\infty}^{+\infty}xf_X(x)\mathrm{d}x + \int_{-\infty}^{+\infty}yf_Y(y)\mathrm{d}y = E(X) + E(Y)
$$

Q.E.D.

Theorem 5.1.4

$$
E\left\{\sum_{i=1}^{n}X_i\right\} = E(X_1) + E(X_2) + \ldots + E(X_n) = \sum_{i=1}^{n}E(X_i) \tag{5.22}
$$

Proof The proof can be given by mathematical induction as follows. Let

$$
Y = X_1 + X_2
$$

Then, by Theorem 5.1.3, we have the following equation:

$$
E(Y) = E(X_1 + X_2) = E(X_1) + E(X_2)
$$

For $n = 3$, we have $X_1 + X_2 + X_3 = Y + X_3$. Therefore, we have the following equation:

$$
E(X_1 + X_2 + X_3) = E(Y + X_3)
$$

Applying Theorem 5.1.3 to the above equation again, we obtain the following equation:

$$
E(X_1 + X_2 + X_3) = E(Y) + E(X_3) = E(X_1) + E(X_2) + E(X_3)
$$

Continuing, thus, we can prove the theorem.

$$E(X_1 + X_2 + \ldots + X_n) = E(X_1) + E(X_2) + \ldots + E(X_n)$$

Theorem 5.1.5 If X and Y are independent,

$$E(XY) = E(X)E(Y)$$

Proof By (5.20), we have the following equation:

$$E(XY) = \int_{-\infty}^{+\infty} \int_{-\infty}^{+\infty} xyf_{XY}(x, y)\mathrm{d}x\mathrm{d}y$$

Substituting (4.82b) into the above equation, we obtain the following equation:

$$\begin{aligned} E(XY) &= \int_{-\infty}^{+\infty} \int_{-\infty}^{+\infty} xyf_X(x)f_Y(y)\mathrm{d}x\mathrm{d}y \\ &= \int_{-\infty}^{+\infty} xf_X(x)\mathrm{d}x \int_{-\infty}^{+\infty} yf_Y(y)\mathrm{d}y = E(X)E(Y) \end{aligned}$$

Q.E.D.

Theorem 5.1.6 If X and Y are independent

$$E[g(X)h(Y)] = E[g(X)]E[h(Y)]$$

Proof Let

$$W = G(X, Y) = g(X)h(Y) = E[g(X)]E[h(Y)]$$

By (5.20), we have the following equation:

$$E[g(X)h(Y)] = E[G(X, Y)] = \int_{-\infty}^{+\infty} \int_{-\infty}^{+\infty} g(X)h(Y)f_{XY}(x, y)\mathrm{d}x\mathrm{d}y$$

Substituting (4.82b) into the above equation, we obtain the following equation:

$$\begin{aligned} E[g(X)h(Y)] &= \int_{-\infty}^{+\infty} \int_{-\infty}^{+\infty} g(X)h(Y)f_X(x)f_Y(y)\mathrm{d}x\mathrm{d}y \\ &= \int_{-\infty}^{+\infty} g(X)f_X(x)\mathrm{d}x \int_{-\infty}^{+\infty} h(Y)f_Y(y)\mathrm{d}y = E[g(X)E[h(Y)] \end{aligned}$$

Q.E.D.

5.2 Variance

The variance of an RV X is a measure of the variability of X. The variance of X is written Var(X), which is commonly denoted by σ^2. The variance of RV X is defined as the expected value of the square of the difference between X and its mean, $(X-\mu)^2$, as follows:

$$\mathrm{Var}(X) = \sigma^2 \triangleq E\left[(X-\mu)^2\right] = \int_{-\infty}^{+\infty} (X-\mu)^2 f_X(x)\mathrm{d}x \quad (5.23)$$

$(X-\mu)$ is an RV representing the random deviation of X from its mean. The difference is squared to make the negative and positive difference the same for the same magnitude.

For a discrete RV X, substituting (4.25) into (5.23), we obtain the following equation:

$$\mathrm{Var}(X) \triangleq \sum_i (x_i - \mu)^2 \, p_i \quad (5.24)$$

In estimation theory, the difference between the RV and its mean is called the error. Because the error is squared, the variance can be used as a measure of the magnitude of the error regardless of the sign of the difference. The square root of the variance is called the standard deviation, sometimes abbreviated as S.D., and is denoted by σ

$$\text{S.D.of } X = \sigma \triangleq \sqrt{\sigma^2} \quad (5.25)$$

Theorem 5.2.1

$$\mathrm{Var}\ (X) \triangleq E\left[(X-\mu)^2\right] = E\left(X^2\right) - \mu^2 \quad (5.26)$$

Proof

$$\begin{aligned}\mathrm{Var}\ (X) &= E\left[(X-\mu)^2\right] = E\left[X^2 - 2\mu X + \mu^2\right] \\ &= E\left(X^2\right) - 2\mu^2 + \mu^2 = E\left(X^2\right) - \mu^2\end{aligned}$$

Q.E.D.

In (5.26), the expected value of X^2, the first term on the right-hand side, is called the second moment of the RV X. If X has zero mean, the variance is equal to the second moment.

Variance of Complex RV

The variance of a complex RV X is defined as the following expected value:

$$\sigma_X{}^2 = \mathrm{Var}(X) \triangleq E\left[(X-\mu)(X-\mu)^*\right] \quad (5.27)$$

In the above definition, the product inside the expectation operator is formed between the error and its complex conjugate so that the product becomes, by (2.15), the square of the magnitude of the error. Therefore, the definition of the variance of a complex RV given by the above equation is equivalent to the following definition:

$$\sigma_X^2 \triangleq E\left[(X - \mu_X)^2\right] \tag{5.28}$$

By comparing (5.28) and (5.23), we see that the definition of the variance is consistent for the real and complex RV. Substituting (4.2) and (5.8) into (5.28), we obtain the following equation:

$$\begin{aligned} {\sigma_X}^2 &= E\left[\left(\left|X_r + jX_i - \mu_{X_r} - j\mu_{X_i}\right|\right)^2\right] = E\left[\left(\left|(X_r - \mu_{X_r}) + j(X_i - \mu_{X_i})\right|\right)^2\right] \\ &= E\left\{(X_r - \mu_{X_r})^2 + (X_i - \mu_{X_i})^2\right\} = E\left\{(X_r - \mu_{X_r})^2\right\} + E\left\{(X_i - \mu_{X_i})^2\right\} \\ &= {\sigma_{X_r}}^2 + {\sigma_{X_i}}^2 \end{aligned} \tag{5.29}$$

where, by (5.26), we have the following equations:

$${\sigma_{X_r}}^2 = E\left\{(X_r - \mu_{X_r})^2\right\} = E\{{X_r}^2\} - \mu_{X_r}^2 = \int_{-\infty}^{+\infty} x^2 f_{X_r}(x)\mathrm{d}x - \mu_{X_r}^2 \tag{5.30}$$

$${\sigma_{X_i}}^2 = E\left\{(X_i - \mu_{X_i})^2\right\} = E\{{X_i}^2\} - \mu_{X_i}^2 = \int_{-\infty}^{+\infty} x^2 f_{X_i}(x)\mathrm{d}x - \mu_{X_i}^2 \tag{5.31}$$

and $f_{X_r}(x)$ and $f_{X_i}(x)$ are the *pdf*s of the real and imaginary components of X. Equation (5.29) shows that the variance of a complex RV is the sum of the variances of its real and imaginary components.

Since the variance is a measure of variability, a fixed value of RV X, that is, $X = x'$, or a constant α should have no variability as follows:

$$\mathrm{Var}(x') = 0 \tag{5.32a}$$

$$\mathrm{Var}(\alpha) = 0 \tag{5.32b}$$

The above rule is consistent with the definition of the variance as follows:

$$\mathrm{Var}(\alpha) = E\left[(\alpha - E(\alpha))^2\right] = E\left[(\alpha - \alpha)^2\right] = 0$$

Q.E.D.

5.3 Covariance and Correlation Coefficient of Two Random Variables

Let the means, the variances and the standard deviations of RVs X and Y be $\mu_X, \sigma_X^2, \sigma_X$ and $\mu_Y, \sigma_Y^2, \sigma_Y$, respectively. The covariance of X and Y, denoted by c_{XY}, is defined by the following equation:

$$\text{Cov}\,(X, Y) = c_{XY} \triangleq E\{(X - \mu_X)\{(Y - \mu_Y)\} \tag{5.33}$$

Expanding the above equation, we obtain the following equation:

$$c_{XY} = E(XY - \mu_X Y - X\mu_Y + \mu_X\mu_Y) = E(XY) - \mu_X\mu_Y \tag{5.34}$$

Correlation Coefficient
The correlation coefficient of two RVs X and Y, denoted by ρ_{XY}, is defined by the following equation:

$$\rho_{XY} \triangleq \frac{c_{XY}}{\sqrt{\sigma_X^2\sigma_Y^2}} = \frac{c_{XY}}{\sigma_X\sigma_Y} \tag{5.35}$$

The correlation coefficient provides the normalized value of the covariance and its magnitude is less than or equal to 1.

Covariance of Complex RV
Given two complex RVs

$$X = X_r + jX_i \qquad Y = Y_r + jY_i \tag{5.36}$$

their covariance is defined by

$$c_{XY} \triangleq E\{(X - \mu_X)(Y - \mu_Y)^*\} \tag{5.37}$$

Using (2.10), we move the complex conjugate operation inside the parentheses and rewrite the above equation as follows:

$$\begin{aligned} c_{XY} &= E\{(X - \mu_X)(Y^* - \mu_Y^*)\} = E(XY^* - \mu_X Y^* - X\mu_Y^* + \mu_X\mu_Y^*) \\ &= E(XY^*) - \mu_X\mu_Y^* \end{aligned} \tag{5.38}$$

Using (5.8) and its complex conjugate, we obtain the following equation:

$$\mu_X\mu_Y^* = (\mu_{X_r} + j\mu_{X_i})(\mu_{Y_r} - j\mu_{Y_i}) = \mu_{X_r}\mu_{Y_r} + \mu_{X_i}\mu_{Y_i} + j\mu_{X_i}\mu_{Y_r} - j\mu_{X_r}\mu_{Y_i} \tag{5.39}$$

Substituting the complex RVs defined by (5.36), we evaluate the following expectation operation:

$$\begin{aligned} E(XY^*) &= E\{(X_r + jX_i)(Y_r - jY_i)\} = E(X_rY_r + X_iY_i + jX_iY_r - jX_rY_i) \\ &= E(X_rY_r) + E(X_iY_i) + jE(X_iY_r) - jE(X_rY_i) \end{aligned} \tag{5.40}$$

Substituting (5.39) and (5.40) into (5.38), we obtain the following equation for the covariance of complex RVs X and Y:

$$\begin{aligned} c_{XY} &= E(X_rY_r) + E(X_iY_i) + jE(X_iY_r) - jE(X_rY_i) - \mu_{X_r}\mu_{Y_r} - \mu_{X_i}\mu_{Y_i} - j\mu_{X_i}\mu_{Y_r} + j\mu_{X_r}\mu_{Y_i} \\ &= \{E(X_rY_r) - \mu_{X_r}\mu_{Y_r}\} + \{E(X_iY_i) - \mu_{X_i}\mu_{Y_i}\} + j\{E(X_iY_r) - \mu_{X_i}\mu_{Y_r}\} \\ &\quad -j\{E(X_rY_i) - \mu_{X_r}\mu_{Y_i}\} \end{aligned}$$

Considering (5.38), we see that the quantities inside the braces in the above equation are the covariances between the real and imaginary components of X and Y and rewrite the above equation as follows:

$$\begin{aligned} c_{XY} &= c_{X_rY_r} + c_{X_iY_i} + jc_{X_iY_r} - jc_{X_rY_i} = (c_{X_rY_r} + c_{X_iY_i}) + j(c_{X_iY_r} - c_{X_rY_i}) \\ &= c_{XY}^r + jc_{XY}^i \end{aligned} \tag{5.41}$$

where c_{XY}^r and c_{XY}^i are the real and imaginary components of the covariance of X and Y, c_{XY}, given by

$$c_{XY}^r = c_{X_rY_r} + c_{X_iY_i} \tag{5.42a}$$

$$c_{XY}^i = c_{X_iY_r} - c_{X_rY_i} \tag{5.42b}$$

where

$$c_{X_rY_r} = E(X_rY_r) - \mu_{X_r}\mu_{Y_r} = \int_{-\infty}^{+\infty}\int_{-\infty}^{+\infty} xyf_{X_rY_r}(x,y)\mathrm{d}x\mathrm{d}y - \mu_{X_r}\mu_{Y_r} \tag{5.42c}$$

$$c_{X_iY_i} = E(X_iY_i) - \mu_{X_i}\mu_{Y_i} = \int_{-\infty}^{+\infty}\int_{-\infty}^{+\infty} xyf_{X_iY_i}(x,y)\mathrm{d}x\mathrm{d}y - \mu_{X_i}\mu_{Y_i} \tag{5.42d}$$

$$c_{X_iY_r} = E(X_iY_r) - \mu_{X_i}\mu_{Y_r} = \int_{-\infty}^{+\infty}\int_{-\infty}^{+\infty} xyf_{X_iY_r}(x,y)\mathrm{d}x\mathrm{d}y - \mu_{X_i}\mu_{Y_r} \tag{5.43a}$$

$$c_{X_rY_i} = E(X_rY_i) - \mu_{X_r}\mu_{Y_i} = \int_{-\infty}^{+\infty}\int_{-\infty}^{+\infty} xyf_{X_rY_i}(x,y)\mathrm{d}x\mathrm{d}y - \mu_{X_r}\mu_{Y_i} \tag{5.43b}$$

As can be seen in the above equations, determination of the covariance between two complex RVs X and Y requires four joint *pdfs*, $f_{X_rY_r}, f_{X_iY_i}, f_{X_iY_r}$, and $f_{X_rY_i}$.

The correlation coefficient of a complex RV is defined by the same Eq. (5.35) with the above covariance and is a complex number as follows:

$$\rho_{XY} \triangleq \frac{c_{XY}}{\sigma_X \sigma_Y} = \frac{c_{XY}^r}{\sigma_X \sigma_Y} + j\frac{c_{XY}^i}{\sigma_X \sigma_Y}$$

where, by (5.25), the S.D. of X and Y is given by

$$\sigma_X = \sqrt{\sigma_X^2} \triangleq \sqrt{E\left[(| X - \mu_X |)^2\right]}$$

$$\sigma_Y = \sqrt{\sigma_Y^2} \triangleq \sqrt{E\left[(| Y - \mu_Y |)^2\right]}$$

Uncorrelated RVs

Two RVs X and Y are said to be uncorrelated if $c_{XY} = 0$, that is, $\rho_{XY} = 0$. By (5.33), c_{XY} would be zero if and only if

$$E(XY) = \mu_X \mu_Y = E(X)E(Y) \Rightarrow c_{XY} = 0 \tag{5.44}$$

$$c_{XY} = 0 \Rightarrow E(XY) = \mu_X \mu_Y = E(X)E(Y) \tag{5.45}$$

Interpretation of Covariance and Correlation Coefficient

Both the covariance and the correlation coefficient of two RVs are measures of how the two RVs vary jointly in a correlated manner. A positive correlation ($\rho > 0$) indicates that the two RVs vary jointly in the same direction such that, if X increases, Y increases, or, if X decreases, Y decreases and *vice versa*. A negative correlation ($\rho < 0$) indicates that the two RVs vary jointly in the opposite directions such that if X increases, Y decreases, or, if X decreases, Y increases and *vice versa*. If $\rho = 0$, the two RVs are uncorrelated.

As shown by (5.33), the covariance is based on the expected value or the mean of the product of X and Y. By taking the product of X and Y, if the deviations of the two RVs from their respective means are in the same direction, that is, both pluses or both minuses, the product XY would contribute to the covariance positively. On the other hand, if the deviations of the two RVs from their respective means are in the opposite directions, that is, one plus and the other minus or vice versa, the product XY would contribute to the covariance negatively. The mean or expected value of these contributions result in the covariance. Figure 5.3 illustrates the random data of (x_i, y_j) pairs for the three cases of correlation. The plots show how x_i and y_j move together in a correlated manner.

A degenerate case of the covariance is the variance, where Y is replaced with X in C_{XY} to yield $\text{Var}(X) = E\{(X-\mu)(X-\mu)\} = E(X^2)-\mu^2$.

Theorem 5.3.1 If an RV X is magnified or reduced by a constant multiplier α, the original variance of the RV is magnified or reduced by the square of the constant multiplier as follows:

$$\text{Var}\ (\alpha X) = \alpha^2\ \text{Var}(X) \tag{5.46}$$

Proof Let $\mu = E(X)$. Then

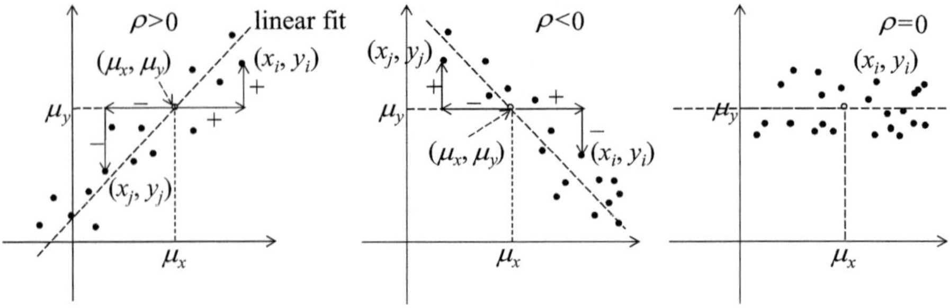

Fig. 5.3 Illustration of correlation coefficient ρ

$$\begin{aligned}\text{Var}\,(\alpha X) &= E\Big[\{\alpha X - E(\alpha X)\}^2\Big] = E\Big[\{\alpha X - \alpha\mu\}^2\Big] = E\Big[\alpha^2\{X-\mu\}^2\Big]\\ &= \alpha^2 E\Big[\{X-\mu\}^2\Big] = \alpha^2\ \text{Var}(X)\end{aligned}$$

Q.E.D.

Theorem 5.3.2

$$-1 \le \rho \le 1 \tag{5.47}$$

Proof Form a positive definite quadratic form of an arbitrary parameter λ as follows:

$$\{(X-\mu_X)\lambda + (Y-\mu_Y)\}^2 = (X-\mu_X)^2\lambda^2 + 2(X-\mu_X)(Y-\mu_Y)\lambda + (Y-\mu_Y)^2 \tag{5.48}$$

Take the expected value of the right-hand side of (5.48) as follows:

$$\begin{aligned}&E\Big\{(X-\mu_X)^2\lambda^2 + 2(X-\mu_X)(Y-\mu_Y)\lambda + (Y-\mu_Y)^2\Big\} = E\Big\{(X-\mu_X)^2\Big\}\lambda^2\\ &\quad +2E\{(X-\mu_X)(Y-\mu_Y)\} + E\Big\{(Y-\mu_Y)^2\Big\} = \sigma_X^2\lambda^2 + 2\sigma_{XY}\lambda + \sigma_Y^2\end{aligned} \tag{5.49}$$

Since the last expression of the above equation is a positive definite quadratic form, its discriminant is given by the following expression:

$$4{c_{XY}}^2 - 4\,\sigma_X^2\sigma_Y^2 \le 0 \tag{5.50}$$

or

$$-\sigma_X\sigma_Y \le c_{XY} \le \sigma_X\sigma_Y \tag{5.51}$$

Substituting (5.51) into (5.35), we obtain (5.47).

Q.E.D.

Theorem 5.3.3

$$\mathrm{Var}(X+Y) = \mathrm{Var}(X) + \mathrm{Var}(Y) + 2\mathrm{Cov}(X,Y) \tag{5.52}$$

Proof Let

$$\mu_X = E(X) \qquad \mu_Y = E(Y) \qquad \mathrm{Var}(X) = \sigma_X^2 \qquad \mathrm{Var}(Y) = \sigma_Y^2$$

Then

$$\begin{aligned}
\mathrm{Var}(X+Y) &= E\left[\{(X+Y) - E(X+Y)\}^2\right] = E\left[\{(X+Y) - \mu_X - \mu_Y\}^2\right] \\
&= E\left[(X-\mu_X)^2 + 2(X-\mu_X)(Y-\mu_Y) + (Y-\mu_Y)^2\right] \\
&= E\left[(X-\mu_X)^2\right] + 2E\left[(X-\mu_X)(Y-\mu_Y)\right] + E\left[(Y-\mu_Y)^2\right] \\
&= \sigma_X^2 + 2\mathrm{Cov}(X,Y) + \sigma_Y^2
\end{aligned}$$

Q.E.D.

Theorem 5.3.4 If two RVs X and Y are independent, X and Y are uncorrelated, that is,

$$E(XY) = E(X)E(Y) \tag{5.53}$$

However, the converse is not necessarily true. The uncorrelatedness is derived from the independence condition. In other words, the independence condition is a stronger condition than the uncorrelatedness condition.

Proof

$$E(XY) = \int_{-\infty}^{+\infty}\int_{-\infty}^{+\infty} xyf_{XY}(x,y)\mathrm{d}x\mathrm{d}y \tag{5.54}$$

Since X and Y are independent, substituting (4.82b) into the above equation, we obtain the following equation:

$$E(XY) = \int_{-\infty}^{+\infty} xf_X(x)\mathrm{d}x \int_{-\infty}^{+\infty} yf_Y(y)\mathrm{d}y = E(X)E(Y)$$

Q.E.D.

Theorem 5.3.5 If two RVs X and Y are independent,

$$\mathrm{Var}(X+Y) = \mathrm{Var}(X) + \mathrm{Var}(Y) \tag{5.55}$$

Proof By (5.53), if X and Y are independent, they are uncorrelated and $c_{XY} = 0$. Setting $c_{XY} = 0$ in (5.52), we obtain (5.55).

Q.E.D.

Equation (5.55) can be generalized as follows. If X_is are independent, the following equation holds true:

$$\mathrm{Var}(X_1 + X_2 + \ldots + X_i + \ldots X_n) = \mathrm{Var}(X_1) + \mathrm{Var}(X_2) + \cdots + \mathrm{Var}(X_i) + \ldots + \mathrm{Var}(X_n) \quad (5.56)$$

5.4 Example Distributions

The following five important distributions are discussed in this book: uniform, binomial, exponential, Gaussian, and Poisson distributions. The first three distributions are discussed in this section and the Gaussian and Poisson distributions, in the sections dealing with their applications later.

5.4.1 Uniform Distribution

Consider an RV X that takes on the values in the interval from x_1 to x_2 on the real line R and no values outside of the interval. If X is equally likely to fall on anywhere over this interval, X is said to be uniformly distributed over the interval. For such an RV X, we say that X is a uniformly distributed RV or that X has a uniform distribution and write as

$$X \sim u(x_1, x_2) \quad (5.57)$$

The space of X is $\Omega = \{x_1 \leq X \leq x_2\}$. The length of this interval is $(x_2 - x_1)$ (Fig. 5.4).

To obtain the CDF of a uniformly distributed RV X, consider the probability of the event $\{X \leq x\}$ as x is varied from $-\infty$ to $+\infty$.

For $x \leq x_1$, $\{X \leq x\} = \{\varnothing\}$,

$$F_X(x) = P[\{X \leq x\}] = P[\{\varnothing\}] = 0 \quad (5.58)$$

For $x_1 < x \leq x_2$, the set $\{X \leq x\}$ is the interval of length $(x - x_1)$ from x_1 to x. The probability that X will fall in this subinterval is proportional to its length and is

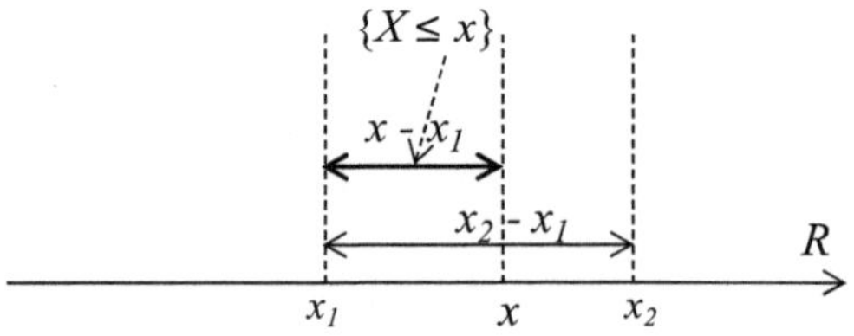

Fig. 5.4 Interval length of $\{X \leq x\}$, $(x - x_1)$, and interval length of the space, $(x_2 - x_1)$

equal to the ratio of the length of the subinterval to that of the total interval, that is, the space Ω, as follows:

$$F_X(x) = P[\{X \le x\}] = \frac{x - x_1}{x_2 - x_1} \tag{5.59}$$

For $x_2 < x$, $\{X \le x\} = \Omega$,

$$F_X(x) = P[\{X \le x\}] = P[\Omega] = 1 \tag{5.60}$$

Putting (5.58) through (5.60) together, we obtain the CDF of a uniformly distributed RV X as follows:

$$F_X(x) = \begin{cases} 0, & \text{if } x \le x_1 \\ \dfrac{x - x_1}{x_2 - x_1} & \text{if } x_1 < x \le x_2 \\ 1, & \text{if } x > x_2 \end{cases} \tag{5.61}$$

The uniform CDF is shown in Fig. 5.5. The CDF is zero for $x < x_1$, monotonically increases as x varies from x_1 to x_2 and stays at 1 for $x > x_2$. This graph agrees with the properties of the CDF, that is, the CDF is bounded by 0 and 1 and nondecreasing.

The uniform *pdf* of X can be obtained either by taking the derivative of the CDF as given by (4.17) or directly from the definition of the *pdf* given by (4.15). For the illustration purposes, the *pdf* is derived by both methods. By taking the derivative of $F_X(x)$, the *pdf* of RV X is given by the following:

$$f_X(x) = \begin{cases} 0 & \text{if } x \le x_1 \\ \dfrac{1}{x_2 - x_1} & \text{if } x_1 < x \le x_2 \\ 0 & \text{if } x > x_2 \end{cases} \tag{5.62}$$

The *pdf* is shown in Fig. 5.5. The *pdf* is a rectangle with the height $1/(x_2 - x_1)$ and the base $(x_2 - x_1)$ so that the area is equal to 1. This satisfies the condition of a *pdf* that the total area under the *pdf* curve be equal to 1.

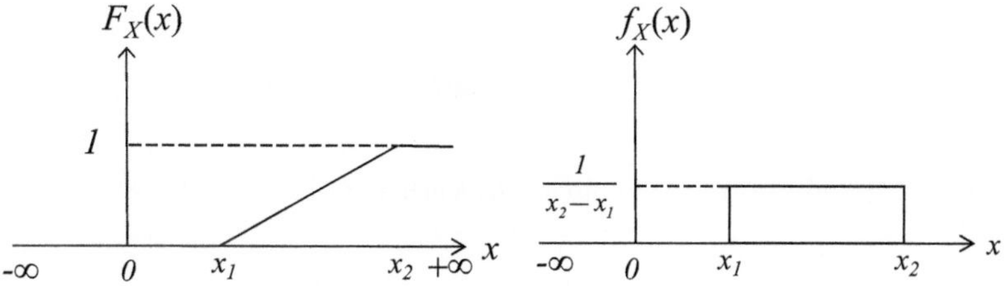

Fig. 5.5 CDF and *pdf* of a uniform distribution

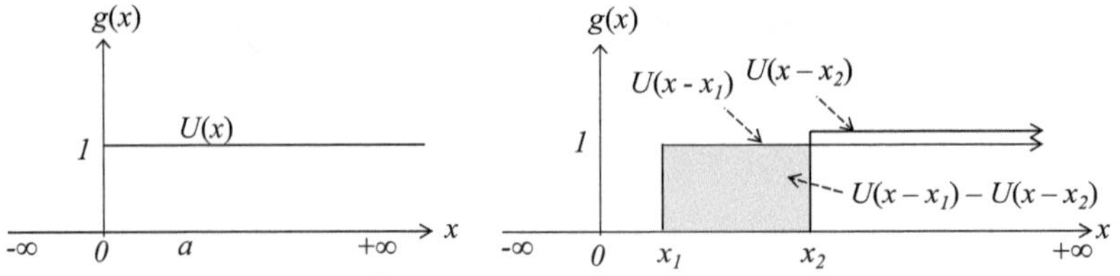

Fig. 5.6 Unit step functions

It is convenient to express the three equations of (5.62) by a single equation using the unit step functions shifted to the right by x_1 and x_2, $U(x-x_1)$ and $U(x-x_2)$, respectively, as follows. As illustrated in Fig. 5.6, the difference between these two shifted unit step functions divided by $x_2 - x_1$ yields the *pdf*.

$$f_X(x) = \frac{1}{x_2 - x_1}\{U(x-x_1) - U(x-x_2)\} \tag{5.63}$$

Now, we derive the *pdf* directly from the definition given by (4.15). Consider the interval from x_1 to x_2 and the probability of X falling in a small interval of length Δx within this interval as follows:

$$P[\{x < X \leq x + \Delta x\}] = \frac{\Delta x}{x_2 - x_1} \tag{5.64}$$

Substituting (5.64) into the definition of the *pdf* given by (4.15), we obtain the following equation:

$$\begin{aligned} f_X(x) &= \lim_{\Delta x \to 0} \frac{P[\{x < X \leq x + \Delta x\}]}{\Delta x} \\ &= \lim_{\Delta x \to 0} \frac{1}{\Delta x}\frac{\Delta x}{x_2 - x_1} = \frac{1}{x_2 - x_1} \end{aligned}$$

We have shown that, by deriving the *pdf* directly from the definition of the *pdf*, we arrive at the same result as (5.62).

The mean and the variance of a uniformly distributed RV X are obtained as follows:

$$E(X) = \int_{x_1}^{x_2} \frac{x\mathrm{d}x}{x_2 - x_1} = \frac{1}{2}\frac{1}{x_2 - x_1}\left(x_2^2 - x_1^2\right) = \frac{1}{2}(x_1 + x_2) \tag{5.65}$$

Using (5.26) and (5.65), we evaluate the variance of X as follows:

$$\begin{aligned}\mathrm{Var}(X) &= E\left(X^2\right) - E(X)^2 = \int_{x_1}^{x_2} \frac{x^2 \mathrm{d}x}{x_2 - x_1} - \left[\frac{1}{2}(x_1 + x_2)\right]^2 \\ &= \frac{1}{3}\frac{\left(x_2^3 - x_1^3\right)}{(x_2 - x_1)} - \frac{1}{4}(x_1 + x_2)^2 = \frac{1}{3}\frac{(x_2 - x_1)\left(x_2^2 + x_2 x_1 + x_1^2\right)}{(x_2 - x_1)} - \frac{1}{4}(x_1 + x_2)^2 \\ &= \frac{1}{12}(x_2 - x_1)^2\end{aligned} \tag{5.66}$$

5.4.2 Binomial Distribution

Let us start with an example to illustrate this important distribution. In digital communications, the information to be transmitted is coded at the source into a string of 0 s and 1 s known as the binary bits. These bits are then transmitted over a communications channel to the destination. The bits received at the destination are decoded to produce the original information sent from the source. Because of the channel noise, however, some of the bits arrive at the destination in error, that is, 0 turned to 1 and *vice versa*. The probability that a bit will arrive at the destination in error is referred to as the bit error rate, denoted by p, and the probability that a bit will arrive correctly, $q = 1-p$.

To continue with the illustration, suppose that a 10-bit string coded into 1010110100 is transmitted from the source. Ideally, 1010110100 should arrive at the destination unaltered. Assume that the bit error rate is 10^{-5}. What is the probability that the fourth and the fifth bits are in error so that the string will arrive as 1010101100? This probability is

$$q \times q \times q \times p \times p \times q \times q \times q \times q \times q = p^2 q^8 = \left(10^{-5}\right)^2 \times \left(1 - 10^{-5}\right)^8$$

This probability remains the same $p^2 q^8$ if the number of errored bits considered is two regardless of which two bits. There are 10-combination 2 ways, that is, $\binom{10}{2}$ in which two errored bits can occur in any order. Hence, the probability of two errors in 10 bits in any order is given by $\binom{10}{2} p^2 q^8$.

Let us now generalize this example as follows. n binary bits are transmitted over a communications channel with a bit error rate p. Let the RV X be the number of bit errors in any order. The probability of k bits in error in any order is given by the following equation:

$$P[\{X = k\}] = \binom{n}{k} p^k (1 - p)^{n-k} \tag{5.67}$$

Using the Dirac delta function, the binomial *pdf* of X can be expressed as follows:

$$f_X(x) = \sum_{k=0}^{n} \binom{n}{k} p^k (1-p)^{n-k} \delta(x-k) \tag{5.68}$$

By integrating the above *pdf*, the CDF of the RV X is given by

$$F_X(x) = \sum_{k=0}^{m} \binom{n}{k} p^k (1-p)^{n-k} \quad \text{for } m \le x \le m+1 \tag{5.69}$$

where

$$\binom{n}{k} = \frac{n!}{k!(n-k)!} \tag{5.70}$$

We say that the RV X has a binomial distribution with parameters n and p if its *pdf* and CDF are given by (5.68) and (5.69) and write as follows:

$$X \sim B(n, p) \tag{5.71}$$

Let us look at the bit error case from the perspective of random experiment as follows. Consider the transmission of each individual bit as a single trial of the same experiment. Each bit can either arrive at the destination correctly, that is, "success (S)," or in error, that is, "failure (F)." The transmission of n bits can be viewed as n independent trials of the same experiment. In general, a random experiment with binary outcomes, "success (S)" or "failure (F)," is referred to as a Bernoulli trial or binomial trial.

The space for this experiment contains two elements, "success (S)" and "failure (F)."

$$\Omega = \{S, F\} \tag{5.72}$$

The probability of the elementary events in this space are as follows:

$$P[\{S\}] = p \qquad\qquad P[\{F\}] = 1 - p \tag{5.73}$$

Suppose that the experiment Ω is tried n times and let the RV X be the total number of successes in any order at the end of the n trials. Let X_i be the RV defined for the ith trial as follows:

$$X_i = \begin{cases} 1 & \text{if outcome of } i\text{th trial is a success } (S) \\ 0 & \text{if outcome of } i\text{th trial is a failure } (F) \end{cases} \tag{5.74}$$

The total number of successes in n trials is given by

$$X = X_1 + X_2 + \ldots + X_i + \ldots + X_n \tag{5.75}$$

Then, the RV X is binomially distributed

$$X \sim B(n, p)$$

The mean and the variance of a binomially distributed RV X are obtained as follows. Taking the expected value of (5.74), we obtain the following equation:

$$E(X_i) = 1 \times P[\{S\}] = p \tag{5.76}$$

$$\text{Var}(X_i) = E\left(X_i^2\right) - [E(X_i)]^2 = p - p^2 = p - p^2 \tag{5.77}$$

Applying (5.21) to (5.75) multiple times and using (5.76), we obtain the following equation for the mean of RV X:

$$\begin{aligned} E(X) &= E(X_1 + X_2 + \cdots + X_i + \cdots + X_n) \\ &= E(X_1) + E(X_2) + \cdots + E(X_i) + \cdots + E(X_n) = np \end{aligned} \tag{5.78}$$

Since X_is are independent, applying (5.56) to (5.75) and using (5.77), we obtain the following equation for the variance of X:

$$\begin{aligned} \text{Var}(X) &= \text{Var}(X_1 + X_2 + \cdots + X_i + \cdots + X_n) \\ &= \text{Var}(X_1) + \text{Var}(X_2) + \cdots + \text{Var}(X_n) = npq \end{aligned} \tag{5.79}$$

In the above formulation, the same experiment is repeated n times, each producing one outcome, and the sequence of the outcomes after the n trials is used in determining the probability. Another way of formulating the same problem is to combine the n repeated trials of the same experiment into a single new experiment in which an outcome is defined as a sequence of n outcomes of the original experiment. In this new "combined" experiment, therefore, a single trial is completed when the original experiment is tried n times. The space of this combined experiment, denoted by Ω_c, is the Cartesian product of the n individual spaces Ω as follows:

$$\begin{aligned} \Omega_c &= \Omega \times \Omega \times \ldots \times \Omega = \{S, F\} \times \{S, F\} \times \ldots \times \{S, F\} \\ &= \{(SSS \ldots SS), (SSS \ldots SF), \ldots, (FFF \ldots FF)\} \end{aligned} \tag{5.80}$$

Ω_c contains 2^n elementary events, which are ordered n-tuples of S and F. For example, for $n = 2$, $\Omega_c = \{(SS), (SF), (FS), (FF)\}$ and the elementary events are $\{(SS)\}$, $\{(SF)\}$, $\{(FS)\}$, and $\{(FF)\}$.

The probabilities of the elementary events in Ω_c are as follows:

$$P[\{(SSS \ldots SS)\}] = P[\{S\} \times \ldots \times \{S\} \times \{S\}] = P[\{S\}] \ldots P[\{S\}]P[\{S\}] = p^n$$

$$\begin{aligned}P[\{(SSS\ldots SF)\}] &= P[\{S\}\times\ldots\times\{S\}\times\{F\}] = P[\{S\}]\ldots P[\{S\}]P[\{F\}]\\&= p^{n-1}(1-p)^1\\P[\{(SSS\ldots FF)\}] &= P[\{S\}\times\ldots\times\{F\}\times\{F\}] = P[\{S\}]\ldots P[\{F\}]P[\{F\}]\\&= p^{n-2}(1-p)^2\end{aligned}$$

....

$$\begin{aligned}P[\{(FFF\ldots FF)\}] &= P[\{S\}\times\ldots\times\{F\}\times\{F\}]\\&= P[\{F\}]\ldots P[\{F\}]P[\{F\}] = (1-p)^n\end{aligned} \tag{5.81}$$

There is one unique elementary event corresponding to k successes in a specific order. The probability of this elementary event is $p^k(1-p)^{n-k}$. There are $\binom{n}{k}$ elementary events with k successes in any order. Since the elementary events are mutually exclusive, the probability of k successes in any order is equal to the sum of the probabilities of these k elementary events. Hence, the RV X is binomially distributed with the *pdf* and the CDF given by (5.68) and (5.69), respectively.

The following examples illustrate the experiments that belong to the class of the Bernoulli trial.

Example 5.4.1
Consider a digital connection with three links. The probability of an error on each link is assumed to be p. A digital bit 0 is transmitted from the source to the destination over the three links. Find the end-to-end probability that the bit 0 will arrive as bit 0.

$$\Omega = \{S, F\} \qquad P[\{S\}] = p \qquad P[\{F\}] = q = 1-p \tag{5.82}$$

The transmission of the bit over each link constitutes a trial of the experiment Ω. The space of the combined experiment of the three identical experiments is the Cartesian product given by

$$\Omega_c = \{(SSS), (SSF), (SFS), (SFF), (FSS), (FSF), (FFS), (FFF)\} \tag{5.83}$$

The event of interest is the event that "bit 0 arrives as 0." Denote this event by A. Referring to Fig. 5.7, we see that this event consists of the following elementary events, which are shown in bold in the figure

$$A = \{(SSS), (SFF), (FSF), (FFS)\} \tag{5.84}$$

The probability of the event A is given by

$$\begin{aligned}P(A) &= P[\{(SSS), (SFF), (FSF), (FFS)\}]\\&= (1-p)^3 + (1-p)p^2 + (1-p)p^2 + (1-p)p^2\end{aligned} \tag{5.85}$$

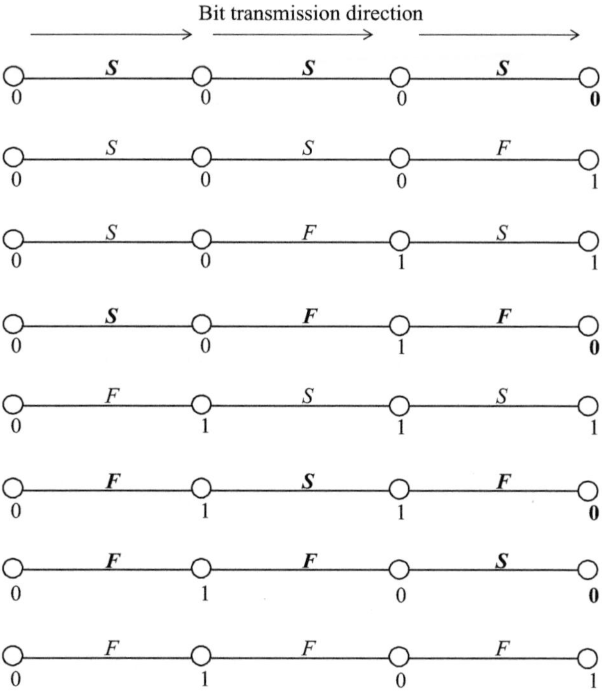

Fig. 5.7 All possible outcomes of Example 5.4.1

Example 5.4.2
Toss a coin n times and determine the probability of k heads in any sequence. Here, let the event be $\{S\}$ if the coin throwing shows the heads and $\{F\}$, if the coin throwing shows the tails.

Example 5.4.3
Suppose that we are interested in analyzing the number of people arriving at a bus station from 9:00 A.M. to 10 A.M. and, for our analysis, we have collected a large amount of data on the arrival times during this time interval. Assume that the arrival times are uniformly distributed over the one-hour interval. This assumption may be unrealistic considering that where and when people leave to go to the bus station could influence their arrival times. Nevertheless, we will consider this example under this assumption for illustration purposes. We will treat this type of arrival in more detail in queueing theory in a later chapter.

Referring to Fig. 5.8, suppose that the total number of people arriving in this one-hour interval is n. Consider the subinterval of from 9:20 A.M. to 9:30 A.M. and the event that a person arrives in the subinterval a success $\{S\}$, and the event of arrival outside of the interval, $\{F\}$. Determine the probability that k people would arrive in this 10-minute subinterval.

The probability that a person would arrive in the 10-minute subinterval, that is, $P[\{S\}] = p$, is 10 min/60 min = 1/6 and $P[\{F\}] = 1-p = 5/6$. Let the RV X be the number of arrivals in the 10-min subinterval. The RV X is binomially distributed with the parameters n and p as discussed earlier.

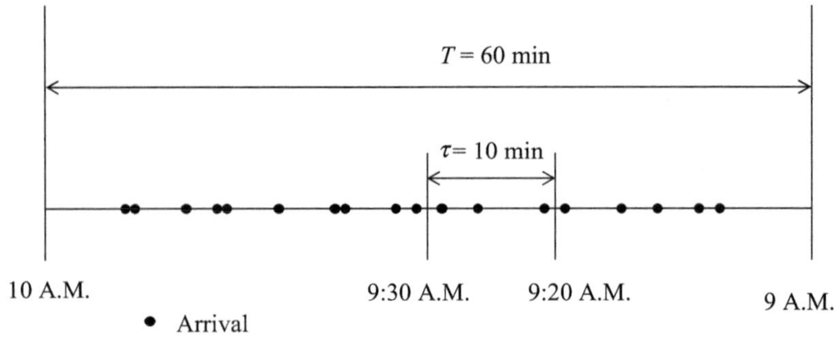

Fig. 5.8 Arrivals in the period from 9 A.M to 10 A.M

5.4.3 Exponential Distribution

The *pdf* of an exponentially distributed RV X with parameter α is given by

$$f_X(x) = \begin{cases} \alpha e^{-\alpha x}, & x > 0 \\ 0, & x \leq 0 \end{cases} \tag{5.86}$$

By integrating (5.86), we obtain the CDF of an exponentially distributed RV X given by

$$F_X(x) = \begin{cases} 1 - e^{-\alpha x}, & x > 0 \\ 0, & x \leq 0 \end{cases} \tag{5.87}$$

Equation (5.87) satisfies the condition for the CDF that its value is equal to 1 at $x = \infty$. This also means that the area under the exponential curve of (5.86) is equal to 1. Figure 5.9 shows exponential pdf and CDF.

The mean and variance of the exponentially distributed RV X are obtained as follows:

$$E(X) = \int_0^{+\infty} x f_X(x) \mathrm{d}x = \int_0^{+\infty} x\alpha e^{-\alpha x} \mathrm{d}x = -[xe^{-\alpha x}]_0^{+\infty} - \left[\frac{1}{\alpha} e^{-\alpha x}\right]_0^{+\infty} \tag{5.88}$$

Applying the l'Hopital's rule to the first term of (5.88), we obtain the following result

$$\lim_{x \to \infty} xe^{-\alpha x} = \lim_{x \to \infty} \frac{x}{e^{\alpha x}} = \lim_{x \to \infty} \frac{\frac{\mathrm{d}}{\mathrm{d}x} x}{\frac{\mathrm{d}}{\mathrm{d}x} e^{\alpha x}} = \lim_{x \to \infty} \frac{1}{\alpha e^{\alpha x}} = 0 \tag{5.89}$$

Substituting (5.89) into 5.88), we obtain the following result for the mean of X:

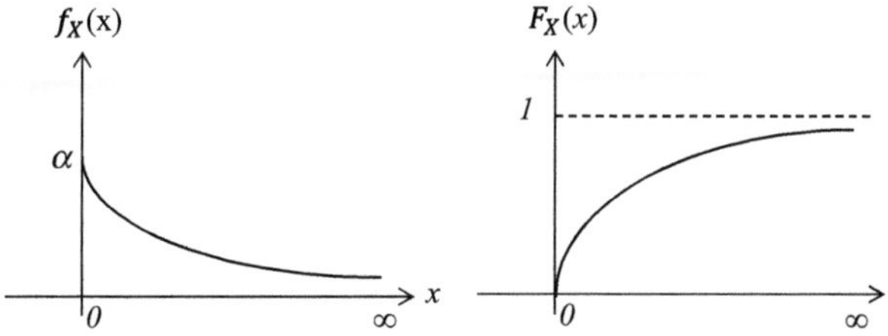

Fig. 5.9 The *pdf* and CDF of an exponential distribution

$$\mu_X = E(X) = \frac{1}{\alpha} \tag{5.90}$$

To find the variance, first obtain the following equation:

$$E(X^2) = \int_0^{+\infty} x^2 \alpha e^{-\alpha x} \mathrm{d}x = \left[-x^2 e^{-\alpha x} - \frac{2}{\alpha} x e^{-\alpha x} + \frac{2}{\alpha^2} e^{-\alpha x} \right]_0^{\infty} = \frac{2}{\alpha^2} \tag{5.91}$$

Substituting (5.90) and (5.91) into the equation below, we obtain the variance of X:

$$\mathrm{Var}(X) = E(X^2) - [E(X)]^2 = \frac{2}{\alpha^2} - \frac{1}{\alpha^2} = \frac{1}{\alpha^2} \tag{5.92}$$

The exponential distribution is defined by a single parameter α, which is $\alpha = \frac{1}{\mu_X}$. In terms of the mean, (5.86) can be expressed as follows:

$$f_X(x) = \frac{1}{\mu_X} e^{-\frac{1}{\mu_X} x}, \quad x > 0 \tag{5.93}$$

Chapter 6
Stochastic Process

This chapter defines a stochastic process and presents its statistical characterizations. It also discusses the stationarity and the ergodicity of the stochastic process and the parameters of the stochastic process.

6.1 Definition of Stochastic Process

Consider the definition of the RV X given by (4.1), which is reproduced below for convenience:

$$X = f(s) \tag{6.1}$$

for which the underlying experiment is specified by the space Ω consisting of all possible outcomes s and the probabilities of the elementary events given by the following equations:

$$P[\{s_i\}] = p_i \qquad s_i \in \Omega \qquad i = 1, 2, \ldots, n \tag{6.2}$$

Suppose that we conduct the experiment defined by (6.2) at two different time points, $t_1$and t_2, and that the probabilities of the elementary events are different at these two time points although the space Ω of all possible outcomes remains the same. Denote the probabilities of the elementary events at the two time points as follows:

$$P[\{s_i(t_1)\}] = p_i(t_1) \qquad P[\{s_i(t_2)\}] = p_i(t_2) \qquad s_i \in \Omega \tag{6.3}$$

Although the RV X is still defined by the function f of the outcome s given by (6.1), we need to distinguish the RVs at the two time points because the probabilities of the elementary events have changed between the two time points. We can

K.I. Park, *Fundamentals of Probability and Stochastic Processes with Applications to Communications*, https://doi.org/10.1007/978-3-319-68075-0_6

use two different letters, e.g., Y and Z, to denote the two RVs at $t_1$and t_2. We can also keep the original letter X for the RV defined by the function f and distinguish the RVs at the two time points by using time as the argument as follows:

$$X(t_1) = f(t_1, s) \qquad X(t_2) = f(t_2, s) \qquad s \in \Omega \tag{6.4}$$

The advantage of the latter method is that, by using the same letter X with time as the argument, we know that the two RVs represent the same random phenomenon even though they are two different RVs.

Consider the RV of Example 4.1.3, where the function maps the temperature readings to the same real numbers, that is, RV X as the temperature readings. Denote the temperature readings at n different time points and define the RVs at the n time points as follows:

$$X(t_1, s), X(t_2, s), \ldots, X(t_i, s), \ldots X(t_n, s) \tag{6.5}$$

In this example, where the outcome is a real number and is mapped to the same real number, that is, $X = s$, s is suppressed for simplicity as follows:

$$X(t_1), X(t_2), \ldots, X(t_i), \ldots X(t_n) \tag{6.6}$$

Discrete Stochastic Process

A discrete stochastic process is defined as a collection of RVs for discrete time points as follows:

$$\{X(t_1), X(t_2), \ldots, X(t_i), \ldots X(t_n)\} = \{X(t_i), t_i \in R\} \tag{6.7}$$

Continuous Stochastic Process

A continuous stochastic process is defined as a collection of RVs for a continuum of time as follows:

$$\{X(t), t \in R\} \tag{6.8}$$

In (6.7) and (6.8), the time points are fixed points and the X's are the RVs for these fixed time points. By taking the time points as variables, the stochastic processes defined by (6.7) and (6.8) are represented by

$$X(t_i) \quad \text{discrete process; } t_i \text{ variable; } X(t) \quad \text{continuous process; } t \text{ variable} \tag{6.9}$$

With the argument t or t_i as the variable, $X(t)$ or $X(t_i)$ is called a stochastics process, a random process, or, simply, a process. One concept that is helpful for the analysis of a stochastic process is as follows. $X(t)$ is a process if the argument t is treated as a variable. Once the variable t of $X(t)$ is fixed at a specific value t^*, the process $X(t)$ reduces to an RV $X(t^*)$. In other words, $X(t)$ is a process, if t is treated as a variable, but, if t is considered fixed at t, $X(t)$ is an RV.

Example 6.1.1

To illustrate the concept of a stochastic process, consider the following experiment. We are interested in the temperature distribution over a one-year period from January 1 to December 31 in a certain geographic area. To determine the distribution, a temperature measurement device is installed at n different locations throughout the area and the measurements are taken over the one-year period.

Figure 6.1a shows the graphs of the measurements collected from the n different locations. Each individual graph is called a sample path and shows the data from a single measurement location plotted over time. The cross section of the sample paths at time t consists of n data points, one per sample path, distributed over the real line R. $X(t)$ is the RV representing the distribution of data at this cross section of the sample paths at time t. The RV $X(t)$ is also called the state of the process at time t. In a stochastic process, the collection of these n data points is sometimes referred to as the ensemble of data at time t.

Keeping the above notion in mind, we note the following pattern of analysis of a stochastic process throughout this chapter. For a given process $X(t)$, first, we treat $X(t)$ as an RV by fixing t. Once we have an RV $X(t)$, we can apply the definitions, the theorems, and the properties associated with the RV obtained in earlier chapters to the RV $X(t)$ to derive its CDF, *pdf*, mean, variance, etc. These quantities will have a constant t as an argument. Once these quantities are derived, however, by treating the argument t as a variable, these quantities can be treated as a function of time. The focus of analysis of a stochastic process is then the dynamic behavior of these quantities in time.

By fixing the time at multiple points, e.g., at t_1 and t_2, we generate multiple RVs, $X(t_1)$ and $X(t_2)$, from the same process and analyze their joint behaviors varying in time, e.g., their covariance. Throughout the discussion in earlier chapters, however, RVs are static and have no time variation, and, therefore, the joint behavior such as the covariance is considered only between RVs representing two different random phenomena, e.g., X and Y. In the earlier chapters, because X is static, no "covariance" of X is defined, and only the variance of X is defined. The variance of X may be considered as the degenerate case of the covariance, which is the covariance between X and itself.

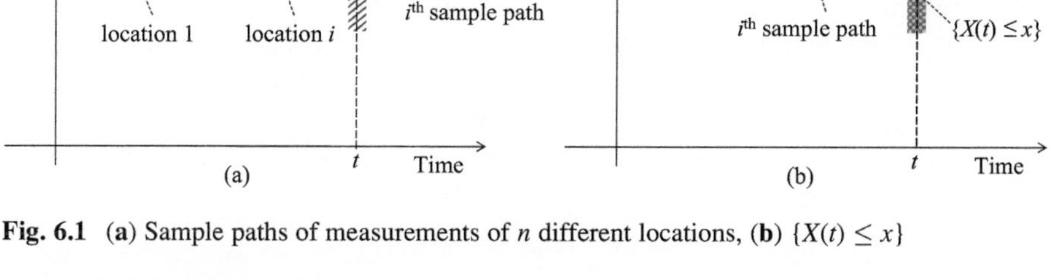

Fig. 6.1 (**a**) Sample paths of measurements of n different locations, (**b**) $\{X(t) \leq x\}$

General Definition of a Stochastic Process
In the above two definitions, the stochastic process is defined as the collection of RVs defined at specific time points, discrete or continuous. In the general definition, however, a stochastic process is simply a collection of RVs, $X(i)$, where i is an arbitrary index identifying the RVs, which happens to be the time in the case of the previous two definitions. Most of the processes discussed in this book are the processes in time. We will see an example of this general definition later in conjunction with a point process.

Definition of a Complex Stochastic Process
Unless otherwise stated, the RV and the process in this book are real. The complex process $X(t)$ is defined by

$$X(t) = X_r(t) + jX_i(t) \tag{6.10}$$

where $X_r(t)$ and $X_i(t)$ are real processes.

The conjugate process $X(t)^*$ is given by

$$X(t)^* = X_r(t) - jX_i(t) \tag{6.11}$$

6.2 Statistical Characterization of a Stochastic Process

We have already discussed that a stochastic process may be thought of as an RV moving in time and that the characterization of a stochastic process deals with analyzing the RVs defined at multiple time points. Since our analysis involves the dynamic behavior of the RVs, we need to determine how their statistical properties change as the selected time points are shifted in time.

Consider n selected time points. The n RVs defined at these time points can be analyzed singly as well as jointly. The number of time points treated together defines the order of the statistical properties. For example, the first-order properties are those of the n RVs treated singly; the second-order properties, those of the n RVs treated jointly two at a time; and the nth-order properties, those of the n RVs treated all together. When two processes are considered, the interactions between the two processes are characterized by analyzing the joint behavior between the RVs at n and m time points selected for the two processes, respectively.

The properties derived for the selected time points are applicable for the fixed time points, and, therefore, may be considered as the static properties. On the other hand, once the static properties are derived for selected time points, we can consider the selected time points as variables and treat the properties as a function of time. The dynamic properties of the processes can be analyzed by analyzing how the properties vary as the selected time points vary.

The static properties are discussed first and the dynamic properties, later.

6.2.1 First-Order Distributions

First-Order CDF
In Fig. 6.1b, the crosshatched vertical strip at time t shows the set $\{X(t) \leq x\}$. By (4.4), the first-order CDF of $X(t)$ is defined as follows:

$$F_{X(t)}(x,t) \triangleq P[\{X(t) \leq x\}] \tag{6.12}$$

The notations used in the above equation are consistent with those used in (4.4) with one change, the inclusion of time t as an argument. The argument t indicates that the equation applies at the specific time t. Since the fixed time t is arbitrary, (6.12) applies for any t. Therefore, the argument t can be treated as a variable and the CDF can be treated as a function of time. The subscript also shows t as an argument, which allows RVs defined at different specific time points to be distinguished. When there is no confusion, however, the argument t will be dropped in the subscript X for simplicity.

First-Order *pdf* By (4.15), the first-order *pdf* of $X(t)$ is defined as follows:

$$f_X(x,t) \triangleq \lim_{\Delta x \to 0} \frac{P[\{x \leq X(t) \leq (x + \Delta x)\}]}{\Delta x} \tag{6.13}$$

By (4.17) and (4.20), the first-order *pdf* and CDF of $X(t)$ can be obtained by

$$f_X(x,t) = \frac{\partial}{\partial x} F_X(x,t) \tag{6.14}$$

$$F_X(x,t) = \int_{-\infty}^{x} f_X(\lambda,t)d\lambda \tag{6.15}$$

The partial derivative is taken with respect to x with the CDF function fixed at t because (4.17) applies for RV $X(t)$ while t is fixed.

The statistical properties of complex $X(t)$ are determined by the statistical properties of its real and imaginary components and the joint statistics between the real and imaginary components. Therefore, the joint distribution of the real and imaginary components of $X(t)$ completely characterizes the complex $X(t)$, which is defined as follows:

$$F_{X(t)}(x_r, x_i, t) \triangleq P[\{X_r(t) \leq x_r, X_i(t) \leq x_i\}] \tag{6.16}$$

If the above joint distribution is given, the first-order distributions of the real and imaginary components of $X(t)$ are obtained as the marginal distributions of the joint distribution as follows;

$$F_{X_r(t)}(x,t) \triangleq P[\{X_r(t) \leq x\}] \tag{6.17a}$$

$$F_{X_i(t)}(x,t)\triangleq P[\{X_i(t)\leq x\}] \tag{6.17b}$$

The joint *pdf* of the real and imaginary components of a complex RV is obtained by

$$f_{X(t)}(x_r,x_i,t) = \frac{\partial^2}{\partial x_r \partial x_i} F_{X(t)}(x_r,x_i,t) \tag{6.18}$$

Its marginal *pdf*s are obtained as follows:

$$f_{X_r(t)}(x_r,t) = \int_{-\infty}^{+\infty} f_{X(t)}(x_r,x_i,t)dx_i \tag{6.19a}$$

$$f_{X_i(t)}(x_i,t) = \int_{-\infty}^{+\infty} f_{X(t)}(x_r,x_i,t)dx_r \tag{6.19b}$$

6.2.2 Second-Order Distributions

Second-Order CDF

In Fig. 6.2a, the crosshatched strips at t_1 and t_2 show the sets defining the two events of the second-order CDF of $X(t)$, $\{X(t_1)\leq x_1\}$, and $\{X(t_2)\leq x_2\}$. By (4.31), the joint CDF of $X(t_1)$ and $X(t_2)$ is defined by

$$F_X(x_1,x_2,t_1,t_2)\triangleq P[\{X(t_1)\leq x_1, X(t_2)\leq x_2\}] \tag{6.20}$$

By (4.31) and as shown in Fig. 6.2b, the joint CDF of $X(t_1)$ and $Y(t_2)$ is defined by

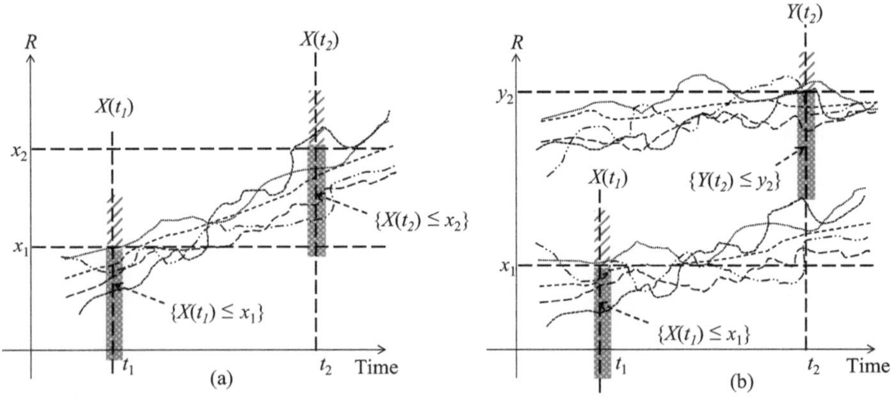

Fig. 6.2 (**a**) RVs $X(t_1)$ and $X(t_2)$, (**b**) RVs $X(t_1)$ and $Y(t_2)$

$$F_{XY}(x, y, t_1, t_2) \triangleq P[\{X(t_1) \leq x, Y(t_2) \leq y\}] \tag{6.21}$$

Second-Order *pdf*

By (4.50), the second-order *pdf* of *X*(*t*) is defined by

$$f_X(x_1, x_2, t_1, t_2) \triangleq \lim_{\Delta x \to 0} \lim_{\Delta y \to 0} \frac{1}{\Delta x \Delta y} P[\{x_1 < X(t_1) < (x_1 + \Delta x), x_2 < X(t_2) < (x_2 + \Delta x)\}] \tag{6.22}$$

By (4.52), the second-order *pdf*s can be obtained by

$$\begin{aligned} f_X(x_1, x_2, t_1, t_2) &= \frac{\partial}{\partial x_2}\left[\frac{\partial}{\partial x_1} F_X(x_1, x_2, t_1, t_2)\right] \\ &= \frac{\partial^2}{\partial x_1 \partial x_2} F_X(x_1, x_2, t_1, t_2) \end{aligned} \tag{6.23a}$$

$$\begin{aligned} f_{XY}(x, y, t_1, t_2) &= \frac{\partial}{\partial y}\left[\frac{\partial}{\partial x} F_{XY}(x, y, t_1, t_2)\right] \\ &= \frac{\partial^2}{\partial x \partial y} F_{XY}(x, y, t_1, t_2) \end{aligned} \tag{6.23b}$$

By (4.57), the second-order CDF of *X*(*t*) can be obtained by

$$F_X(x_1, x_2, t_1, t_2) = \int_{-\infty}^{x_2} \int_{-\infty}^{x_1} f_X(\lambda, \delta, t_1, t_2) \mathrm{d}\lambda \, \mathrm{d}\delta \tag{6.24}$$

By (4.60), the marginal *pdf*s can be obtained by

$$f_X(x_1, t_1, t_2) = \int_{-\infty}^{+\infty} f_X(x_1, x_2, t_1, t_2) dx_2 \tag{6.25}$$

For complex *X*(*t*) and *Y*(*t*), the second-order statistics of *X*(*t*) and the joint second-order statistics of *X*(*t*) and *Y*(*t*) involve four RVs, the real and imaginary components at the two time points of *X*(*t*) and the real and imaginary components of *X*(*t*) and *Y*(*t*) at one time point each. The CDFs and the pdfs are given by the following equations:

$$F_{XY}(x_r, x_i, y_r, y_i, t_1, t_2) \triangleq P[\{X_r(t_1) \leq x_r, X_i(t_1) \leq x_i, Y_r(t_2) \leq y_r, Y_i(t_2) \leq y_i\}] \tag{6.26}$$

$$F_{X(t)}(x_{r1},x_{i1},x_{r2},x_{i2},t_1,t_2) \triangleq P[\{X_r(t_1) \le x_{r1}, X_i(t_1) \le x_{i1}, X_r(t_2) \le x_{r2}, X_i(t_2) \le x_{i2}\}] \tag{6.27}$$

$$f_{X(t)}(x_{r1},x_{i1},x_{r2},x_{i2},t_1,t_2) = \frac{\partial^4 F_X(x_{r1},x_{i1},x_{r2},x_{i2},t_1,t_2)}{\partial x_{r1}\partial x_{r2}\partial x_{i1}\partial x_{i2}} \tag{6.28}$$

$$f_{XY}(x_r,x_i,y_r,y_i,t_1,t_2) = \frac{\partial^4 F_{XY}(x_r,x_i,y_r,y_i,t_1,t_2)}{\partial x_r \partial x_i \partial y_r \partial y_i} \tag{6.29}$$

Uncorrelated Process

Two processes $X(t)$ and $Y(t)$ are defined to be uncorrelated, if the cross-covariance between the two processes is zero at all pairs of t_1 and t_2:

$$c_{XY}(t_1,t_2) = 0 \tag{6.30}$$

Orthogonal Process

Two processes $X(t)$ and $Y(t)$ are called orthogonal if their cross-correlation is zero:

$$R_{XY}(t_1,t_2) = E\{X(t_1)Y^*(t_2)\} = 0 \tag{6.31}$$

6.3 Vector RVs

It is convenient to use vectors to handle multivariate distributions. In this book, vectors are denoted by boldface letters. This section defines vector RVs and multivariate distributions using these vector RVs. This section also explains the concept of complete statistical characterization of these vector RVs.

6.3.1 Definition of Vector RVs

Figure 6.3 shows two processes $X(t)$ and $Y(t)$ and n and m time points selected for $X(t)$ and $Y(t)$, respectively.

Consider the n and m RVs at these time points, $X(t_i), i = 1, 2, \ldots, n$ and $Y\left(t'_j\right)$, $j = 1, 2, \ldots, m$. Form an n- and m-dimensional column vectors $\boldsymbol{X}$ and $\boldsymbol{Y}$ with $X(t_i)'$ s and $Y\left(t'_j\right)'$ s as the components, respectively, as follows:

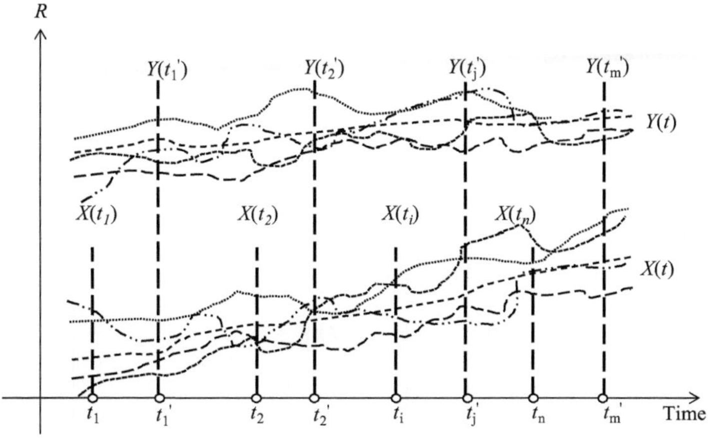

Fig. 6.3 n and m time points selected for two processes $X(t)$ and $Y(t)$

$$\boldsymbol{X} = \boldsymbol{X}(t_1, t_2, \ldots, t_i, \ldots, t_n) = \begin{bmatrix} X_1 \\ X_2 \\ \cdot \\ \cdot \\ X_i \\ \cdot \\ \cdot \\ X_n \end{bmatrix} = \begin{bmatrix} X(t_1) \\ X(t_2) \\ \cdot \\ \cdot \\ X(t_i) \\ \cdot \\ \cdot \\ X(t_n) \end{bmatrix} \tag{6.32a}$$

$$\boldsymbol{Y} = \boldsymbol{Y}\left(t'_1, t'_2, \ldots, t'_j, \ldots, t'_m\right) = \begin{bmatrix} Y_1 \\ Y_2 \\ \cdot \\ \cdot \\ Y_j \\ \cdot \\ \cdot \\ Y_m \end{bmatrix} = \begin{bmatrix} Y(t'_1) \\ Y(t'_2) \\ \cdot \\ \cdot \\ Y(t'_j) \\ \cdot \\ \cdot \\ Y(t'_m) \end{bmatrix} \tag{6.32b}$$

Concatenating $\boldsymbol{X}$ and $\boldsymbol{Y}$ into a single vector, we form a new vector $\boldsymbol{Z}_{XY}$ as follows with the subscript $\boldsymbol{XY}$ to keep track of the component vectors:

$$\boldsymbol{Z}_{XY} = \boldsymbol{Z}_{XY}\left(t_1, .., t_n, t'_1, .., t'_m\right) = \begin{bmatrix} \boldsymbol{X} \\ - \\ \boldsymbol{Y} \end{bmatrix} = \begin{bmatrix} X_1 \\ \cdot \\ X_n \\ - \\ Y_1 \\ \cdot \\ Y_m \end{bmatrix} = \begin{bmatrix} X(t_1) \\ \cdot \\ X(t_n) \\ -- \\ Y(t'_1) \\ \cdot \\ Y(t'_m) \end{bmatrix} \tag{6.33}$$

where the dashed lines inside the matrices show the partitions of the matrices. The spaces for $\boldsymbol{X}, \boldsymbol{Y}$, and $\boldsymbol{Z}_{XY}$ are the following Cartesian products:

$$\begin{aligned}\Omega_X &= R_{X_1} \times \ldots \times R_{X_n} \\ \Omega_Y &= R_{Y_1} \times \ldots \times R_{Y_m} \\ \Omega_{Z_{XY}} &= \Omega_X \times \Omega_Y = R_{X_1} \times \ldots \times R_{X_n} \times R_{Y_1} \times \ldots \times R_{Y_m}\end{aligned} \tag{6.34}$$

The vectors of the specific values that the three vector RVs take are denoted by the lowercase letters as follows:

$$\boldsymbol{x} = \begin{bmatrix} x_1 \\ x_2 \\ \cdot \\ x_i \\ \cdot \\ \cdot \\ x_n \end{bmatrix} \qquad \boldsymbol{y} = \begin{bmatrix} y_1 \\ y_2 \\ \cdot \\ y_i \\ \cdot \\ \cdot \\ y_m \end{bmatrix} \qquad \boldsymbol{z}_{XY} = \begin{bmatrix} \boldsymbol{x} \\ - \\ \boldsymbol{y} \end{bmatrix} = \begin{bmatrix} x_1 \\ \cdot \\ x_n \\ - \\ y_1 \\ \cdot \\ y_m \end{bmatrix} \tag{6.35}$$

The vector complex RVs are defined by the same equations (6.32) and (6.33), if the component RVs are considered to be complex RVs as follows:

$$\boldsymbol{X} = \boldsymbol{X}(t_1, .., t_n) = \begin{bmatrix} X_1 \\ X_2 \\ \cdot \\ \cdot \\ X_i \\ \cdot \\ \cdot \\ X_n \end{bmatrix} = \begin{bmatrix} X(t_1) \\ X(t_2) \\ \cdot \\ \cdot \\ X(t_i) \\ \cdot \\ \cdot \\ X(t_n) \end{bmatrix} = \begin{bmatrix} X_r(t_1) + jX_i(t_1) \\ X_r(t_2) + jX_i(t_2) \\ \cdot \\ \cdot \\ X_r(t_i) + jX_i(t_i) \\ \cdot \\ \cdot \\ X_r(t_n) + jX_i(t_n) \end{bmatrix} \tag{6.36}$$

$$\boldsymbol{X}^* = \boldsymbol{X}^*(t_1, .., t_n) = \begin{bmatrix} X_1^* \\ X_2^* \\ \cdot \\ \cdot \\ X_i^* \\ \cdot \\ \cdot \\ X_n^* \end{bmatrix} = \begin{bmatrix} X_1^*(t_1) \\ X_2^*(t_2) \\ \cdot \\ \cdot \\ X_i^*(t_i) \\ \cdot \\ \cdot \\ X_n^*(t_n) \end{bmatrix} = \begin{bmatrix} X_r(t_1) - jX_i(t_1) \\ X_r(t_2) - jX_i(t_2) \\ \cdot \\ \cdot \\ X_r(t_i) - jX_i(t_i) \\ \cdot \\ \cdot \\ X_r(t_n) - jX_i(t_n) \end{bmatrix} \tag{6.37}$$

$$\boldsymbol{Y} = \boldsymbol{Y}(t'_1.., t'_m) = \begin{bmatrix} Y_1 \\ Y_2 \\ \cdot \\ \cdot \\ Y_j \\ \cdot \\ \cdot \\ Y_m \end{bmatrix} = \begin{bmatrix} Y(t'_1) \\ Y(t'_2) \\ \cdot \\ \cdot \\ Y(t'_j) \\ \cdot \\ \cdot \\ Y(t'_m) \end{bmatrix} = \begin{bmatrix} Y_r(t'_1) + jY_i(t'_1) \\ Y_r(t'_2) + jY_i(t'_2) \\ \cdot \\ \cdot \\ Y_r(t'_j) + jY_i(t'_j) \\ \cdot \\ \cdot \\ Y_r(t'_m) + jY_i(t'_m) \end{bmatrix} \tag{6.38a}$$

$$
\boldsymbol{Y}^* = \boldsymbol{Y}^*(t_1^{\prime}..,t_m^{\prime}) = \begin{bmatrix} Y_1^* \\ Y_2^* \\ \cdot \\ \cdot \\ Y_j^* \\ \cdot \\ \cdot \\ Y_m^* \end{bmatrix} = \begin{bmatrix} Y^*(t_1^{\prime}) \\ Y^*(t_2^{\prime}) \\ \cdot \\ \cdot \\ Y^*(t_j^{\prime}) \\ \cdot \\ \cdot \\ Y^*(t_m^{\prime}) \end{bmatrix} = \begin{bmatrix} Y_r(t_1^{\prime}) - jY_i(t_1^{\prime}) \\ Y_r(t_2^{\prime}) - jY_i(t_2^{\prime}) \\ \cdot \\ \cdot \\ Y_r(t_j^{\prime}) - jY_i(t_j^{\prime}) \\ \cdot \\ \cdot \\ Y_r(t_m^{\prime}) - jY_i(t_m^{\prime}) \end{bmatrix} \quad (6.38b)
$$

$$
\boldsymbol{Z}_{XY} = \boldsymbol{Z}_{XY}(t_1,..,t_n,t_1^{\prime},..,t_m^{\prime}) = \begin{bmatrix} \boldsymbol{X} \\ - \\ \boldsymbol{Y} \end{bmatrix} = \begin{bmatrix} X_1 \\ \cdot \\ X_n \\ - \\ Y_1 \\ \cdot \\ Y_m \end{bmatrix} = \begin{bmatrix} X(t_1) \\ \cdot \\ X(t_n) \\ -- \\ Y(t_1^{\prime}) \\ \cdot \\ Y(t_m^{\prime}) \end{bmatrix}
$$

$$
= \begin{bmatrix} X_r(t_1) + jX_i(t_1) \\ \cdot \\ X_r(t_n) + jX_i(t_n) \\ -- \\ Y_r(t_1^{\prime}) + jY_i(t_1^{\prime}) \\ \cdot \\ Y_r(t_m^{\prime}) + jY_i(t_m^{\prime}) \end{bmatrix} \quad (6.39a)
$$

$$
\boldsymbol{Z}_{XY}^* = \boldsymbol{Z}_{XY}^*(t_1,..,t_n,t_1^{\prime},..,t_m^{\prime}) = \begin{bmatrix} \boldsymbol{X}^* \\ - \\ \boldsymbol{Y}^* \end{bmatrix} = \begin{bmatrix} X_1^* \\ \cdot \\ X_n^* \\ - \\ Y_1^* \\ \cdot \\ Y_m^* \end{bmatrix} = \begin{bmatrix} X^*(t_1) \\ \cdot \\ X^*(t_n) \\ -- \\ Y^*(t_1^{\prime}) \\ \cdot \\ Y^*(t_m^{\prime}) \end{bmatrix}
$$

$$
= \begin{bmatrix} X_r(t_1) - jX_i(t_1) \\ \cdot \\ X_r(t_n) - jX_i(t_n) \\ -- \\ Y_r(t_1^{\prime}) - jY_i(t_1^{\prime}) \\ \cdot \\ Y_r(t_m^{\prime}) - jY_i(t_m^{\prime}) \end{bmatrix} \quad (6.39b)
$$

The spaces for complex vector RVs $\boldsymbol{X}, \boldsymbol{Y}$, and $\boldsymbol{Z}_{XY}$ are the following Cartesian products:

$$
\begin{aligned}
\Omega_X &= \Omega_{X_1} \times \ldots \times \Omega_{X_n} \\
\Omega_Y &= \Omega_{Y_1} \times \ldots \times \Omega_{Y_m} \\
\Omega_{Z_{XY}} &= \Omega_{X_1} \times \ldots \times \Omega_{X_n} \times \Omega_{Y_1} \times \ldots \times \Omega_{Y_m} \\
\Omega_{X_i} &= R_{X_{ri}} \times R_{X_{ii}} \; , i = 1, \ldots, n \\
\Omega_{Y_j} &= R_{Y_{rj}} \times R_{Y_{ij}}, j = 1, \ldots, m
\end{aligned}
\tag{6.40}
$$

The vectors of the specific values that the three complex vector RVs take are denoted by the lowercase letters as follows:

$$
\boldsymbol{x} = \begin{bmatrix} x_{r1} + jx_{i1} \\ x_{r2} + jx_{i2} \\ \cdot \\ x_{ri} + jx_{ii} \\ \cdot \\ \cdot \\ x_{rn} + jx_{in} \end{bmatrix} \quad
\boldsymbol{y} = \begin{bmatrix} y_{r1} + jy_{i1} \\ y_{r2} + jy_{i2} \\ \cdot \\ y_{rj} + jy_{ij} \\ \cdot \\ \cdot \\ y_{rn} + jy_{in} \end{bmatrix} \quad
z_{XY} = \begin{bmatrix} \boldsymbol{x} \\ - \\ \boldsymbol{y} \end{bmatrix} = \begin{bmatrix} x_{r1} + jx_{i1} \\ \cdot \\ x_{rn} + jx_{in} \\ ----- \\ y_{r1} + jy_{i1} \\ \cdot \\ y_{rn} + jy_{in} \end{bmatrix}
\tag{6.41}
$$

6.3.2 Multivariate Distributions

The *n*th- and *m*th-order CDFs and *pdf*s of $X(t)$ and $Y(t)$ for the n and m time points are defined as follows, respectively:

$$
F_{X(t_1,..,t_n)}(x_1, x_2, \ldots, x_n; t_1, t_2, \ldots, t_n) \triangleq P[\{X(t_1) \leq x_1, X(t_2) \leq x_2, \ldots, X(t_n) \leq x_n\}]
\tag{6.42}
$$

$$
f_{X(t_1,..,t_n)}(x_1, x_2, \ldots, x_n; t_1, t_2, \ldots, t_n) = \frac{\partial^n}{\partial x_1 \ldots \partial x_n} F_X(x_1, x_2, \ldots, x_n, t_1, t_2, \ldots, t_n)
\tag{6.43}
$$

$$
\begin{aligned}
&F_{Y(t'_1,..,t'_m)}\left(y_1, y_2, \ldots, y_j, \ldots, y_m; t'_1, t'_2, \ldots, t'_m\right) \triangleq \\
&\qquad P\left[\left\{Y(t'_1) \leq y_1, Y(t'_2) \leq y_2, \ldots, Y(t'_m) \leq y_m\right\}\right]
\end{aligned}
\tag{6.44}
$$

$$
\begin{aligned}
&f_{Y(t'_1,..,t'_m)}\left(y_1, y_2, \ldots, y_j, \ldots, y_m; t'_1, t'_2, \ldots, t'_m\right) \\
&\quad = \frac{\partial^n}{\partial y_1 \ldots \partial y_n} F_X\left(y_1, y_2, \ldots, y_j, \ldots, y_m, t'_1, t'_2, \ldots, t'_m\right)
\end{aligned}
\tag{6.45}
$$

The $(n + m)$th-order joint CDF and the *pdf* of $X(t)$ and $Y(t)$ are given by

$$
\begin{aligned}
&F_{Z_{XY}\left(t_1,\ldots,t_n,t'_1,\ldots,t'_m\right)}\left(x_1, \ldots, x_n; y_1, \ldots, y_m; t_1, \ldots, t_n; t'_1, \ldots, t'_m\right) \\
&\triangleq P\left[\left\{X(t_1) \leq x_1, \ldots, X(t_n) \leq x_n, Y(t'_1) \leq y_1, \ldots, Y(t'_m) \leq y_m\right\}\right]
\end{aligned}
\tag{6.46}
$$

$$f_{Z_{XY}(t_1,\ldots,t_n,t'_1,\ldots,t'_m)}(x_1,\ldots,x_n;y_1,\ldots,y_m;t_1,\ldots,t_n;t'_1,\ldots,t'_m)$$
$$= \frac{\partial^n}{\partial x_1\ldots\partial x_n\partial y_1\ldots\partial y_m}F_{Z_{XY}}(x_1,\ldots,x_n;y_1,\ldots,y_m;t_1,\ldots,t_n;t'_1,\ldots,t'_m) \tag{6.47}$$

The *n*th- and *m*th-order CDFs and *pdf*s of complex *X*(*t*) and *Y*(*t*) for the *n* and *m* time points are defined as follows, respectively, involving 2*n* and 2*m* RVs:

$$F_{X(t_1,\ldots,t_n)}(x_{r_1},x_{i_1},\ldots,x_{r_n},x_{i_n},t_1,\ldots,t_n)$$
$$\triangleq P[\{X_r(t_1)\le x_{r_1},X_i(t_1)\le x_{r_i}\ldots X_r(t_n)\le x_{r_n},X_i(t_n)\le x_{r_n}\}] \tag{6.48}$$

$$f_{X(t_1,\ldots,t_n)}(x_{r_1},x_{i_1},\ldots,x_{r_n},x_{i_n},t_1,\ldots,t_n)$$
$$= \frac{\partial^{2n}F_X(x_{r_1},x_{i_1},\ldots,x_{r_n},x_{i_n},t_1,\ldots,t_n,t_1,t_2,\ldots,t_n)}{\partial x_{r_1}\ldots\partial x_{r_1}\partial x_{i_1}\ldots\partial x_{i_n}} \tag{6.49}$$

$$F_{Y(t'_1,\ldots,t'_m)}(y_{r_1},y_{i_1},\ldots,y_{r_m},y_{i_m};t'_1,t'_2,\ldots,t'_m)$$
$$\triangleq P[\{Y_r(t'_1)\le y_{r_1},Y_i(t'_1)\le y_{i_1},\ldots,Y_r(t'_m)\le y_{r_m},Y_i(t'_m)\le y_{i_m}\}] \tag{6.50}$$

$$f_{Y(t'_1,\ldots,t'_m)}(y_{r_1},y_{i_1},\ldots,y_{r_m},y_{i_m};t'_1,t'_2,\ldots,t'_m)$$
$$= \frac{\partial^{2m}F_Y(y_{r_1},y_{i_1},\ldots,y_{r_m},y_{i_m};t'_1,t'_2,\ldots,t'_m)}{\partial x_{r_1}\ldots\partial x_{r_1}\partial x_{i_1}\ldots\partial x_{i_m}} \tag{6.51}$$

The (*n* + *m*)th-order joint CDF and the *pdf* of *X*(*t*) and *Y*(*t*) are given by

$$F_{Z_{XY}(t_1,\ldots,t_n,t'_1,\ldots,t'_m)}(x_{r_1},x_{i_1},\ldots,x_{r_n};y_{r_1},y_{i_1},\ldots,y_{r_m},y_{i_m};t_1,..,t_n;t'_1,..,t'_m)$$
$$\triangleq P\{X_r(t_1)\le x_{r_1},X_i(t_1)\le,\ldots,X_r(t_n)\le x_{r_n},X_i(t_n)\le x_{r_n},$$
$$Y_r(t'_1)\le y_{r_1},Y_i(t'_1)\le y_{i_1},\ldots,Y_r(t'_m)\le y_{r_m},Y_i(t'_m)\} \tag{6.52}$$

$$f_{Z_{XY}(t_1,\ldots,t_n,t'_1,\ldots,t'_1)}(x_{r_1},x_{i_1},\ldots,x_{r_n};y_{r_1},y_{i_1},\ldots,y_{r_m},y_{i_m};t_1,..,t_n;t'_1,..,t'_m)$$
$$= \frac{\partial^{2n+2m}F_{Z_{XY}}(x_{r_1},x_{i_1},\ldots,x_{r_n};y_{r_1},y_{i_1},\ldots,y_{r_m},y_{i_m};t_1,..,t_n;t'_1,..,t'_m)}{\partial x_{r_1}\ldots\partial x_{r_1}\partial x_{r_1}\ldots\partial x_{i_n}\partial x_{r_1}\ldots\partial x_{r_1}\partial x_{i_1}\ldots\partial x_{i_m}} \tag{6.53}$$

6.3.3 Complete Statistical Characterization

Complete characterization of the statistical properties of a real stochastic process *X* (*t*) and the joint characteristics with another process *Y*(*t*) require determining the *n*th-order distributions and the (*n* + *m*)th-order joint distributions given by (6.42) through (6.47) for an infinitely large *n* and *m* and for any values of the time points, t_is and t'_1s. This may be possible if the processes are known, for example, by theory, in the form of analytical expressions such as the normal process.

Empirically, however, "complete" characterization of the statistical properties can be determined only approximately. The more time points are selected, the closer the characterization would approach complete characterization.

Complete characterization of the statistical properties of a complex stochastic process $X(t)$ requires complete characterization of the two real processes $X_r(t)$ and $X_i(t)$ and the joint statistics of $X_r(t)$ and $X_i(t)$. Complete characterization of the joint statistical properties of two complex stochastic process $X(t)$ and $Y(t)$ requires characterization of joint behaviors between the two real processes of $X(t)$ and those of $Y(t)$, a total of four pairs of joint behaviors, (X_r, Y_r), (X_r, Y_i), (X_i, Y_r) and (X_i, Y_i).

6.4 Characteristic Function

This section defines the characteristic function of an RV first for a scalar RV and then for a vector RV. This section also discusses the concept of independent stochastic processes.

6.4.1 Characteristic Function of a Scalar RV

One important transformation that is useful in analyzing the properties of RVs is the characteristic function. For a real RV X, its characteristic function is defined by the following expected value:

For continuous RV X,

$$\Psi_X(\omega) \triangleq E\{e^{j\omega X}\} = \int_{-\infty}^{+\infty} e^{j\omega x} f_X(x) \mathrm{d}x \tag{6.54}$$

For discrete RV X,

$$\psi_X(\omega) \triangleq E\{e^{j\omega X}\} = \sum_n e^{j\omega x_n} p_X(x_n) \tag{6.55}$$

where

$$p_X(x_n) = P[\{X = x_n\}] \tag{6.56}$$

The following properties are consequences of the above definition:

$$\psi_X(0) = E\{e^{j0X}\} = \int_{-\infty}^{+\infty} 1 \times f_X(x) \mathrm{d}x = 1 \tag{6.57}$$

Theorem 6.4.1

$$|\Psi_X(\omega)| \le 1 \tag{6.58}$$

Proof

$$\begin{aligned}|\Psi_X(\omega)| &= |E\{e^{j\omega X}\}| = |E(\cos \omega x) + jE(\sin \omega x)| \\ &= \sqrt{\{E(\cos \omega x)\}^2 + \{E(\sin \omega x)\}^2} \le \sqrt{E\left\{(\cos \omega x)^2\right\} + E\left\{(\sin \omega x)^2\right\}} \\ &= \sqrt{E\left\{(\cos \omega x)^2 + (\sin \omega x)^2\right\}} = \sqrt{E\{1\}} = 1\end{aligned}$$

Q.E.D.

The *pdf* of an RV X can be obtained by the following inverse transformation of its characteristic function:

$$\frac{1}{2\pi}\int_{-\infty}^{+\infty} \psi_X(\omega) e^{-j\omega x} d\omega$$

Substituting (6.54) into the above equation with a different variable z to distinguish it with the current variable x, we obtain the following equation:

$$\begin{aligned}\frac{1}{2\pi}\int_{-\infty}^{+\infty} \Psi_X(\omega) e^{-j\omega x} d\omega &= \frac{1}{2\pi}\int_{-\infty}^{+\infty}\left\{\int_{-\infty}^{+\infty} e^{j\omega z} f_X(z) dz\right\} e^{-j\omega x} d\omega \\ &= \int_{-\infty}^{+\infty}\left\{\frac{1}{2\pi}\int_{-\infty}^{+\infty} e^{j\omega(z-x)} d\omega\right\} f_X(z) dz \\ &= \int_{-\infty}^{+\infty} \delta(z-x) f_X(z) dz = f_X(x)\end{aligned} \tag{6.59}$$

where the integral inside the braces is the Dirac delta function as follows:

$$\frac{1}{2\pi}\int_{-\infty}^{+\infty} e^{j\omega(z-x)} d\omega = \delta(z-x) \tag{6.60}$$

Hence, $\Psi_X(\omega)$ *and* $f_X(x)$ are a Fourier transform pair:

$$\psi_X(\omega) \Leftrightarrow f_X(x) \tag{6.61}$$

6.4.2 Characteristic Function of a Vector RV

The characteristic function of a vector RV is defined by the following equation:

$$\Psi_X(\omega_1,\ldots,\omega_n) \triangleq E\left\{e^{j\omega^T X}\right\} = E\left\{e^{j(\omega_1 X_1+\cdots+\omega_n X_n)}\right\}$$
$$= \int_{-\infty}^{+\infty}\cdots\int_{-\infty}^{+\infty} e^{j(\omega_1 x_1+\ldots+\omega_n x_n)} f_X(x_1,\ldots,x_n)\mathrm{d}x_1\ldots\mathrm{d}x_n \tag{6.62}$$

By the similar derivation as the scalar case given by (6.59), the *pdf* of a vector RV is obtained by the following inverse transformation:

$$\frac{1}{(2\pi)^n}\int_{-\infty}^{+\infty}\cdots\int_{-\infty}^{+\infty} e^{-j(\omega_1 x_1+\omega_2 x_2+\ldots+\omega_n x_n)}\Psi_X(\omega_1,\omega_2,\ldots,\omega_n)\mathrm{d}\omega_1\ldots\mathrm{d}\omega_n$$

Substituting (6.62) into the above equation with a different variable z to distinguish it with the current variable x, we obtain the following equation:

$$\frac{1}{(2\pi)^n}\int_{-\infty}^{+\infty}\cdots\int_{-\infty}^{+\infty} e^{-j(\omega_1 x_1+\omega_2 x_2+\ldots+\omega_n x_n)}\Psi_X(\omega_1,\omega_2,\ldots,\omega_n)\mathrm{d}\omega_1\ldots\mathrm{d}\omega_n$$
$$= \frac{1}{(2\pi)^n}\int_{-\infty}^{+\infty}\cdots\int_{-\infty}^{+\infty} e^{-j(\omega_1 x_1+\ldots+\omega_n x_n)}\left\{\int_{-\infty}^{+\infty}\cdots\int_{-\infty}^{+\infty} e^{j\omega_1 z_1}\ldots e^{j\omega_n z_n} f_X(z_1,\ldots,z_n)\mathrm{d}z_1\ldots\mathrm{d}z_n\right\}\mathrm{d}\omega_1\ldots\mathrm{d}\omega_n$$
$$= \int_{-\infty}^{+\infty}\cdots\int_{-\infty}^{+\infty}\left\{\frac{1}{(2\pi)^n}\int_{-\infty}^{+\infty}\cdots\int_{-\infty}^{+\infty} e^{j\omega_1(z_1-x_1)}\ldots e^{j\omega_n(z_n-x_n)}\mathrm{d}\omega_1\ldots\mathrm{d}\omega_n\right\} f_X(z_1,\ldots,z_n)\mathrm{d}z_1\ldots\mathrm{d}z_n$$
$$= \int_{-\infty}^{+\infty}\cdots\int_{-\infty}^{+\infty}\left\{\frac{1}{(2\pi)^n}\int_{-\infty}^{+\infty}\cdots\int_{-\infty}^{+\infty} e^{j\sum_{i=1}^{n}\omega_1(z_1-x_1)}\mathrm{d}\omega_1\ldots\mathrm{d}\omega_n\right\} f_X(z_1,\ldots,z_n)\mathrm{d}z_1\ldots\mathrm{d}z_n$$
$$= \int_{-\infty}^{+\infty}\cdots\int_{-\infty}^{+\infty}\left\{\prod_{i=1}^{n}\delta(z_i-x_i)\right\} f_X(z_1,\ldots,z_n)\mathrm{d}z_1\ldots\mathrm{d}z_n$$
$$= \int_{-\infty}^{+\infty}\cdots\int_{-\infty}^{+\infty}\delta(z_i-x_i)\ldots\delta(z_n-x_n) f_X(z_1,\ldots,z_n)\mathrm{d}z_1\ldots\mathrm{d}z_n = f_X(x_1,x_2,\ldots,x_n) \tag{6.63}$$

Note that, between the fourth and fifth lines of the above equation, the following substitution is made:

$$\frac{1}{(2\pi)^n}\int_{-\infty}^{+\infty}\cdots\int_{-\infty}^{+\infty} e^{j\sum_{i=1}^{n}\omega_1(z_1-x_1)}\mathrm{d}\omega_1\ldots\mathrm{d}\omega_n = \prod_{i=1}^{n}\delta(z_i-x_i) \tag{6.64}$$

The characteristic function and the *pdf* are a Fourier pair:

$$\Psi_X(\omega_1,\omega_2,\ldots,\omega_n) \quad \Leftrightarrow \quad f_X(x_1,x_2,\ldots,x_n) \tag{6.65}$$

6.4.3 Independent Process

Two processes $X(t)$ and $Y(t)$ are said to be mutually independent *iff* the two vector RVs $\boldsymbol{X}$ and $\boldsymbol{Y}$ defined by (6.32) are mutually independent for any n and m and for any $(n + m)$ time points. $\boldsymbol{X}$ and $\boldsymbol{Y}$ are mutually independent *iff* the multivariate distributions of the concatenated vector RV $\boldsymbol{Z_{XY}}$ defined by (6.46) and (6.47) are the products of the marginal distributions as follows:

$$\begin{aligned}
&F_{\boldsymbol{Z_{XY}}\left(t_1,\ldots,t_n,t'_1,\ldots,t'_m\right)}\left(x_1,\ldots,x_n,y_1,\ldots,y_m;t_1,\ldots,t_n,t'_1,\ldots,t'_m\right)\\
&\quad = F_{\boldsymbol{X}(t_1,\ldots,t_n)}(x_1,\ldots,x_n;t_1,\ldots,t_n)\ F_{\boldsymbol{Y}\left(t'_1,t'_2,\ldots,t'_m\right)}\left(y_1,\ldots,y_m;t'_1,\ldots,t'_m\right)\\
&f_{\boldsymbol{Z_{XY}}\left(t_1,\ldots,t_n,t'_1,\ldots,t'_m\right)}\left(x_1,\ldots,x_n,y_1,\ldots,y_m;t_1,\ldots,t_n,t'_1,\ldots,t'_m\right)\\
&\quad = f_{\boldsymbol{X}(t_1,\ldots,t_n)}(x_1,\ldots,x_n;t_1,\ldots,t_n)\ f_{\boldsymbol{Y}\left(t'_1,t'_2,\ldots,t'_m\right)}\left(y_1,\ldots,y_m;t'_1,\ldots,t'_m\right)
\end{aligned} \tag{6.66}$$

Theorem 6.4.2 If two processes $X(t)$ and $Y(t)$ are mutually independent, the two processes are uncorrelated.

Proof By the definition of the cross-covariance,

$$\begin{aligned}
c_{XY}(t_1,t_2) &= E\{X(t_1)Y(t_2)\} - \mu_X(t_1)\mu_Y(t_2)\\
&= \int_{-\infty}^{+\infty}\int_{-\infty}^{+\infty} xyf_{XY}(x,y,t_1,t_2)\mathrm{d}x\mathrm{d}y - \mu_X(t_1)\mu_Y(t_2)
\end{aligned} \tag{6.67}$$

Since $X(t)$ and $Y(t)$ are independent, by (4.82b), we have

$$f_{XY}(x,y,t_1,t_2) = f_X(x,t_1)f_Y(y,t_2)$$

and, thus, the double integral of (6.67) becomes the following:

$$\int_{-\infty}^{+\infty}\int_{-\infty}^{+\infty} xyf_{XY}(x,y,t_1,t_2)\mathrm{d}x\mathrm{d}y = \int_{-\infty}^{+\infty} xf_X(x,t_1)dx \int_{-\infty}^{+\infty} yf_Y(y,t_2)\mathrm{d}y = \mu_X(t_1)\mu_Y(t_2) \tag{6.68}$$

Substituting (6.68) into (6.67), we have

$$c_{XY}(t_1,t_2) = 0$$

By (6.30), $X(t)$ and $Y(t)$ are uncorrelated.

Q.E.D.

6.5 Stationarity

To analyze the dynamic properties of a stochastic process, we consider that the selected time points are moving to the right and examine how the static properties change as the time points move. One important concept in describing the dynamic properties of a stochastic process is its stationarity. Unless the stationarity is established for the stochastic process under consideration, the analysis of the process often becomes intractable.

The analysis of stationarity is much more complicated, if the process is complex rather than real. We will discuss the stationarity for a real process first, and, then, for a complex process.

If the static properties associated with the selected time points remain unchanged as the time points are shifted, the process is considered stationary with respect to the properties. For example, if the first-order properties do not change as the single selected time point is varied, the process is considered to be first-order stationary. Similarly, if the second-order properties derived for two selected time points do not change as the two time points are shifted concurrently with the interval length between the time points kept unchanged, the process is taken to be second-order stationary.

6.5.1 *n*th-Order Stationarity

Given a stochastic process $X(t)$, consider the vector RV $\boldsymbol{X}$ defined by (6.32a) for n arbitrary time points $t_1, t_2, \ldots, t_i, \ldots, t_n$. Shift the n time points by the same amount τ to n new time points, $t_1+\tau, t_2+\tau, \ldots, t_i+\tau, \ldots, t_n+\tau$, and consider the corresponding vector RV at these new time points. To distinguish the two vectors, we will use time as the argument, $\boldsymbol{X}(t_1, .., t_n)$ and $\boldsymbol{X}(t_1+\tau, .., t_n+\tau)$.

Figure 6.4a shows the original n time points and the associated n RVs and the new n time points obtained by shifting the original time points by the same amountτ and the associated new n RVs. The process is nth-order stationary, if its nth-order distribution for arbitrarily selected time points defined by (6.42) and (6.43) and all lower kth-order distributions remain unchanged at the two sets of time points as follows for all $k \leq n$:

$$F_{X(t_1,\ldots,t_k)}(x_1,\ldots,x_k,t_1,\ldots,t_k) = F_{X(t_1+\tau\ldots,t_k+\tau)}(x_1,\ldots,x_k,t_1+\tau\ldots,t_k+\tau),\quad \text{for } k=1,2,\ldots,n \tag{6.69}$$

$$f_{X(t_1,\ldots,t_k)}(x_1,\ldots,x_k,t_1,\ldots,t_k) = f_{X(t_1+\tau\ldots,t_k+\tau)}(x_1,\ldots,x_k,t_1+\tau\ldots,t_k+\tau),\quad \text{for } k=1,2,\ldots,n \tag{6.70}$$

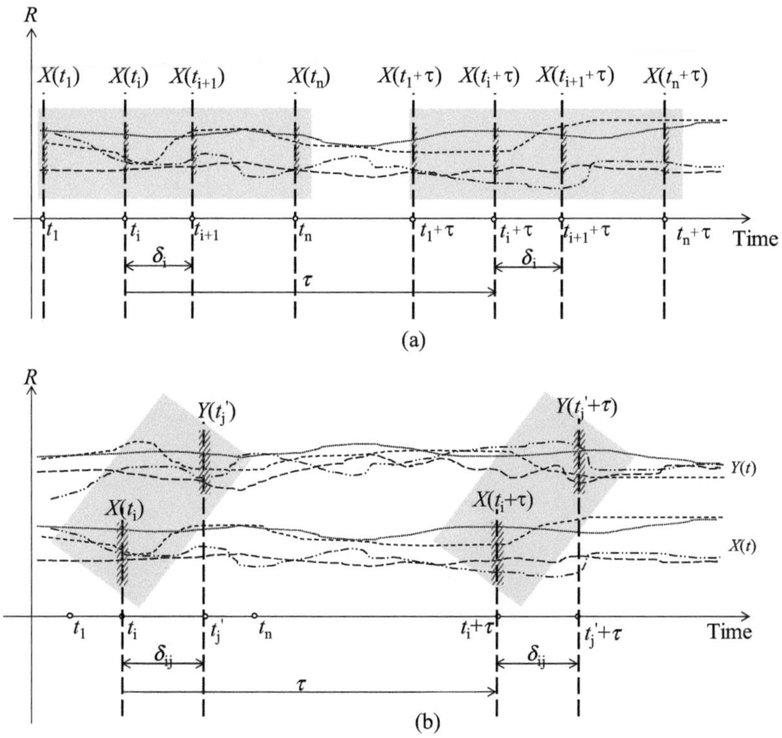

Fig. 6.4 (**a**) Shifting n time points on $X(t)$ by the same amount τ, (**b**) Shifting n time points on $X(t)$ and m time points on $Y(t)$ by the same amount τ

By this definition, if a process is nth-order stationary, it is kth-order stationary for $k < n$.

One parameter that affects the distributions is the distance between the two adjacent time points, which we will call the interval length, as follows:

$$\delta_i = t_{i+1} - t_i, i = 1, \ldots, n-1 \tag{6.71}$$

Since all original time points are shifted by the same amount τ, the interval lengths remain unchanged with the new time points as follows:

$$(t_{i+1} + \tau) - (t_i + \tau) = t_{i+1} - t_i = \delta_i, i = 1, \ldots, n-1 \tag{6.72}$$

The nth-order stationarity defined by (6.69) and (6.70) may, therefore, be stated as the following theorem.

Theorem 6.5.1 If a process $X(t)$ is nth-order stationary, its nth-order distribution remains unchanged for any n time points provided the interval lengths, δ_i's, between the n time points are kept the same. Stated alternatively, if a process $X(t)$ is nth-order stationary, its nth-order distribution may be expressed as a function of the relative interval lengths, δ_i's, instead of the absolute time points as follows:

$$F_{X(t_1,\ldots,t_n)}(x_1,\ldots,x_k,t_1,\ldots,t_k) = F_{X(t_1+\tau,\ldots,t_n+\tau)}(x_1,x_2,\ldots,x_n,\delta_1,\delta_2,\ldots,\delta_n) \tag{6.73}$$

$$f_{X(t_1,\ldots,t_n)}(x_1,x_2,\ldots,x_n,t_1,t_2,\ldots,t_n) = f_{X(t_1+\tau,\ldots,t_n+\tau)}(x_1,x_2,\ldots,x_n,\delta_1,\delta_2,\ldots,\delta_n) \tag{6.74}$$

Similarly, if $X(t)$ and $Y(t)$ are jointly stationary, the $(n+m)$th-order joint distribution may be expressed as a function of the relative interval lengths, δ_i's and δ_{ij}"s, instead of the $(n+m)$ absolute time points as follows:

$$\begin{aligned} &F_{Z_{XY}(t_1,\ldots,t_n,t'_1,\ldots,t'_m)}(x_1,\ldots,x_n;y_1,\ldots,y_m;t_1,\ldots,t_n,t'_1,\ldots,t'_m) \\ &\quad = F_{Z_{XY}(t_1+\tau,\ldots,t_n+\tau;\ t'_1+\tau,\ldots,t'_m+\tau)}(x_1,\ldots,x_n;y_1,\ldots,y_m;\delta_1,\delta_2,\ldots,\delta_n;\delta'_{11},\delta'_{12},\ldots,\delta'_{nm}) \end{aligned} \tag{6.75}$$

$$\begin{aligned} &f_{Z_{XY}(t_1,\ldots,t_n,t'_1,\ldots,t'_m)}(x_1,\ldots,x_n;y_1,\ldots,y_m;t_1,\ldots,t_n,t'_1,\ldots,t'_m) \\ &\quad = f_{Z_{XY}(t_1+\tau,\ldots,t_n+\tau;\ t'_1+\tau,\ldots,t'_m+\tau)}(x_1,\ldots,x_n;y_1,\ldots,y_m;\delta_1,\delta_2,\ldots,\delta_n;\delta'_{11},\delta'_{12},\ldots,\delta'_{nm}) \end{aligned} \tag{6.76}$$

where δ_i's and δ_{ij}"s are the interval lengths between two time points given by (6.71).

For the poof of the theorem, see the discussion leading up to (6.71) and (6.72).

We now consider the stationarity of a complex stochastic process. A complex stochastic process $X(t)$ defined by (6.10) is called stationary if its real and imaginary parts, $X_r(t)$ and $X_i(t)$, are jointly stationary in accordance with the definition of the joint stationarity given by (6.87) and (6.88) to be discussed a little later. Given a complex process $X(t)$, consider the vector complex RV $\boldsymbol{X}(t_1,\ldots,t_n)$ defined by (6.36) for n arbitrary time points $t_1, t_2, \ldots, t_i, \ldots, t_n$. Shift the original n time points by the same amount τ to n new time points, $t_1+\tau, t_2+\tau, \ldots, t_i+\tau, \ldots, t_n+\tau$, and consider the corresponding vector complex RV $\boldsymbol{X}(t_1+\tau,\ldots,t_n+\tau)$. If the statistical properties of the two vector complex RVs are the same, the complex process $X(t)$ is nth-order stationary.

For the complex process $X(t)$, for each time point t_i, the corresponding complex RV $X(t_i)$ is determined by two real RVs, $X_r(t_i)$ and $X_i(t_i)$, and, thus, the kth-order characterization of $X(t)$ at k time points involves $2k$ RVs and their multivariate joint distributions. The complex process $X(t)$ is nth-order stationary, if its nth-order distribution for arbitrarily selected time points defined by (6.48) and (6.49) and all lower kth-order distributions remain unchanged at the two sets of time points for all $k \le n$ as follows:

$$\begin{aligned} &F_{X(t_1,\ldots,t_k)}(x_{r1},x_{i1},\ldots,x_{rk},x_{ik},t_1,\ldots,t_k) \\ &\quad = F_{X(t_1+\tau\ldots,t_k+\tau)}(x_{r1},x_{i1},\ldots,x_{rk},x_{ik},t_1+\tau,\ldots,t_k+\tau) \\ &\quad = F_{X(t_1+\tau\ldots,t_k+\tau)}(x_{r1},x_{i1},\ldots,x_{rk},x_{ik},t_1+\tau,\ldots,t_k+\tau), \text{for } k=1,2,\ldots,n \end{aligned}$$

$$\begin{aligned} & f_{X(t_1,\ldots,t_k)}(x_{r1}, x_{i1}, \ldots, x_{rk}, x_{ik}, t_1, \ldots, t_k) \\ & \quad = f_{X(t_1+\tau\ldots,t_k+\tau)}(x_{r1}, x_{i1}, \ldots, x_{rk}, x_{ik}, t_1+\tau\ldots, t_k+\tau) \\ & \quad = f_{X(t_1+\tau\ldots,t_k+\tau)}(x_{r1}, x_{i1}, \ldots, x_{rk}, x_{ik}, t_1+\tau, \ldots, t_k+\tau), \text{for } k = 1, 2, \ldots, n \end{aligned} \tag{6.77}$$

By this definition, if a process is nth-order stationary, it is kth-order stationary for $k < n$.

6.5.2 Strict Sense Stationarity

Strict sense stationarity (SSS) is the generalization of the nth-order stationarity defined by (6.69) and (6.70). A stochastic process $X(t)$ is called strict sense stationary if the two vector RVs $\boldsymbol{X}(t_1, \ldots, t_n)$ and $\boldsymbol{X}(t_1+\tau, \ldots, t_n+\tau)$ have the same nth-order distribution for any arbitrarily selected n time points $t_1, t_2, \ldots, t_n$ and for any arbitrary value of τ. The arbitrary nth-order distribution of $X(t)$ determined in this manner is called the "complete" statistical characterization of $X(t)$ as discussed in Sect. 6.3.3.

The definition of the SSS for a complex process is same as that for a real process. The complete statistical properties of the process should remain unchanged with a shift of time. Section 6.3.3 discusses what it means to completely characterize the statistical properties of a complex process.

6.5.3 First-Order Stationarity

In the definition of the nth-order stationarity given by (6.69) and (6.70), let $n = 1$ and select a single time point t. If the CDF and the *pdf* remain unchanged as the time point is moved from t to $(t + \tau)$, the process is first-order stationary. If the process is first-order stationary, (6.69) and (6.70) yield the following equations:

$$\begin{aligned} F_{X(t)}(x, t) &= F_{X(t+\tau)}(x, t+\tau) \\ f_{X(t)}(x, t) &= f_{X(t+\tau)}(x, t+\tau) \end{aligned}$$

The above equations show that, if the process is first-order stationary, the distribution does not change as time varies, e.g., from t to $(t + \tau)$, that is, its CDF and *pdf* are independent of t as follows:

$$F_{X(t)}(x, t) = F_X(x) \tag{6.78a}$$

$$f_{X(t)}(x, t) = f_X(x) \tag{6.78b}$$

A converse statement is also true as follows. If the distribution of the RV X at an arbitrary time point is independent of time, the process $X(t)$ is first-order stationary.

This statement is true by the definition of stationarity. In fact, by the definition of stationarity, we can state that, if the distribution of the RV X at an arbitrary time point is independent of time, the process $X(t)$ is nth-order stationary for any n.

Similarly, for the complex process, if the process is first-order stationary, its CDF and *pdf* are independent of t and vice versa as follows:

$$F_{X(t)}(x_r, x_i, t) = F_{X(t+\tau)}(x_r, x_i; t+\tau) = F_X(x_r, x_i) \tag{6.79a}$$

$$f_{X(t)}(x_r, x_i, t) = f_{X(t+\tau)}(x_r, x_i; t+\tau) = f_X(x_r, x_i) \tag{6.79b}$$

The marginal distributions should be independent of time as follows:

$$F_{X_r}(x_r, t) = F_{X_r(t+\tau)}(x_r; t+\tau) = F_{X_r}(x_r) \tag{6.80a}$$

$$f_{X_r}(x_r, t) = f_{X_r(t+\tau)}(x_r; t+\tau) = f_{X_r}(x_r) \tag{6.80b}$$

$$F_{X_i}(x_i, t) = F_{X_r(t+\tau)}(x_i; t+\tau) = F_{X_i}(x_i) \tag{6.80c}$$

$$f_{X_i}(x_i, t) = f_{X_i(t+\tau)}(x_i; t+\tau) = f_{X_i}(x_i) \tag{6.80d}$$

The weakest case of the stationarity of a complex process requires that the joint distribution between the real and imaginary components of the process at a single time point should be invariant under the shift of time.

Fig. 6.5 (a) illustrates a first-order stationary process and (b), a non-stationary process.

6.5.4 Second-Order Stationarity

Refer to Fig. 6.6a for a discussion of the second-order stationarity. If a process is second-order stationary, (6.69) and (6.70) yield the following equations, which show that the second-order distribution remains unchanged, if the two time points are shifted by the same amount τ:

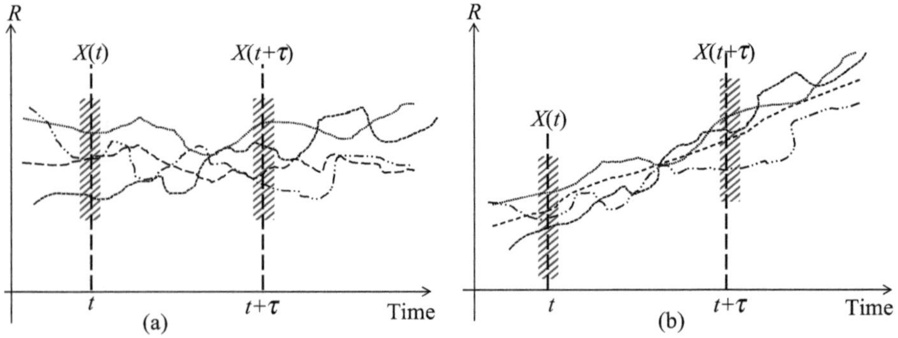

Fig. 6.5 (**a**) Illustration of a first-order stationary process, (**b**) Illustration of a non- stationary process

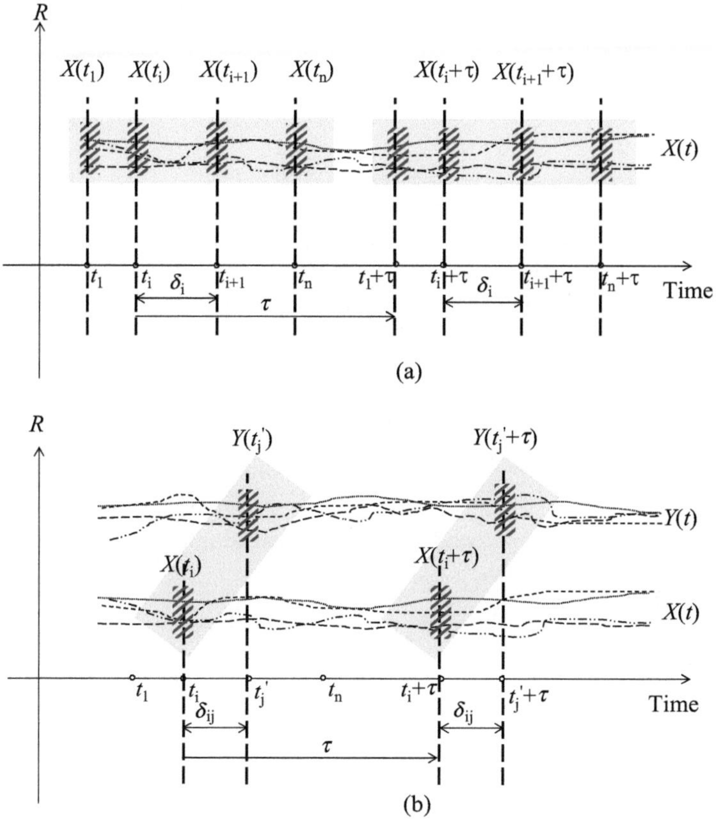

Fig. 6.6 (a) Illustration of of second order stationarity of $X(t)$, (**b**) Illustration of of second order stationarity of $X(t)$ and $Y(t)$

$$F_{X(t_1,t_2)}(x_1, x_2, t_1, t_2) = F_{X(t_1+\tau,t_2+\tau)}(x_1, x_2, t_1+\tau, t_2+\tau) \tag{6.81a}$$

$$f_{X(t_1,t_2)}(x_1, x_2, t_1, t_2) = f_{X(t_1+\tau,t_2+\tau)}(x_1, x_2, t_1+\tau, t_2+\tau) \tag{6.81b}$$

A converse statement is also true as follows. If the second-order distribution of the RVs at two arbitrary time points are equal to that at the two new time points obtained by shifting the original time points by the same amount τ, the process is second-order stationary.

If a process is second-order stationary, (6.73) and (6.74) yield the following equations, which show that the second-order distribution depends on the interval length δ between the two time points only and not on the specific time points:

$$F_{X(t_1,t_2)}(x_1, x_2, t_1, t_2) = F_{X(t_1+\tau,t_2+\tau)}(x_1, x_2, t_1+\tau, +t_1+\tau) = F_{X(t_1,t_2)}(x_1, x_2, \delta) \tag{6.82a}$$

$$f_{X(t_1,t_2)}(x_1,x_2,t_1,t_2) = f_{X(t_1+\tau,t_2+\tau)}(x_1,x_2,t_1+\tau,+t_1+\tau) = f_{X(t_1,t_2)}(x_1,x_2,\delta) \tag{6.82b}$$

where

$$\delta = t_2 - t_1 \; or \; t_2 = \delta + t_1 \tag{6.83}$$

A converse statement of the above statement is also true.

If a complex process is second-order stationary, the second-order distributions, which are defined by (6.48) and (6.49) with $n = 2$, remain unchanged with a shift of time as follows:

$$\begin{aligned} F_{X(t_1,t_2)}(x_{r1},x_{i1},x_{r2},x_{i2},t_1,t_2) &= F_{X(t_1+\tau,t_2+\tau)}(x_{r1},x_{i1},x_{r2},x_{i2},t_1+\tau,t_2+\tau) \\ f_{X(t_1,t_2)}(x_{r1},x_{i1},x_{r2},x_{i2},t_1,t_2) &= f_{X(t_1+\tau,t_2+\tau)}(x_{r1},x_{i1},x_{r2},x_{i2},t_1+\tau,t_2+\tau) \end{aligned} \tag{6.84}$$

By a same derivation as that for the real process as given by (6.73) and (6.74), the second-order distributions depend on the interval length δ only and not on the specific time points as follows:

$$\begin{aligned} F_{X(t_1,t_2)}(x_{r1},x_{i1},x_{r2},x_{i2},t_1,t_2) &= F_{X(t_1+\tau,t_2+\tau)}(x_{r1},x_{i1},x_{r2},x_{i2},t_1+\tau,\delta+t_1+\tau) \\ &= F_{X(t_1,t_2)}(x_{r1},x_{i1},x_{r2},x_{i2},\delta) \\ f_{X(t_1,t_2)}(x_{r1},x_{i1},x_{r2},x_{i2},t_1,t_2) &= f_{X(t_1+\tau,t_2+\tau)}(x_{r1},x_{i1},x_{r2},x_{i2},t_1+\tau,\delta+t_1+\tau) \\ &= f_{X(t_1,t_2)}(x_{r1},x_{i1},x_{r2},x_{i2},\delta) \end{aligned} \tag{6.85}$$

where

$$\delta = t_2 - t_1 \; or \; t_2 = \delta + t_1 \tag{6.86}$$

We will see in a later section that, with the above distribution, the autocorrelation function of a second-order stationary process depends only on δ.

By the definition of the general nth-order stationarity given by (6.77), the nth-order stationarity guarantees the kth-order stationarity for all $k \le n$. Therefore, if a complex process is second-order stationary, the process is also first-order stationary, and, thus, its first-order distribution is independent of time and its mean is constant.

6.5.5 Wide Sense Stationarity (WSS)

We will see later that, if the process is second-order stationary, the autocorrelation function depends only on the interval length δ between two time points. By the definition of the general nth-order stationarity given by (6.69) and (6.70), the nth-order stationarity guarantees the kth-order stationarity for all $k \leq n$. Therefore, if a process is second-order stationary, the process is also first-order stationary, and, thus, its first-order distribution is independent of time and its mean is constant. In conclusion, if a process is at least second-order stationary, its mean is constant and its autocorrelation depends on δ only. The stationarity in which mean is constant and autocorrelation depends on the interval length only is called the WSS. Therefore, the second-order stationarity implies the WSS. However, the converse is not necessarily true except for certain special cases. This will be discussed further later in conjunction with autocorrelation.

6.5.6 (n + m)th-Order Joint Stationarity

Referring to Fig. 6.4b, two processes $X(t)$ and $Y(t)$ are jointly stationary, if their $(n + m)$th joint distributions defined by (6.46) and (6.47) remain unchanged with a shift of the time points by the same amount τ as follows:

$$F_{Z_{XY}(t_1,..,t_n,t'_1,..,t'_m)}(x_1,..,x_n,y_1,..,y_m;t_1,..,t_n,t'_1,..,t'_m)$$
$$= F_{Z_{XY}(\ t_1+\tau,..,t_n+\tau;\ t'_1+\tau,..,t'_m+\tau)}(x_1,\ldots,x_k,y_1,\ldots,y_m;t_1+\tau,..,t_n+\tau;\ t'_1+\tau,..,t'_m+\tau) \tag{6.87}$$

$$f_{Z_{XY}(t_1,..,t_n,t'_1,..,t'_m)}(x_1,\ldots,x_k,y_1,\ldots,y_m;t_1,..,t_n;t'_1,..,t'_m)$$
$$= f_{Z_{XY}(\ t_1+\tau,..,t_n+\tau;\ t'_1+\tau,..,t'_m+\tau)}(x_1,\ldots,x_k,y_1,\ldots,y_m;t_1+\tau,..,t_n+\tau;\ t'_1+\tau,..,t'_m+\tau) \tag{6.88}$$

The joint stationarity between two processes, however, does not necessarily guarantee the stationarity of individual processes, that is, even if two processes are individually non-stationary, their joint behavior can be stationary.

Two complex processes $X(t)$ and $Y(t)$ are jointly stationary, if their $(n + m)$th joint distributions defined by (6.52) and (6.53) remain unchanged with a shift of time points as follows:

$$F_{Z_{XY}(t_1,\ldots,t_n,t'_1,\ldots,t'_m)}(x_{r1},x_{i1},\ldots,x_{rn},x_{in},y_{r1},y_{i1},\ldots,y_{rm},y_{im};t_1,\ldots,t_n,t'_1,\ldots,t'_m)$$
$$= F_{Z_{XY}(t_1+\tau,\ldots,t_n+\tau;\ t'_1+\tau,\ldots,t'_m+\tau)}(x_{r1}x_{i1},\ldots,x_{rn},x_{in},y_{r1},y_{i1},\ldots,y_{rm},y_{im};t_1+\tau,\ldots,t_n+\tau;t'_1+\tau,\ldots,t'_m+\tau) \tag{6.89}$$

$$f_{Z_{XY}(t_1,\ldots,t_n,t'_1,\ldots,t'_m)}(x_{r1},x_{i1},\ldots,x_{rn},x_{in},y_{r1},y_{i1},\ldots,y_{rm},y_{im};t_1,\ldots,t_n,t'_1,..,t'_m)$$
$$= f_{Z_{XY}(t_1+\tau,\ldots,t_n+\tau;\ t'_1+\tau,\ldots,t'_m+\tau)}(x_1,\ldots,x_k,y_1,\ldots,y_m;t_1+\tau,\ldots,t_n+\tau;\ t'_1+\tau,\ldots,t'_m+\tau) \tag{6.90}$$

The joint stationarity does not necessarily guarantee the stationarity of individual processes.

6.5.7 Joint Second-Order Stationarity

As shown in Fig. 6.6b, the second-order joint stationarity is defined between two time points, one for $X(t)$ and another for $Y(t)$, in terms of the distribution of the concatenated vector RVs defined by (6.33) as follows:

$$F_{Z_{XY}(t_1,t_1')}(x_1,y_1;t_1,t_1') = F_{Z_{XY}(t_1+\tau,t_1'+\tau)}(x_1,y_1;t_1+\tau,t_1'+\tau) = F_{Z_{XY}(t_1+\tau,t_1'+\tau)}(x_1,y_1,\delta) \tag{6.91a}$$

$$f_{Z_{XY}(t_1,t_1')}(x_1,y_1;t_1,t_1') = f_{Z_{XY}(t_1+\tau,t_2+\tau)}(x_1,y_1;t_1+\tau,t_2+\tau) = f_{Z_{XY}(t_1+\tau,t_2+\tau)}(x_1,y_1,\delta) \tag{6.91b}$$

In the above expressions, since we only need one time point for $X(t)$ and $Y(t)$, we can replace x_1 , y_1, and t_1' with x , y , and t_2 and rewrite the equation as follows for simplicity:

$$F_{Z_{XY}(t_1,t_2)}(x,y;t_1,t_2) = F_{Z_{XY}(t_1+\tau,t_2+\tau)}(x,y;t_1+\tau,t_2+\tau) = F_{Z_{XY}(t_1+\tau,t_2+\tau)}(x,y,\delta) \tag{6.92a}$$

$$f_{Z_{XY}(t_1,t_2)}(x,y;t_1,t_2) = f_{Z_{XY}(t_1+\tau,t_2+\tau)}(x,y;t_1+\tau,t_2+\tau) = f_{Z_{XY}(t_1+\tau,t_2+\tau)}(x,y,\delta) \tag{6.92b}$$

Note that the second-order joint stationarity provides the "relative" stationarity as the two processes move together maintaining the interval length between the two RVs under consideration constant, but does not guarantee the stationarity of the individual processes. For example, even if the statistical properties of both processes may change with the shift of time, they may change together in such a way that their relative second-order statistical properties stay the same with the shift of time, satisfying the condition for the joint second-order stationary. Therefore, as we will see later, if two processes are second-order joint stationary, their autocorrelation depends only on the interval length δ between two time points, but their respective means are not necessarily constant. Therefore, for the joint statistics between two different processes, the joint second-order stationarity and the joint WSS are not the same.

Theorem 6.5.2 If $X(t)$ and $Y(t)$ are jointly stationary, the $(n+m)$th joint distributions may be expressed as a function of the relative interval lengths, δ_i's and δ_{ij}"s, instead of the $(n+m)$ absolute time points as follows:

$$\begin{aligned} &F_{Z_{XY}(t_1,\ldots,t_n,t_1',\ldots,t_m')}(x_1,\ldots,x_n;y_1,\ldots,y_m;t_1,\ldots,t_n,t_1',\ldots,t_m') \\ &= F_{Z_{XY}(t_1+\tau,\ldots,t_n+\tau;\ t_1'+\tau,\ldots,t_m'+\tau)}(x_1,\ldots,x_n;y_1,\ldots,y_m;\delta_1,\delta_2,\ldots,\delta_n;\delta_{11}',\delta_{12}',\ldots,\delta_{nm}') \end{aligned} \tag{6.93}$$

$$
\begin{aligned}
&f_{Z_{XY}\left(t_1,\ldots,t_n,t'_1,\ldots,t'_m\right)}\left(x_1,\ldots,x_n;y_1,\ldots,y_m;t_1,\ldots,t_n,t'_1,\ldots,t'_m\right)\\
&=f_{Z_{XY}\left(t_1+\tau,\ldots,t_n+\tau;\; t'_1+\tau,\ldots,t'_m+\tau\right)}\left(x_1,\ldots,x_n;y_1,\ldots,y_m;\delta_1,\delta_2,\ldots,\delta_n;\delta'_{11},\delta'_{12},\ldots,\delta'_{nm}\right)
\end{aligned}
\tag{6.94}
$$

where δ_i's and δ_{ij}"s are the interval lengths given by (6.71).

For the proof of the theorem, see the discussion leading up to (6.71) and (6.72).

Now, consider the complex process. The joint second-order stationarity of the complex process is defined between two time points, one for *X*(*t*) and another for *Y*(*t*), in terms of the distribution of the concatenated vector RVs defined by (6.33) as follows:

$$
\begin{aligned}
F_{Z_{XY}\left(t_1,t'_1\right)}\left(x_r,x_i,y_r,y_i;t_1,t'_1\right) &= F_{Z_{XY}\left(t_1+\tau,t'_1+\tau\right)}\left(x_r,x_i,y_r,y_i;t_1+\tau,t'_1+\tau\right)\\
&= F_{Z_{XY}\left(t_1+\tau,t'_1+\tau\right)}\left(x_r,x_i,y_r,y_i,\delta\right)
\end{aligned}
\tag{6.95}
$$

$$
\begin{aligned}
f_{Z_{XY}\left(t_1,t'_1\right)}\left(x_r,x_i,y_r,y_i;t_1,t'_1\right) &= f_{Z_{XY}\left(t_1+\tau,t'_1+\tau\right)}\left(x_r,x_i,y_r,y_i;t_1+\tau,t'_1+\tau\right)\\
&= f_{Z_{XY}\left(t_1+\tau,t'_1+\tau\right)}\left(x_r,x_i,y_r,y_i,\delta\right)
\end{aligned}
\tag{6.96}
$$

where $\delta = t'_1 - t_1$.

6.5.8 Jointly WSS Process

Note that WSS was defined earlier as a noun, "Wide Sense Staionarity." As in this section, WSS is sometimes used as an adjective, "Wide Sense Staionary." Two processes *X*(*t*) and *Y*(*t*) are jointly WSS, if the two processes are individually WSS and the cross-correlation between the two processes, which will be discussed later, depends only on the interval length between two time points.

Two complex processes *X*(*t*) and *Y*(*t*) are jointly WSS, if the two processes are individually WSS and the cross-correlation between the two processes depends only on the interval length between the two time points. Complex processes *X*(*t*) and *Y*(*t*) are individually WSS, if their real and imaginary components $X_r(t)$ and $X_i(t)$ and $Y_r(t)$ and $Y_i(t)$ are, respectively, jointly WSS.

6.6 Ergodicity

The ergodicity is another important dynamic property of a stochastic process, related to the stationarity. By the general definition of the ergodicity, a process is said to be ergodic with respect to a statistical parameter such as the mean, if the parameter characterized over time along a randomly selected sample path is equivalent to the same parameter characterized across the ensemble of the sample paths at a fixed time. For example, a process is said to be mean-ergodic, if the mean of the process taken along a sample path is equal to the mean of the process across the ensemble at any time t or, simply stated, if the time mean is same as the ensemble mean.

It would not be possible to assume the ergodicity for a process without establishing the required stationarity for the process. Certain statistical conditions must be satisfied to determine whether a process has the ergodicity property.

Consider the measurements taken over a period of length T on multiple sample paths of a process $X(t)$ and let

n = total number of sample paths
m = total number of data taken on each sample path over a period of length T
x_i^j = value of $X(t_i)$ of the jth sample path

The mean of the process at time t_i, $\mu_X(t_i)$, is estimated by the average of n data points of the ensemble of n sample paths. This average is called the ensemble average, which is given by

$$\text{Ensemble average} \qquad \widehat{\mu_X(t_i)} = \frac{1}{n}\sum_{j=1}^{n} x_i^j \tag{6.97}$$

Another kind of average of a stochastic process is obtained by selecting a sample path randomly and taking the average of m data points taken on that sample path over the measurement period T. This average is called the time average and is given by

$$\text{Time average} \qquad \widehat{\mu_X^j} = \frac{1}{m}\sum_{i=1}^{m} x_i^j \tag{6.98}$$

A process is called mean-ergodic if the time average approaches the ensemble average for a large T.

Example 6.6.1
Figure 6.7 illustrates the concept of the mean-ergodicity and its application for estimating the mean of a stochastic process.

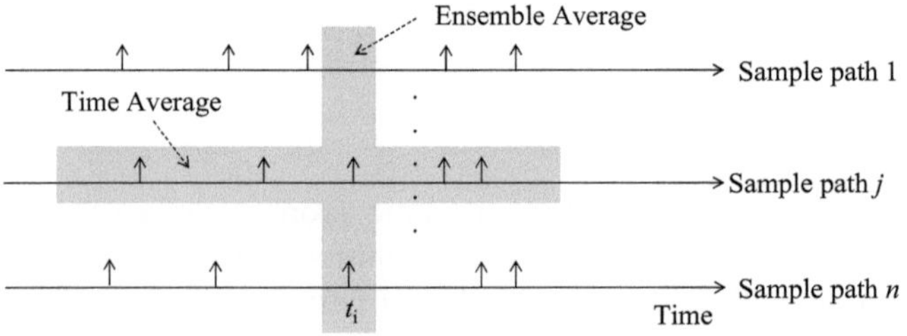

Fig. 6.7 Illustration of the concept of the mean-ergodicity

6.7 Parameters of a Stochastic Process

This section discusses the following parameters of a stochastic process: mean, variance, autocorrelation, autocovariance, cross-correlation, cross-covariance, and correlation coefficient. We will discuss the parameters for a complex process first. The parameters for the real process are then derived simply by setting the imaginary components to zero.

6.7.1 Mean and Variance

The mean of a complex RV is defined by (5.8) and (5.9). The mean of the complex process is defined by the same equations for the RV $X(t)$ at an arbitrary time point t of the process except that the mean of the stochastic process is shown as a function of time t. By the definition given by (5.8) and (5.9) and using the marginal *pdf*s given by (6.19), we obtain the following equation for the mean of the complex $X(t)$:

$$\mu_X(t) = \mu_{X_r}(t) + j\mu_{X_i}(t) \tag{6.99}$$

where

$$\mu_{X_r}(t) = \int_{-\infty}^{+\infty} x_r f_{X(t)}(x_r, t)\mathrm{d}x_r \tag{6.100a}$$

$$\mu_{X_i}(t) = \int_{-\infty}^{+\infty} x_i f_{X_i}(x_i)\mathrm{d}x_i \tag{6.100b}$$

Using (5.29) through (5.31), we obtain the variance of $X(t)$ in terms of the variances of the real and imaginary components of $X(t)$ as follows:

$$\sigma_X(t)^2 = \mathrm{Var}\{X(t)\} \triangleq E\left[\{X(t) - \mu_X(t)\}\{X(t) - \mu_X(t)\}^*\right] = \sigma_{X_r}(t)^2 + \sigma_{X_i}(t)^2 \tag{6.101}$$

where

$$\sigma_{X_r}(t)^2 = E\left\{X_r(t)^2\right\} - \mu_{X_r}(t)^2 = \int_{-\infty}^{+\infty} x^2 f_{X_r}(x, t)\mathrm{d}x - \mu_{X_r}(t)^2 \tag{6.102}$$

$$\sigma_{X_i}(t)^2 = E\left\{X_i(t)^2\right\} - \mu_{X_i}(t)^2 = \int_{-\infty}^{+\infty} x^2 f_{X_i}(x, t)\mathrm{d}x - \mu_{X_i}(t)^2 \tag{6.103}$$

and $f_{X_r}(x,t)$ and $f_{X_i}(x,t)$ are the pdf's of the real and imaginary components of *X*(*t*). The above equations show that the variance of the complex *X*(*t*) is the sum of the variances of its real and imaginary components.

For a real process, the mean and variance are obtained by (6.99) and (6.101) by setting $\mu_{X_i} = 0$ and $\sigma_{X_i}(t)^2 = 0$:

$$\mu_X(t) = \mu_{X_r}(t) \tag{6.104a}$$

$$\sigma_X(t)^2 = \sigma_{X_r}(t)^2 \tag{6.104b}$$

If a process is first-order stationary, its mean is constant as follows. Using the marginal *pdfs* given by (6.80), we obtain the following constant mean:

$$\begin{aligned}\mu_X(t) &= E\{X_r(t)\} + jE\{X_i(t)\} = \int_{-\infty}^{+\infty} x_r f_{X_r}(x_r)\mathrm{d}x_r + j\int_{-\infty}^{+\infty} x_i f_{X_i}(x_i)\mathrm{d}x_i \\ &= \mu_{X_r} + j\mu_{X_i} = \mu_X\end{aligned} \tag{6.105}$$

However, the converse is not necessarily true, that is, if the mean is constant, the process is not necessarily first-order stationary. To illustrate this point, consider the means of a real process at two different time points as follows:

$$\mu_X(t) = \int_{-\infty}^{+\infty} x f_X(x,t)\mathrm{d}x$$

$$\mu_X(t+\tau) = \int_{-\infty}^{+\infty} x f_X(x,t+\tau)\mathrm{d}x$$

Suppose that the two means are both equal to a constant α and equate the two integrals as follows:

$$\int_{-\infty}^{+\infty} x f_X(x,t)\mathrm{d}x = \int_{-\infty}^{+\infty} x f_X(x,t+\tau)\mathrm{d}x = \alpha$$

The above equality does not imply that the two *pdfs* must be equal. For the given *pdf* on the left-hand side that produces the constant mean α, there is an infinite possibility of the *pdf* on the right that can produce the same mean. If, for example, the *pdf* is symmetric with respect to $x = \alpha$, the two *pdfs* in the above equation can give the same mean α if the symmetry with respect to $x = \alpha$ is maintained regardless of whether the shape of one *pdf* is broader than the other.

If a process is first-order stationary, by substituting the constant marginal *pdfs* given by (6.80) into (6.101) through (6.103), we obtain a constant variance as follows:

$$\sigma_X(t)^2 = \mathrm{Var}\{X(t)\} \triangleq E\left[\{X(t) - \mu_X(t)\}\{X(t) - \mu_X(t)\}^*\right] = \sigma_{X_r}(t)^2 + \sigma_{X_i}(t)^2 = \sigma_X{}^2 \tag{6.106}$$

where

$$\sigma_{X_r}(t)^2 = \int_{-\infty}^{+\infty} x^2 f_{X_r}(x,t)\mathrm{d}x - \mu_{X_r}(t)^2 = \int_{-\infty}^{+\infty} x^2 f_{X_r}(x)\mathrm{d}x - {\mu_{X_r}}^2 = {\sigma_{X_r}}^2 \tag{6.107}$$

$$\sigma_{X_i}(t)^2 = \int_{-\infty}^{+\infty} x^2 f_{X_i}(x,t)\mathrm{d}x - \mu_{X_i}(t)^2 = \int_{-\infty}^{+\infty} x^2 f_{X_i}(x)\mathrm{d}x - {\mu_{X_i}}^2 = {\sigma_{X_i}}^2 \tag{6.108}$$

The converse is not necessarily true, that is, if the variance is constant, the process is not necessarily first-order stationary.

6.7.2 Autocorrelation

The autocorrelation of $X(t)$, denoted by $r_{XX}(t_1, t_2)$, is defined as the expected value of the product of the values of the process at two different time points as follows:

$$r_{XX}(t_1, t_2) \triangleq E\{X(t_1)X^*(t_2)\} \tag{6.109}$$

The prefix "auto" in autocorrelation refers to the fact that the correlation is between the RVs defined for the same process $X(t)$ whereas the cross-correlation, to be defined later, refers to the correlation between two different processes. We did not need this distinction in earlier chapters where we discussed the RVs without time t as an argument. The complex conjugate in the above definition assures that the product becomes the square of the magnitude of the second moment of $X(t)$ when the two time points coincide, $t_1 = t_2$. By substituting $X(t_1)$ and $X^*(t_2)$ into the above definition and expanding it, we obtain the following expression:

$$\begin{aligned} r_{XX}(t_1, t_2) &= E[\{X_r(t_1) + jX_i(t_1)\}\{X_r(t_2) - jX_i(t_2)\}] \\ &= E[\{X_r(t_1)X_r(t_2) + X_i(t_1)X_i(t_2)\} + j\{X_i(t_1)X_r(t_2) - X_r(t_1)X_i(t_2)\}] \\ &= [E\{X_r(t_1)X_r(t_2)\} + E\{X_i(t_1)X_i(t_2)\}] + j[E\{X_i(t_1)X_r(t_2)\} - E\{X_r(t_1)X_i(t_2)\}] \end{aligned}$$

The first two expectation operations in the above equation are the autocorrelations defined by (6.109) for the real and imaginary components of $X(t)$. The third and fourth expectation operations in the above equation are the cross-correlations between the real and imaginary components of $X(t)$, which will be defined later by (6.132). The above equation can be written in terms of these autocorrelation and cross-correlation functions as follows:

$$\begin{aligned} r_{XX}(t_1, t_2) &= [r_{X_rX_r}(t_1, t_2) + r_{X_iX_i}(t_1, t_2)] + j[r_{X_iX_r}(t_1, t_2) - r_{X_rX_i}(t_1, t_2)] \\ &= r_{XX}^r(t_1, t_2) + j r_{XX}^i(t_1, t_2) \end{aligned} \tag{6.110}$$

where the superscripts r and i denote the real and imaginary components of the autocorrelation of $X(t)$:

$$r_{XX}^r(t_1, t_2) = r_{X_rX_r}(t_1, t_2) + r_{X_iX_i}(t_1, t_2) \tag{6.111}$$

$$r_{XX}^{i}(t_1,t_2) = r_{X_iX_r}(t_1,t_2) - r_{X_rX_i}(t_1,t_2) \tag{6.112}$$

$$r_{X_rX_r}(t_1,t_2) = E\{X_r(t_1)X_r(t_2)\} = \int_{-\infty}^{+\infty}\int_{-\infty}^{+\infty} x_{r_1}x_{r_2}f_{X_rX_r}(x_{r_1}x_{r_2},t_1,t_2)\mathrm{d}x_{r_1}\mathrm{d}x_{r_2} \tag{6.113}$$

$$r_{X_iX_i}(t_1,t_2) = E\{X_i(t_1)X_i(t_2)\} = \int_{-\infty}^{+\infty}\int_{-\infty}^{+\infty} x_{i_1}x_{i_2}f_{X_iX_i}(x_{i_1}x_{i_2},t_1,t_2)\mathrm{d}x_{i_1}\mathrm{d}x_{i_2} \tag{6.114}$$

$$r_{X_iX_r}(t_1,t_2) = E\{X_r(t_2)X_i(t_1)\} = \int_{-\infty}^{+\infty}\int_{-\infty}^{+\infty} x_ix_rf_{X_rX_i}(x_i,x_r,t_1,t_2)\mathrm{d}x_i\mathrm{d}x_r \tag{6.115}$$

$$r_{X_rX_i}(t_1,t_2) = E\{X_r(t_1)X_i(t_2)\} = \int_{-\infty}^{+\infty}\int_{-\infty}^{+\infty} x_rx_if_{X_rX_i}(x_r,x_i,t_1,t_2)\mathrm{d}x_r\mathrm{d}x_i \tag{6.116}$$

The above equations show that the autocorrelation function of a complex process is given by the two autocorrelation functions of the real and imaginary components of the process and the two cross-correlation functions between the real and imaginary components of the process.

For a real process, since $r_{X_iX_i}(t_1,t_2) = r_{X_iX_r}(t_1,t_2) = r_{X_rX_i}(t_1,t_2) = 0$, the autocorrelation defined by (6.110) reduces to the following equation:

$$r_{XX}(t_1,t_2) = r_{X_rX_r}(t_1,t_2) \tag{6.117}$$

If a process is at least second-order stationary, its autocorrelation depends only on the interval length δ between two time points. By integrating the second-order *pdf* given by (6.85), we obtain the following four marginal *pdf*s needed for computing the autocorrelation function:

$$f_{\mathbf{X}(t_1,t_2)}(x_{r_1}x_{r_2},\delta) = \int_{-\infty}^{+\infty}\int_{-\infty}^{+\infty} f_{\mathbf{X}(t_1,t_2)}(x_{r_1},x_{i_1},x_{r_2},x_{i_2},\delta)\mathrm{d}x_{i_1}\mathrm{d}x_{i_2}$$

$$f_{\mathbf{X}(t_1,t_2)}(x_{i_1}x_{i_2},\delta) = \int_{-\infty}^{+\infty}\int_{-\infty}^{+\infty} f_{\mathbf{X}(t_1,t_2)}(x_{r_1},x_{i_1},x_{r_2},x_{i_2},\delta)\mathrm{d}x_{r_1}\mathrm{d}x_{r_2}$$

$$f_{\mathbf{X}(t_1,t_2)}(x_{i1},x_{r2},\delta) = \int_{-\infty}^{+\infty}\int_{-\infty}^{+\infty} f_{\mathbf{X}(t_1,t_2)}(x_{r_1},x_{i_1},x_{r_2},x_{i_2},\delta)\mathrm{d}x_{r_1}\mathrm{d}x_{i_2}$$

$$f_{\mathbf{X}(t_1,t_2)}(x_{r1},x_{i2},\delta) = \int_{-\infty}^{+\infty}\int_{-\infty}^{+\infty} f_{\mathbf{X}(t_1,t_2)}(x_{r1},x_{i1},x_{r2},x_{i2},\delta)\mathrm{d}x_{r_2}\mathrm{d}x_{i_1}$$

Substituting the above marginal *pdf*s, which depend only on the interval length $\delta = t_2 - t_1$, into (6.110) through (6.116), we obtain the following autocorrelation function of a second-order stationary complex process:

$$r_{XX}(t_1,t_2) = r^r_{XX}(t_1,t_2) + jr^i_{XX}(t_1,t_2) = r^r_{XX}(\delta) + jr^i_{XX}(\delta) = r_{XX}(\delta) \quad (6.118)$$

where

$$r^r_{XX}(\delta) = r_{X_rX_r}(\delta) + r_{X_iX_i}(\delta) \quad (6.119a)$$

$$r^i_{XX}(\delta) = r_{X_iX_r}(\delta) - r_{X_rX_i}(\delta) \quad (6.119b)$$

$$r_{X_rX_r}(\delta) = \int_{-\infty}^{+\infty}\int_{-\infty}^{+\infty} x_{r_1}x_{r_2}f_{X(t_1,t_2)}(x_{r_1}x_{r_2},\delta)\mathrm{d}x_{r_1}\mathrm{d}x_{r_2} \quad (6.120)$$

$$r_{X_iX_i}(\delta) = \int_{-\infty}^{+\infty}\int_{-\infty}^{+\infty} x_{i_1}x_{i_2}f_{X(t_1,t_2)}(x_{i_1}x_{i_2},\delta)\mathrm{d}x_{i_1}\mathrm{d}x_{i_2} \quad (6.121)$$

$$r_{X_iX_r}(\delta) = \int_{-\infty}^{+\infty}\int_{-\infty}^{+\infty} x_ix_rf_{X(t_1,t_2)}(x_{i1},x_{r2},\delta)\mathrm{d}x_i\mathrm{d}x_r \quad (6.122)$$

$$r_{X_rX_i}(\delta) = \int_{-\infty}^{+\infty}\int_{-\infty}^{+\infty} x_rx_if_{X(t_1,t_2)}(x_{r1},x_{i2},\delta)\mathrm{d}x_r\mathrm{d}x_i \quad (6.123)$$

If a process is real and second-order stationary, by (6.117) and (6.120), we obtain its autocorrelation as a function of the interval length δ only as follows:

$$r_{XX}(t_1,t_2) = r_{X_rX_r}(t_1,t_2) = r_{X_rX_r}(\delta) = r_{XX}(\delta) \quad (6.124)$$

The wide sense stationarity (WSS) requires that the process have the constant mean, and the autocorrelation depends on the interval length δ only. As shown above, if the process is at least second-order stationary, it is WSS.

However, the converse is not necessarily true, that is, if a process is WSS, the process is not necessarily second-order stationary. The WSS is a weaker condition than the second-order stationarity because the latter requires the *pdf* as a function of δ only while the former, only the autocorrelation as a function of δ only. To illustrate this point, consider two autocorrelation functions given by the following double integrals:

$$r_{XX}(t_1,t_2) = \int_{-\infty}^{+\infty}\int_{-\infty}^{+\infty} x_1x_2f_{X(t_1,t_2)}(x_1,x_2,t_1,t_2)\mathrm{d}x_1\,\mathrm{d}x_2$$

$$r_{XX}(t_1+,t_2+\tau) = \int_{-\infty}^{+\infty}\int_{-\infty}^{+\infty} x_1x_2f_{X(t_1,t_2)}(x_1,x_2,t_1+\tau,t_2+\tau)\mathrm{d}x_1\,\mathrm{d}x_2$$

Supposing that the two double integrals produce the same value of the autocorrelation function because the interval length is the same, $(t_1+\tau)-(t_2+\tau) = t_1-t_2 = \delta$, we equate the two double integrals as follows:

$$\int_{-\infty}^{+\infty}\int_{-\infty}^{+\infty} x_1x_2f_{X(t_1,t_2)}(x_1,x_2,t_1,t_2)\mathrm{d}x_1\,\mathrm{d}x_2 = \int_{-\infty}^{+\infty}\int_{-\infty}^{+\infty} x_1x_2f_{X(t_1,t_2)}(x_1,x_2,t_1+\tau,t_2+\tau)\mathrm{d}x_1\,\mathrm{d}x_2$$

Given the left-hand side that produces the autocorrelation, there is an infinite possibility of the *pdf* in the right-hand side that can produce the same autocorrelation. For the sake of making the point, suppose that the two RVs are independent so that we can write the double integral as the product of two integrals, each of which gives the mean, as follows:

$$\int_{-\infty}^{+\infty} x_1 f_{X(t_1)}(x_1, t_1)dx_1 \int_{-\infty}^{+\infty} x_2 f_{X(t_2)}(x_2, t_2)\mathrm{d}x_2$$
$$= \int_{-\infty}^{+\infty} x_1 f_{X(t_1)}(x_1, t_1+\tau)dx_1 \int_{-\infty}^{+\infty} x_2 f_{X(t_2)}(x_2, t_2+\tau)\mathrm{d}x_2$$

As discussed under the discussion on the mean in Sect. 6.7.1, the above equality does not imply that the *pdf*s on both sides of the equation must be the same.

6.7.3 Autocovariance

The autocovariance of $X(t)$ is defined as the covariance of two RVs $X(t_1)$ and $X(t_2)$ representing the process at two arbitrary time points,t_1 and t_2, by the same equation (5.33), which defines the covariance of two RVs, except that the autocovariance is shown as a function of two time points, t_1 and t_2, as follows:

$$c_{XX}(t_1, t_2) \triangleq E\left[\{X(t_1) - \mu_X(t_1)\}\{X(t_1) - \mu_X(t_1)\}^*\right]$$

By expanding the above equation, we have the following equation:

$$\begin{aligned} c_{XX}(t_1, t_2) &= E\left[\{X(t_1) - \mu_X(t_1)\}\{X^*(t_2) - \mu_X^*(t_2)\}\right] \\ &= E\{X(t_1)X^*(t_2)\} \quad - \mu_X(t_1)\mu_X^*(t_2) \\ &= r_{XX}(t_1, t_2) - \mu_X(t_1)\mu_X^*(t_2) \end{aligned} \tag{6.125}$$

which is a function of the mean and the autocorrelation given by (6.99) and (6.110), respectively. The prefix "auto" in autocovariance refers to the fact that the covariance is between two RVs defined for the same process $X(t)$.

The correlation coefficient of the process $X(t)$ is defined by the same equation (5.35) as follows:

$$\rho_{XX}(t_1, t_2) = \frac{Cov\,\{X(t_1), X(t_2)\}}{\sigma_X(t_1)\sigma_X(t_2)} = \frac{c_{XX}(t_1, t_2)}{\sigma_X(t_1)\sigma_X(t_2)} \tag{6.126}$$

If the process is second-order stationary, its autocovariance depends only on the interval length δ between two time points. By the definition of the nth-order stationarity given by (6.69) and (6.70), the second-order stationarity guarantees the first-order stationarity, and thus, if a process is second-order stationary, its mean is a constant. Therefore, for a second-order stationary process $X(t)$, its autocovariance is given by the following equation:

$$c_{XX}(t_1, t_2) = r_{XX}(\delta) - \mu_X \mu_X^* \tag{6.127}$$

By (5.35), the correlation coefficient of the second-order stationary process $X(t)$ is given by

$$\rho_{XX}(t_1, t_2) = \frac{Cov\,\{X(t_1), X(t_2)\}}{\sigma_X(t_1)\sigma_X(t_2)} = \frac{c_{XX}(t_1, t_2)}{\sigma_X(t_1)\sigma_X(t_2)} = \frac{c_{XX}(\delta)}{\sigma_X^2} \tag{6.128}$$

Theorem 6.7.1

$$c_{XX}(t_2, t_1) = c_{XX}^*(t_1, t_2) \tag{6.129}$$

Proof By (6.148) to be shown later, we have

$$\begin{aligned} r_{XX}(t_2, t_1) &= r_{XX}^*(t_1, t_2) \\ c_{XX}(t_2, t_1) &= r_{XX}(t_2, t_1) - \mu_X(t_2)\mu_X^*(t_1) \end{aligned} \tag{6.130}$$

Consider

$$\begin{aligned} \mu_X(t_1)\mu_X^*(t_2) &= \{\mu_{X_r}(t_1) + j\mu_{X_i}(t_1)\}\{\mu_{X_r}(t_2) - j\mu_{X_i}(t_2)\} \\ &= \{\mu_{X_r}(t_2)\mu_{X_r}(t_1) + \mu_{X_i}(t_2)\mu_{X_i}(t_1)\} + j\{\mu_{X_i}(t_2)\mu_{X_r}(t_1) - \mu_{X_r}(t_2)\mu_{X_i}(t_1)\} \end{aligned}$$

By comparing the last two equations, we obtain the following equation:

$$\mu_X(t_2)\mu_X^*(t_1) = \{\mu_X(t_1)\mu_X^*(t_2)\}^* \tag{6.131}$$

Substituting (6.148) and the above equation into (6.130), we have

$$\begin{aligned} c_{XX}(t_2, t_1) &= r_{XX}^*(t_1, t_2) - \{\mu_X(t_1)\mu_X^*(t_2)\}^* = \{r_{XX}(t_1, t_2) - \mu_X(t_1)\mu_X^*(t_2)\}^* \\ &= c_{XX}^*(t_1, t_2) \end{aligned}$$

Q.E.D.

6.7.4 Cross-correlation

Figure 6.2b shows two processes $X(t)$ and $Y(t)$. The cross-correlation of $X(t)$ and $Y(t)$ is defined by

$$r_{XY}(t_1, t_2) \triangleq E\{X(t_1)Y^*(t_2)\} \tag{6.132}$$

Expanding the above equation, we obtain the following expression:

$$\begin{aligned} r_{XY}(t_1,t_2) &= E[\{X_r(t_1)+jX_i(t_1)\}\{Y_r(t_2)-jY_i(t_2)\}] \\ &= E[\{X_r(t_1)Y_r(t_2)+X_i(t_1)Y_i(t_2)\}+j\{X_i(t_1)Y_r(t_2)-X_r(t_1)Y_i(t_2)\}] \\ &= [E\{X_r(t_1)Y_r(t_2)\}+E\{X_i(t_1)Y_i(t_2)\}]+j[E\{X_i(t_1)Y_r(t_2)\}-E\{X_r(t_1)Y_i(t_2)\}] \end{aligned}$$

The four expected values in the above equation are the cross-correlations defined by (6.132) for the real and imaginary components of X(t) and the cross-correlation becomes the following equation:

$$\begin{aligned} r_{XY}(t_1,t_2) &= [r_{X_rY_r}(t_1,t_2)+r_{X_iY_i}(t_1,t_2)]+j[r_{X_iY_r}(t_1,t_2)-r_{X_rY_i}(t_1,t_2)] \\ &= r_{XY}^r(t_1,t_2)+jr_{XY}^i(t_1,t_2) \end{aligned} \tag{6.133}$$

where

$$r_{XY}^r(t_1,t_2) = r_{X_rY_r}(t_1,t_2)+r_{X_iY_i}(t_1,t_2) \tag{6.134a}$$

$$r_{XY}^i(t_1,t_2) = r_{X_iY_r}(t_1,t_2)-r_{X_rY_i}(t_1,t_2) \tag{6.134b}$$

$$r_{X_rY_r}(t_1,t_2) = \int_{-\infty}^{+\infty}\int_{-\infty}^{+\infty} x_ry_rf_{\mathbf{Z}_{XY}(t_1,t_1')}(x_r,y_r,t_1,t_2)dx_rdy_r \tag{6.135}$$

$$r_{X_iY_i}(t_1,t_2) = \int_{-\infty}^{+\infty}\int_{-\infty}^{+\infty} x_iy_if_{\mathbf{Z}_{XY}(t_1,t_1')}(x_i,y_i,t_1,t_2)dx_idy_i \tag{6.136}$$

$$r_{X_iY_r}(t_1,t_2) = \int_{-\infty}^{+\infty}\int_{-\infty}^{+\infty} x_iy_rf_{\mathbf{Z}_{XY}(t_1,t_1')}(x_i,y_r,t_1,t_2)dx_idy_r \tag{6.137}$$

$$r_{X_rY_i}(t_1,t_2) = \int_{-\infty}^{+\infty}\int_{-\infty}^{+\infty} x_ry_if_{\mathbf{Z}_{XY}(t_1,t_1')}(x_r,y_i,t_1,t_2)\mathrm{d}x_r\mathrm{d}y_i \tag{6.138}$$

For real $X(t)$ and $Y(t)$, the cross-correlation is obtained by setting $r_{X_iY_i}(t_1,t_2) = r_{X_iY_r}(t_1,t_2) = r_{X_rY_i}(t_1,t_2) = 0$ in (6.133) as follows:

$$r_{XY}(t_1,t_2) = r_{X_rY_r}(t_1,t_2) \tag{6.139}$$

which is determined by (6.135).

If two processes are jointly second-order stationary, their cross-correlation depends only on the interval length δ between two time points. By integrating the second-order *pdf* given by (6.96), we obtain the following four marginal *pdf*s required in (6.135) through (6.138) as follows:

$$\begin{aligned} f_{\mathbf{Z}_{XY}(t_1,t_1')}(x_r,y_r,\delta) &= \int_{-\infty}^{+\infty}\int_{-\infty}^{+\infty} f_{\mathbf{Z}_{XY}(t_1,t_1')}(x_r,x_i,y_r,y_i;\delta)\mathrm{d}x_i\mathrm{d}y_i \\ f_{\mathbf{Z}_{XY}(t_1,t_1')}(x_i,y_i,\delta) &= \int_{-\infty}^{+\infty}\int_{-\infty}^{+\infty} f_{\mathbf{Z}_{XY}(t_1,t_1')}(x_r,x_i,y_r,y_i;\delta)\mathrm{d}x_r\mathrm{d}y_r \\ f_{\mathbf{Z}_{XY}(t_1,t_1')}(x_i,y_r,\delta) &= \int_{-\infty}^{+\infty}\int_{-\infty}^{+\infty} f_{\mathbf{Z}_{XY}(t_1,t_1')}(x_r,x_i,y_r,y_i;\delta)\mathrm{d}x_r\mathrm{d}y_i \\ f_{\mathbf{Z}_{XY}(t_1,t_1')}(x_r,y_i,\delta) &= \int_{-\infty}^{+\infty}\int_{-\infty}^{+\infty} f_{\mathbf{Z}_{XY}(t_1,t_1')}(x_r,x_i,y_r,y_i;\delta)\mathrm{d}x_i\mathrm{d}y_r \end{aligned}$$

Substituting the above four marginal *pdf*s into (6.135) through (6.138), we obtain the cross-correlation of the jointly second-order stationary complex $X(t)$ and $Y(t)$ as a function δ only as follows:

$$r_{XY}(t_1,t_2) = r^r_{XY}(t_1,t_2) + jr^i_{XY}(t_1,t_2) = r^r_{XY}(\delta) + jr^i_{XY}(\delta) = r_{XY}(\delta) \tag{6.140}$$

where

$$r^r_{XY}(t_1,t_2) = r_{X_rY_r}(t_1,t_2) + r_{X_iY_i}(t_1,t_2) \tag{6.141a}$$

$$r^i_{XY}(t_1,t_2) = r_{X_iY_r}(t_1,t_2) - r_{X_rY_i}(t_1,t_2) \tag{6.141b}$$

$$r_{X_rY_r}(\delta) = \int_{-\infty}^{+\infty}\int_{-\infty}^{+\infty} x_r y_r f_{Z_{XY}(t_1,t_1')}(x_r, y_r, \delta)\,dx_r dy_r \tag{6.142}$$

$$r_{X_iY_i}(\delta) = \int_{-\infty}^{+\infty}\int_{-\infty}^{+\infty} x_i y_i f_{Z_{XY}(t_1,t_1')}(x_i, y_i, \delta)\,dx_i dy_i \tag{6.143}$$

$$r_{X_iY_r}(\delta) = \int_{-\infty}^{+\infty}\int_{-\infty}^{+\infty} x_i y_r f_{Z_{XY}(t_1,t_1')}(x_i, y_r, \delta)\,dx_i dy_r \tag{6.144}$$

$$r_{X_rY_i}(\delta) = \int_{-\infty}^{+\infty}\int_{-\infty}^{+\infty} x_r y_i f_{Z_{XY}(t_1,t_1')}(x_r, y_i, \delta)\,dx_r dy_i \tag{6.145}$$

If two real processes are jointly second-order stationary, their cross-correlation is a function of δ as follows:

$$r_{XY}(t_1,t_2) = r_{X_rY_r}(t_1,t_2) = r_{X_rY_r}(\delta) = r_{XY}(\delta) \tag{6.146}$$

The joint second-order stationarity between $X(t)$ and $Y(t)$ does not guarantee the stationarity of the individual processes, $X(t)$ and $Y(t)$. The joint WSS requires that the individual processes are WSS. Therefore, if the two processes are jointly WSS, the cross-correlation between $X(t)$ and $Y(t)$ and the autocorrelations of $X(t)$ and $Y(t)$ all depend on the interval length δ only and the means of $X(t)$ and $Y(t)$ are constants as follows:

$$r_{XY}(t_1,t_2) = r_{XY}(\delta) \quad r_{XX}(t_1,t_2) = r_{XX}(\delta) \quad r_{YY}(t_1,t_2) = r_{YY}(\delta) \quad \mu_X(t) = \mu_{X_r} + j\mu_{X_i} \tag{6.147}$$

Theorem 6.7.2

$$r_{XX}(t_2,t_1) = r^*_{XX}(t_1,t_2) \tag{6.148}$$

Proof By (6.109),

$$\begin{aligned} r_{XX}(t_2,t_1) &= E\{X(t_2)X^*(t_1)\} = E[\{X_r(t_2)+jX_i(t_2)\}\{X_r(t_1)-jX_i(t_1)\}] \\ &= E[\{X_r(t_1)X_r(t_2)+X_i(t_1)X_i(t_2)\}-j\{X_i(t_1)X_r(t_2)-X_r(t_1)X_i(t_2)\}] \\ &= E[\{X_r(t_1)X_r(t_2)+X_i(t_1)X_i(t_2)\}]-[jE\{X_i(t_1)X_r(t_2)-X_r(t_1)X_i(t_2)\}] \\ &= r^r_{XX}(t_1,t_2)-jr^i_{XX}(t_1,t_2) \end{aligned}$$

where the last two expectation operations are substituted by (6.111) and (6.112). By (6.110), we see that the last expression becomes the following equation:

$$r^r_{XX}(t_1,t_2)-jr^i_{XX}(t_1,t_2)=r^*_{XX}(t_1,t_2)$$

Q.E.D.

6.7.5 Cross-covariance

The cross-covariance of $X(t)$ and $Y(t)$ is defined as the covariance between two RVs, $X(t_1)$ and $Y(t_2)$, as follows:

$$\begin{aligned} c_{XY}(t_1,t_2) &= \text{Cov}\ \{X(t_1),Y(t_2)\} \\ &\triangleq E\left[\{X(t_1)-\mu_X(t_1)\}\{Y(t_2)-\mu_Y(t_2)\}^*\right] \\ &= E\{X(t_1)Y^*(t_2)\}-\mu_X(t_1)\mu_Y{}^*(t_2) \\ &= r_{XY}(t_1,t_2)-\mu_X(t_1)\mu_Y{}^*(t_2) \end{aligned} \tag{6.149}$$

If $X(t)$ and $Y(t)$ are jointly second-order stationary, we have the following relationship:

$$c_{XY}(t_1,t_2)=r_{XY}(\delta)-\mu_X(t_1)\mu_Y{}^*(t_2) \tag{6.150}$$

Since a joint stationarity does not guarantee the stationarity of individual processes, the means are not necessarily constants in the above equation.

If the two process are jointly WSS, however, the means of the individual processes are constant and their cross-covariance depends only on δ as follows:

For real process:

$$c_{XY}(t_1,t_2)=r_{XY}(\delta)-\mu_X\mu_Y \tag{6.151a}$$

For complex process:

$$c_{XY}(t_1,t_2)=r_{XY}(\delta)-\mu_X\mu_Y{}^* \tag{6.151b}$$

6.8 Properties of the Autocorrelation of a WSS Process

Theorem 6.8.1

(a) If $X(t)$ is a real WSS process, its autocorrelation function satisfies the following property:

$$r_{XX}(-\tau) = r_{XX}(\tau) \tag{6.152}$$

(b) If the two real processes $X(t)$ and $Y(t)$ are jointly WSS, their cross-correlation function satisfies the following property:

$$r_{XY}(-\tau) = r_{YX}(\tau) \tag{6.153}$$

(c) For a complex process $X(t)$, the autocorrelation function of $X(t)$ satisfies the following property:

$$r_{XX}(-\tau) = r_{XX}^{*}(\tau) \tag{6.154}$$

Proof (a) For a real WSS process $X(t)$, we have

$$\begin{aligned} r_{XX}(-\tau) &= R_{XX}(t_2 - t_1) = E\{X(t_2)X(t_1)\} = E\{X(t_1)X(t_2)\} \\ &= r_{XX}(t_1 - t_2) = r_{XX}(\tau) \end{aligned}$$

Q.E.D.

(b) For real jointly WSS processes $X(t)$ and $Y(t)$, we have

$$\begin{aligned} r_{XY}(-\tau) &= R_{XY}(t_2 - t_1) = E\{X(t_2)Y(t_1)\} \\ &= E\{Y(t_1)X(t_2)\} = r_{YX}(t_1 - t_2) = r_{YX}(\tau) \end{aligned}$$

Q.E.D.

(c) By part (a) of the above theorem, the two autocorrelation functions appearing in the real component of the autocorrelation of complex $X(t)$ given by (6.111) satisfy the following property:

$$r_{X_rX_r}(-\tau) = r_{X_rX_r}(\tau) \tag{6.155}$$

$$r_{X_iX_i}(-\tau) = r_{X_iX_i}(\tau) \tag{6.156}$$

Therefore, the real component satisfies the following property:

$$\begin{aligned} r_{XX}^r(-\tau) &= r_{X_rX_r}(-\tau) + r_{X_iX_i}(-\tau) \\ &= r_{X_rX_r}(\tau) + r_{X_iX_i}(\tau) = r_{XX}^r(\tau) \end{aligned} \tag{6.157}$$

Now, consider the two cross-correlation functions between X_r and X_i, which appear in the imaginary component of the autocorrelation of $X(t)$ given by (6.112). By part (b) of the above theorem, we have

$$r_{X_iX_r}(-\tau) = r_{X_rX_i}(\tau) \tag{6.158}$$

$$r_{X_rX_i}(-\tau) = r_{X_iX_r}(\tau) \tag{6.159}$$

Therefore, the imaginary component satisfies the following property:

$$\begin{aligned} r_{XX}^i(-\tau) &= r_{X_iX_r}(-\tau) - r_{X_rX_i}(-\tau) = r_{X_rX_i}(\tau) - r_{X_iX_r}(\tau) \\ &= -\{r_{X_iX_r}(\tau) - r_{X_rX_i}(\tau)\} = -r_{XX}^i(\tau) \end{aligned} \tag{6.160}$$

Substituting (6.160) and (6.157) into (6.110) with $-\tau$, we have

$$r_{XX}(-\tau) = r_{XX}^r(-\tau) + jr_{XX}^i(-\tau) = r_{XX}^r(\tau) - jr_{XX}^i(\tau) = r_{XX}^*(\tau)$$

Q.E.D.

Theorem 6.8.2 For a WSS process $X(t)$, real or complex, the autocorrelation function satisfies the following property:

$$r_{XX}(\tau) \le r_{XX}(0) \tag{6.161}$$

Proof Frist, consider a real WSS process.

$$\begin{aligned} E\left[\{X(t) + X(t-\tau)\}^2\right] &= E\left[\{X(t)\}^2 + 2X(t)X(t-\tau) + \{X(t-\tau)\}^2\right] \\ &= 2\,r_{XX}(0) + 2\,r_{XX}(\tau) \ge 0 \\ E\left[\{X(t) - X(t-\tau)\}^2\right] &= E\left[\{X(t)\}^2 - 2X(t)X(t-\tau) + \{X(t-\tau)\}^2\right] \\ &= 2\,r_{XX}(0) - 2\,r_{XX}(\tau) \ge 0 \end{aligned}$$

From the above two equations, we have

$$-r_{XX}(0) \le r_{XX}(\tau) \le r_{XX}(0)$$

Because the autocorrelation at $\tau = 0$ is the expected value of a squared term, we have

$$r_{XX}(0) \ge 0$$

From the last two relations, we have

$$|r_{XX}(\tau)| \le r_{XX}(0)$$

This proves the theorem for a real WSS process.

Now, consider a complex WSS process $X(t)$ and its autocorrelation given by

$$r_{XX}(\tau) = r_{XX}^r(\tau) + jr_{XX}^i(\tau) \tag{6.162}$$

For a real process, we have proven above the following relationships:

$$\left|r_{XX}^r(\tau)\right| \leq r_{XX}^r(0) \tag{6.163a}$$

$$\left|r_{XX}^i(\tau)\right| \leq r_{XX}^i(0) \tag{6.163b}$$

The magnitude of the autocorrelation of a complex WSS process is given by

$$|r_{XX}(\tau)|^2 = \{r_{XX}^r(\tau) + jr_{XX}^i(\tau)\}\{r_{XX}^r(\tau) - jr_{XX}^i(\tau)\} = \{r_{XX}^r(\tau)\}^2 + \{r_{XX}^i(\tau)\}^2$$

Substituting (6.163)-(a) and -(b) into the above equation, we obtain the following relationship:

$$|r_{XX}(\tau)|^2 \leq \{r_{XX}^r(0)\}^2 + \{r_{XX}^i(0)\}^2 = |r_{XX}(0)|^2$$

Q.E.D.

6.9 Parameter Vectors and Matrices

This section defines parameter vectors and parameter matrices of vector RVs including the mean vector, the autocorrelation matrix, the autocovariance matrix, and the cross-covariance matrix.

6.9.1 Mean Vectors

The n-, m-, and $(n+m)$-dimensional mean vectors of $\boldsymbol{X}$, $\boldsymbol{Y}$ and $\boldsymbol{Z}_{XY}$ are given by the following equations:

$$\boldsymbol{\mu}_X = \begin{bmatrix} \mu_{X_1} \\ \cdot \\ \cdot \\ \mu_{X_n} \end{bmatrix} = E(\boldsymbol{X}) = \begin{bmatrix} E\{X(t_1)\} \\ \cdot \\ \cdot \\ E\{X(t_n)\} \end{bmatrix} \tag{6.164a}$$

$$\boldsymbol{\mu_Y} = \begin{bmatrix} \mu_{Y_1} \\ \vdots \\ \mu_{Y_m} \end{bmatrix} = E(\boldsymbol{Y}) = \begin{bmatrix} E\{Y(t_1)\} \\ \vdots \\ E\{Y(t_m)\} \end{bmatrix} \tag{6.164b}$$

$$\boldsymbol{\mu_{Z_{XY}}} = \begin{bmatrix} \mu_{Z_{XY1}} \\ \cdot \\ \mu_{Z_{XYi}} \\ \cdot \\ \mu_{Z_{XYn+m}} \end{bmatrix} = E(\boldsymbol{Z_{XY}}) = E\begin{bmatrix} \boldsymbol{X} \\ - \\ \boldsymbol{Y} \end{bmatrix} = \begin{bmatrix} E\{X(t_1)\} \\ \cdot \\ E\{X(t_n)\} \\ ---- \\ E\{Y(t_1')\} \\ \cdot \\ E\{Y(t_m')\} \end{bmatrix} \tag{6.165}$$

For a complex process, the n-, m-, and $(n+m)$-dimensional mean vectors of $\boldsymbol{X}$, $\boldsymbol{Y}$ and $\boldsymbol{Z_{XY}}$ are given by the following equations:

$$\begin{aligned} \boldsymbol{\mu_X} &= \begin{bmatrix} \mu_{X1} \\ \cdot \\ \cdot \\ \mu_{Xn} \end{bmatrix} = E(\boldsymbol{X}) = \begin{bmatrix} E\{X(t_1)\} \\ \cdot \\ \cdot \\ E\{X(t_n)\} \end{bmatrix} \\ &= \begin{bmatrix} E\{X_r(t_1) + jX_i(t_1)\} \\ \cdot \\ \cdot \\ E\{X_r(t_n) + jX_i(t_n)\} \end{bmatrix} = \begin{bmatrix} \mu_{Xr}(t_1) + j\mu_{X_i}(t_1) \\ \cdot \\ \cdot \\ \mu_{Xr}(t_n) + j\mu_{X_i}(t_n) \end{bmatrix} \end{aligned} \tag{6.166a}$$

$$\begin{aligned} \boldsymbol{\mu_Y} &= \begin{bmatrix} \mu_{Y1} \\ \cdot \\ \cdot \\ \mu_{Ym} \end{bmatrix} = E(\boldsymbol{Y}) = \begin{bmatrix} E\{Y(t_1)\} \\ \cdot \\ \cdot \\ E\{Y(t_m)\} \end{bmatrix} \\ &= \begin{bmatrix} E\{Y_r(t_1) + jY_i(t_1)\} \\ \cdot \\ \cdot \\ E\{Y_r(t_m) + jY_i(t_m)\} \end{bmatrix} = \begin{bmatrix} \mu_{Y_r}(t_1) + j\mu_{Y_i}(t_1) \\ \cdot \\ \cdot \\ \mu_{Y_r}(t_m) + j\mu_{Y_i}(t_m) \end{bmatrix} \end{aligned} \tag{6.166b}$$

$$\boldsymbol{\mu_{Z_{XY}}} = \begin{bmatrix} \mu_{Z_{XY1}} \\ \cdot \\ \mu_{Z_{XYi}} \\ \cdot \\ \cdot \\ \mu_{Z_{XYn+m}} \end{bmatrix} = E(\boldsymbol{Z_{XY}}) = E\begin{bmatrix} \boldsymbol{X} \\ - \\ \boldsymbol{Y} \end{bmatrix} = \begin{bmatrix} \mu_{X_r}(t_1) + j\mu_{X_i}(t_1) \\ \cdot \\ \mu_{X_r}(t_n) + j\mu_{X_i}(t_n) \\ --------- \\ \mu_{Y_r}(t_1') + j\mu_{Y_i}(t_1') \\ \cdot \\ \mu_{Y_r}(t_m') + j\mu_{Y_i}(t_m') \end{bmatrix} \tag{6.167}$$

6.9.2 Autocovariance Matrices

Strictly speaking, among RVs, there is no distinction between the autocovariance and the cross-covariance and both types are simply the covariance. The "auto" vs "cross" is significant only when the covariance is referred to in conjunction with the underlying stochastic process. When the covariance is between the RVs defined for the same process such as $X_i's$ for $X(t)$, the covariance is called the autocovariance, and when it is between the RVs from two different processes, e.g., $X_i's$, and $Y_i's$ for $X(t)$ and $Y(t)$, it is called the cross-covariance.

By the definition given by (6.125), the $(n \times n)$- dimensional autocovariance matrix of $\boldsymbol{X}$ is given by the following equation:

$$\mathbb{C}_{XX} \triangleq E\left\{(\boldsymbol{X}-\boldsymbol{\mu}_X)(\boldsymbol{X}-\boldsymbol{\mu}_X)^{\mathrm{T}}\right\}$$
$$= \begin{bmatrix} (X_1-\mu_{X_1})(X_1-\mu_{X_1}) & . & . & . & (X_1-\mu_{X_1})(X_n-\mu_{X_n}) \\ . & . & . & . & . \\ . & . & (X_i-\mu_{X_i})(X_j-\mu_{X_j}) & . & . \\ . & . & . & . & . \\ (X_n-\mu_{X_n})(X_1-\mu_{X_1}) & . & . & . & (X_n-\mu_{X_n})(X_n-\mu_{X_n}) \end{bmatrix}$$
$$= \begin{bmatrix} c_{X_1X_1} & . & . & . & c_{X_1X_n} \\ c_{X_2X_1} & . & . & . & . \\ . & . & c_{X_iX_j} & . & . \\ . & . & . & . & . \\ c_{X_nX_1} & . & . & . & c_{X_nX_n} \end{bmatrix} \tag{6.168}$$

where

$$c_{X_iX_j} = Cov(X_i, X_j) = E\left\{(X_i-\mu_{X_i})(X_j-\mu_{X_j})\right\} = E(X_iX_j) - \mu_{X_i}\mu_{X_j} \tag{6.169}$$

Expanding the matrix multiplication inside the expectation operator, we have

$$\begin{aligned} \mathbb{C}_{XX} &= E\left\{(\boldsymbol{X}-\boldsymbol{\mu}_X)(\boldsymbol{X}-\boldsymbol{\mu}_X)^{\mathrm{T}}\right\} = E\{(\boldsymbol{X}-\boldsymbol{\mu}_X)(\boldsymbol{X}^{\mathrm{T}}-\boldsymbol{\mu}_X^{\mathrm{T}})\} \\ &= E\{\boldsymbol{X}(\boldsymbol{X}^{\mathrm{T}}-\boldsymbol{\mu}_X^{\mathrm{T}}) - \boldsymbol{\mu}_X(\boldsymbol{X}^{\mathrm{T}}-\boldsymbol{\mu}_X^{\mathrm{T}})\} \\ &= E\{\boldsymbol{X}\boldsymbol{X}^{\mathrm{T}} - \boldsymbol{X}\boldsymbol{\mu}_X^{\mathrm{T}} - \boldsymbol{\mu}_X\boldsymbol{X}^{\mathrm{T}} + \boldsymbol{\mu}_X\boldsymbol{\mu}_X^{\mathrm{T}}\} \\ &= E(\boldsymbol{X}\boldsymbol{X}^{\mathrm{T}}) - \boldsymbol{E}(\boldsymbol{X}\boldsymbol{\mu}_X^{\mathrm{T}}) - \boldsymbol{E}(\boldsymbol{\mu}_X\boldsymbol{X}^{\mathrm{T}}) + \boldsymbol{E}(\boldsymbol{\mu}_X\boldsymbol{\mu}_X^{\mathrm{T}}) \\ &= E(\boldsymbol{X}\boldsymbol{X}^{\mathrm{T}}) - E(\boldsymbol{X})\boldsymbol{\mu}_X^{\mathrm{T}} - \boldsymbol{\mu}_X E(\boldsymbol{X}^{\mathrm{T}}) + \boldsymbol{\mu}_X\boldsymbol{\mu}_X^{\mathrm{T}} \\ &= E(\boldsymbol{X}\boldsymbol{X}^{\mathrm{T}}) - \boldsymbol{\mu}_X\boldsymbol{\mu}_X^{\mathrm{T}} - \boldsymbol{\mu}_X\boldsymbol{\mu}_X^{\mathrm{T}} + \boldsymbol{\mu}_X\boldsymbol{\mu}_X^{\mathrm{T}} = E(\boldsymbol{X}\boldsymbol{X}^{\mathrm{T}}) - \boldsymbol{\mu}_X\boldsymbol{\mu}_X^{\mathrm{T}} \end{aligned} \tag{6.170}$$

Similarly, we have the following autocovariance matrix of $\boldsymbol{Y}$:

$$\mathbb{C}_{YY} = \begin{bmatrix} c_{Y_1Y_1} & \cdot & \cdot & \cdot & c_{Y_1Y_n} \\ \cdot & \cdot & \cdot & \cdot & \cdot \\ \cdot & \cdot & c_{Y_iY_j} \cdot & \cdot & \cdot \\ c_{Y_nY_1} & \cdot & \cdot & \cdot & c_{Y_nY_n} \end{bmatrix} \tag{6.171}$$

where

$$c_{Y_iY_j} = \mathrm{Cov}\left(Y_i, Y_j\right) = E\left\{\left(Y_i - \mu_{Y_i}\right)\left(Y_j - \mu_{Y_j}\right)\right\} = E\left(Y_iY_j\right) - \mu_{Y_i}\mu_{Y_j} \tag{6.172}$$

Theorem 6.9.1 If $\boldsymbol{X}$ is real, its autocovariance matrix is symmetric, that is,

$$\mathbb{C}_{XX} = \mathbb{C}_{XX}^{\mathrm{T}} \tag{6.173}$$

Proof Consider the *ji*th element of C_{XX}. By (6.169), we have

$$c_{X_jX_i} = \mathrm{Cov}\left(X_j, X_i\right) = E\left\{\left(X_j - \mu_{X_j}\right)\left(X_i - \mu_i\right)\right\} = \mathrm{Cov}\left(X_j, X_i\right) = c_{X_iX_j}$$

Q.E.D.

For a complex process, by the definition given by (6.125), the $(n \times n)$- dimensional autocovariance matrix of $\boldsymbol{X}$ is as follows:

$$\mathbb{C}_{XX} \triangleq E\left[\{(\boldsymbol{X} - \boldsymbol{\mu}_X)\}\{(\boldsymbol{X} - \boldsymbol{\mu}_X)^*\}^{\mathrm{T}}\right]$$
$$= \begin{bmatrix} \left(X_1 - \mu_{X_1}\right)\left(X_1 - \mu_{X_1}\right)^* & \cdot & \cdot & \cdot & \left(X_1 - \mu_{X_1}\right)\left(X_n - \mu_{X_n}\right)^* \\ \cdot & \cdot & \cdot & \cdot & \cdot \\ \cdot & \cdot & \left(X_i - \mu_{X_i}\right)\left(X_j - \mu_{X_j}\right)^* & \cdot & \cdot \\ \cdot & \cdot & \cdot & \cdot & \cdot \\ \left(X_n - \mu_{X_n}\right)\left(X_1 - \mu_{X_1}\right)^* & \cdot & \cdot & \cdot & \left(X_n - \mu_{X_n}\right)\left(X_n - \mu_{X_n}\right)^* \end{bmatrix}$$
$$= \begin{bmatrix} c_{X_1X_1} & \cdot & \cdot & \cdot & c_{X_1X_n} \\ c_{X_2X_1} & \cdot & \cdot & \cdot & \cdot \\ \cdot & \cdot & c_{X_iX_j} & \cdot & \cdot \\ \cdot & \cdot & \cdot & \cdot & \cdot \\ c_{X_nX_1} & \cdot & \cdot & \cdot & c_{X_nX_n} \end{bmatrix} \tag{6.174}$$

where

$$c_{X_iX_j} = \mathrm{Cov}\left(X_i, X_j\right) = E\left\{\left(X_i - \mu_{X_i}\right)\left(X_j - \mu_{X_j}\right)^*\right\} = E\left(X_iX_j^*\right) - \mu_{X_i}\mu_{X_j}^* \tag{6.175}$$

Expanding the matrix multiplication inside the expectation operator, we have

$$
\begin{aligned}
\mathbb{C}_{XX} &= E\left[\{(\boldsymbol{X}-\boldsymbol{\mu}_X)\}\{(\boldsymbol{X}-\boldsymbol{\mu}_X)^*\}^{\mathrm{T}}\right] \\
&= E\left[\{(\boldsymbol{X}-\boldsymbol{\mu}_X)\}\{(\boldsymbol{X}^*-\boldsymbol{\mu}_X^*)\}^{\mathrm{T}}\right] \\
&= E\{(\boldsymbol{X}-\boldsymbol{\mu}_X)(\boldsymbol{X}^{*\mathrm{T}}-\boldsymbol{\mu}_X^{*\mathrm{T}})\} \\
&= E\{\boldsymbol{X}(\boldsymbol{X}^{*\mathrm{T}}-\boldsymbol{\mu}_X^{*\mathrm{T}})-\boldsymbol{\mu}_X(\boldsymbol{X}^{*\mathrm{T}}-\boldsymbol{\mu}_X^{*\mathrm{T}})\} \\
&= E\{\boldsymbol{X}\boldsymbol{X}^{*\mathrm{T}}-\boldsymbol{X}\boldsymbol{\mu}_X^{*\mathrm{T}}-\boldsymbol{\mu}_X\boldsymbol{X}^{*\mathrm{T}}+\boldsymbol{\mu}_X\boldsymbol{\mu}_X^{*\mathrm{T}}\} \\
&= E(\boldsymbol{X}\boldsymbol{X}^{*\mathrm{T}})-E(\boldsymbol{X}\boldsymbol{\mu}_X^{*\mathrm{T}})-E(\boldsymbol{\mu}_X\boldsymbol{X}^{*\mathrm{T}})+E(\boldsymbol{\mu}_X\boldsymbol{\mu}_X^{*\mathrm{T}}) \\
&= E(\boldsymbol{X}\boldsymbol{X}^{*\mathrm{T}})-\boldsymbol{\mu}_X\boldsymbol{\mu}_X^{*\mathrm{T}}
\end{aligned}
\tag{6.176}
$$

Similarly, we have the following autocovariance matrix of $\boldsymbol{Y}$:

$$
\mathbb{C}_{YY}=\begin{bmatrix} c_{Y_1Y_1} & \cdot & \cdot & \cdot & c_{Y_1Y_n} \\ \cdot & \cdot & \cdot & \cdot & \cdot \\ \cdot & \cdot & c_{Y_iY_j}\cdot & \cdot & \cdot \\ c_{Y_nY_1} & \cdot & \cdot & \cdot & c_{Y_nY_n} \end{bmatrix}
\tag{6.177}
$$

where

$$
c_{Y_i,Y_j}=Cov(Y_i,Y_j)=E\left\{(Y_i-\mu_{Y_i})\left(Y_j-\mu_{Y_j}\right)^*\right\}=E\left(Y_iY_j^*\right)-\mu_{Y_i}\mu_{Y_j}^*
\tag{6.178}
$$

Theorem 6.9.2 The autocovariance matrix of a complex process $X(t)$ is Hermitian, that is,

$$
c_{X_iX_j}=c_{X_jX_i}^* \qquad c_{X_iX_i}=c_{X_iX_i}^*=\text{real}
\tag{6.179}
$$

Proof Consider the *ji*th element of $\mathbb{C}_{XX}$. By (6.175), we have

$$
c_{X_jX_i}=\mathrm{Cov}(X_j,X_i)=E\left\{\left(X_j-\mu_{X_j}\right)(X_i-\mu_{X_i})^*\right\}=E(X_jX_i^*)-\mu_{X_j}\mu_{X_i}^*
$$

By (6.129), we know that

$$
c_{X_jX_i}=c_{X_iX_j}^*
\tag{6.180}
$$

Consider the diagonal elements, *ii*th elements, as follows:

$$
c_{X_iX_i}=\mathrm{Cov}(X_i,X_i)=E\left\{(X_i-\mu_{X_i})(X_i-\mu_{X_i})^*\right\}=E\left\{|X_i-\mu_{X_i}|^2\right\}=\sigma_{X_i}^2
$$

The diagonal elements are real numbers.

Q.E.D.

6.9.3 Cross-covariance Matrix

The $(n \times m)$-dimensional cross-covariance matrix of $\boldsymbol{X}$ and $\boldsymbol{Y}$ is given below:

$$\mathbb{C}_{XY} \triangleq E\left\{(\boldsymbol{X}-\boldsymbol{\mu}_X)(\boldsymbol{Y}-\boldsymbol{\mu}_Y)^{\mathrm{T}}\right\} = \begin{bmatrix} c_{X_1Y_1} & c_{X_1Y_2} & \cdot & c_{X_1Y_j} & \cdot & c_{X_1Y_m} \\ c_{X_2Y_1} & \cdot & \cdot & \cdot & \cdot & \cdot \\ \cdot & \cdot & \cdot & c_{X_iY_j} & \cdot & \cdot \\ \cdot & \cdot & \cdot & \cdot & \cdot & \cdot \\ c_{X_nY_1} & \cdot & \cdot & c_{X_nY_j} & \cdot & c_{X_nY_m} \end{bmatrix} \tag{6.181}$$

where

$$c_{X_iY_j} = Cov(X_i, Y_j) = E\left\{(X_i - \mu_{X_i})\left(Y_j - \mu_{Y_j}\right)\right\} = E(X_iY_j) - \mu_{X_i}\mu_{Y_j} \tag{6.182}$$

Expanding the matrix multiplication inside the expectation operator in (6.181), we obtain the following equation:

$$\mathbb{C}_{XY} = E\left\{(\boldsymbol{X}-\boldsymbol{\mu}_X)(\boldsymbol{Y}-\boldsymbol{\mu}_Y)^{\mathrm{T}}\right\} = E\left(\boldsymbol{X}\boldsymbol{Y}^{\mathrm{T}}\right) - \boldsymbol{\mu}_X\boldsymbol{\mu}_Y^{\mathrm{T}} \tag{6.183}$$

As stated by (6.173), the autocovariance matrix is symmetric if the process is real. We can see that this is true intuitively because $\mathrm{Cov}(X_i, X_j)$ and $\mathrm{Cov}(X_j, X_i)$ are the covariance between the same two real RVs, $X(t_i)$ and $X(t_j)$ so that $c_{X_iX_j} = c_{X_jX_i}$. In other words, in both expressions, we are dealing with the same two RVs. The same is not true in the case of the cross-covariance matrix, that is, $\mathrm{Cov}(X_i, Y_j) \neq \mathrm{Cov}(Y_i, X_j)$. $\mathrm{Cov}(X_i, Y_j)$ is the covariance between $X(t_i)$ and $Y(t_j)$, whereas $\mathrm{Cov}(Y_i, X_j)$ is the covariance between $Y(t_i)$ and $X(t_j)$.

For the complex process, the $(n \times m)$-dimensional cross-covariance matrix of $\boldsymbol{X}$ and $\boldsymbol{Y}$ is given below:

$$\mathbb{C}_{XY} \triangleq E\left[\{(\boldsymbol{X}-\boldsymbol{\mu}_X)\}\{(\boldsymbol{Y}-\boldsymbol{\mu}_Y)^*\}^{\mathrm{T}}\right]$$

$$= \begin{bmatrix} c_{X_1Y_1} & c_{X_1Y_2} & \cdot & c_{X_1Y_j} & \cdot & c_{X_1Y_m} \\ c_{X_2Y_1} & \cdot & \cdot & \cdot & \cdot & \cdot \\ \cdot & \cdot & \cdot & c_{X_iY_j} & \cdot & \cdot \\ \cdot & \cdot & \cdot & \cdot & \cdot & \cdot \\ c_{X_nY_1} & \cdot & \cdot & c_{X_nY_j} & \cdot & c_{X_nY_m} \end{bmatrix} \tag{6.184}$$

where

$$c_{X_iY_j} = \mathrm{Cov}(X_i, Y_j) = E\left\{(X_i - \mu_{X_i})\left(Y_j - \mu_{Y_j}\right)^*\right\} = E\left(X_iY_j^*\right) - \mu_{X_i}\mu_{Y_j}^* \tag{6.185}$$

Expanding the matrix multiplication inside the expectation operator, we obtain the following equation:

$$\mathbb{C}_{XY} = E\left\{(X - \mu_X)(Y - \mu_Y)^{\mathrm{T}}\right\} = E\left(XY^{*\mathrm{T}}\right) - \mu_X \mu_Y^{*\mathrm{T}} \tag{6.186}$$

Similarly,

$$\mathbb{C}_{YX} = E\left\{(Y - \mu_Y)(X - \mu_X)^{\mathrm{T}}\right\} = E\left(YX^{*\mathrm{T}}\right) - \mu_Y \mu_X^{*\mathrm{T}} \tag{6.187}$$

$$c_{Y_jX_i} = \mathrm{Cov}\left(Y_j, X_i\right) = E\left\{\left(Y_j - \mu_{Y_j}\right)\left(X_i - \mu_{X_i}\right)^*\right\} = E\left(Y_j X_i^*\right) - \mu_{Y_j}\mu_{X_i}^* \tag{6.188}$$

6.9.4 Covariance Matrix of a Concatenated Vector RV

Since Z_{XY} is a mixed vector of X and Y, its covariance matrix does not have the qualifying prefix "auto." The covariance matrix of Z_{XY} involves the autocovariance and the cross-covariance matrices of X and Y.

By the same process of (6.170), we obtain the following equation:

$$\mathbb{C}_{Z_{XY}Z_{XY}} \triangleq E\left\{\left(Z_{XY} - \mu_{Z_{XY}}\right)\left(Z_{XY} - \mu_{Z_{XY}}\right)^{\mathrm{T}}\right\} = E\left(Z_{XY}Z_{XY}^{\mathrm{T}}\right) - \mu_{Z_{XY}}\mu_{Z_{XY}}^{\mathrm{T}} \tag{6.189}$$

Consider the matrix multiplication inside the expectation operator as follows. By (2.31), we obtain the following equations:

$$Z_{XY}Z_{XY}^{\mathrm{T}} = \begin{bmatrix} XX^{\mathrm{T}} & | & XY^{\mathrm{T}} \\ -- & - & -- \\ YX^{\mathrm{T}} & | & YY^{\mathrm{T}} \end{bmatrix} \qquad \mu_{Z_{XY}}\mu_{Z_{XY}}^{\mathrm{T}} = \begin{bmatrix} \mu_X\mu_X^{\mathrm{T}} & | & \mu_X\mu_Y^{\mathrm{T}} \\ -- & - & -- \\ \mu_Y\mu_X^{\mathrm{T}} & | & \mu_Y\mu_Y^{\mathrm{T}} \end{bmatrix} \tag{6.190}$$

Substituting (6.190) into (6.189), we obtain the following equation:

$$\begin{aligned} \mathbb{C}_{Z_{XY}Z_{XY}} &= \begin{bmatrix} c_{Z_{XY11}} & \cdot & c_{Z_{XY1j}} & \cdot & c_{Z_{XY1(n+m)}} \\ \cdot & \cdot & \cdot & \cdot & \cdot \\ \cdot & \cdot & c_{Z_{XYij}} & \cdot & \cdot \\ \cdot & \cdot & \cdot & \cdot & \cdot \\ c_{Z_{XY(n+m)1}} & \cdot & \cdot & \cdot & c_{Z_{XY(n+m)(n+m)}} \end{bmatrix} \\ &= \begin{bmatrix} E\left(XX^{\mathrm{T}}\right) - \mu_X\mu_X^{\mathrm{T}} & | & E\left(XY^{\mathrm{T}}\right) - \mu_X\mu_Y^{\mathrm{T}} \\ -------- & - & -------- \\ E\left(YX^{\mathrm{T}}\right) - \mu_Y\mu_X^{\mathrm{T}} & | & E\left(YY^{\mathrm{T}}\right) - \mu_Y\mu_Y^{\mathrm{T}} \end{bmatrix} \\ &= \begin{bmatrix} \mathbb{C}_{XX} & | & \mathbb{C}_{XY} \\ -- & - & -- \\ \mathbb{C}_{YX} & | & \mathbb{C}_{YY} \end{bmatrix} \end{aligned} \tag{6.191}$$

The submatrices of (6.191), $\mathbb{C}_{XX}, \mathbb{C}_{XY}$, and $\mathbb{C}_{YY}$, are given by (6.168), (6.181), and (6.171), respectively, and $\mathbf{C}_{YX}$ is given by the transposition of $\mathbf{C}_{XY}$. The dashed lines inside the matrices show the partitions of the matrices and do not change the matrices in any way.

Since $\boldsymbol{Z}_{XY}$ is a mixed vector of $\boldsymbol{X}$ and $\boldsymbol{Y}$, its covariance matrix does not have the qualifying prefix "auto." The covariance matrix of $\boldsymbol{Z}_{XY}$ involves the autocovariance and the cross-covariance matrices of $\boldsymbol{X}$ and $\boldsymbol{Y}$.

For a complex process, by the same process of (6.170), we have

$$\mathbb{C}_{Z_{XY}Z_{XY}} \triangleq E\left\{\left(\boldsymbol{Z}_{XY}-\boldsymbol{\mu}_{Z_{XY}}\right)\left\{\left(\boldsymbol{Z}_{XY}-\boldsymbol{\mu}_{Z_{XY}}\right)^{*}\right\}^{\mathrm{T}}\right\} = E\left(\boldsymbol{Z}_{XY}\boldsymbol{Z}_{XY}^{*\mathrm{T}}\right) - \boldsymbol{\mu}_{Z_{XY}}\boldsymbol{\mu}_{Z_{XY}}^{*\mathrm{T}} \tag{6.192}$$

Consider the matrix multiplication inside the expectation operator as follows. By (2.31), we have the following equation:

$$\begin{aligned} \boldsymbol{Z}_{XY}\boldsymbol{Z}_{XY}^{*\mathrm{T}} &= \begin{bmatrix} \boldsymbol{X}\boldsymbol{X}^{*\mathrm{T}} & | & \boldsymbol{X}\boldsymbol{Y}^{*\mathrm{T}} \\ -- & - & -- \\ \boldsymbol{Y}\boldsymbol{X}^{*\mathrm{T}} & | & \boldsymbol{Y}\boldsymbol{Y}^{*\mathrm{T}} \end{bmatrix} \\ \boldsymbol{\mu}_{Z_{XY}}\boldsymbol{\mu}_{Z_{XY}}^{\mathrm{T}} &= \begin{bmatrix} \boldsymbol{\mu}_X\boldsymbol{\mu}_X^{*\mathrm{T}} & | & \boldsymbol{\mu}_X\boldsymbol{\mu}_Y^{*\mathrm{T}} \\ -- & - & -- \\ \boldsymbol{\mu}_Y\boldsymbol{\mu}_X^{*\mathrm{T}} & | & \boldsymbol{\mu}_Y\boldsymbol{\mu}_Y^{*\mathrm{T}} \end{bmatrix} \end{aligned} \tag{6.193}$$

Substituting (6.193) into (6.192), we obtain the following equation:

$$\begin{aligned} \mathbb{C}_{Z_{XY}Z_{XY}} &= \begin{bmatrix} c_{Z_{XY11}} & \cdot & c_{Z_{XY1j}} & \cdot & c_{Z_{XY1(n+m)}} \\ \cdot & \cdot & \cdot & \cdot & \cdot \\ \cdot & \cdot & c_{Z_{XYij}} & \cdot & \cdot \\ \cdot & \cdot & \cdot & \cdot & \cdot \\ c_{Z_{XY(n+m)1}} & \cdot & \cdot & \cdot & c_{Z_{XY(n+m)(n+m)}} \end{bmatrix} \\ &= \begin{bmatrix} E\left(\boldsymbol{X}\boldsymbol{X}^{*\mathrm{T}}\right) - \boldsymbol{\mu}_X\boldsymbol{\mu}_X^{*\mathrm{T}} & | & E\left(\boldsymbol{X}\boldsymbol{Y}^{*\mathrm{T}}\right) - \boldsymbol{\mu}_X\boldsymbol{\mu}_Y^{*\mathrm{T}} \\ --------- & - & --------- \\ E\left(\boldsymbol{Y}\boldsymbol{X}^{*\mathrm{T}}\right) - \boldsymbol{\mu}_Y\boldsymbol{\mu}_X^{*\mathrm{T}} & | & E\left(\boldsymbol{Y}\boldsymbol{Y}^{*\mathrm{T}}\right) - \boldsymbol{\mu}_Y\boldsymbol{\mu}_Y^{*\mathrm{T}} \end{bmatrix} \\ &= \begin{bmatrix} \mathbb{C}_{XX} & | & \mathbb{C}_{XY} \\ -- & + & -- \\ \mathbb{C}_{YX} & | & \mathbb{C}_{YY} \end{bmatrix} \end{aligned} \tag{6.194}$$

The submatrices of (6.194), $\mathbb{C}_{XX}, \mathbb{C}_{YY}, \mathbb{C}_{XY}$, and $\mathbb{C}_{YX}$, are given by (6.174), (6.177), (6.186), and (6.187), respectively. The dashed lines inside the matrices show the partitions of the matrices and do not change the matrices in any way.

6.9.5 Linear Combination

Theorem 6.9.3 Consider m linear combinations of n RVs, $X_1, \ldots, X_n$, given by the following equation:

$$\mathbf{W} = \mathbb{B}\boldsymbol{X} \tag{6.195}$$

where

$$\mathbf{W} = \begin{bmatrix} W_1 \\ \cdot \\ \cdot \\ W_m \end{bmatrix} \qquad \boldsymbol{X} = \begin{bmatrix} X_1 \\ \cdot \\ \cdot \\ X_n \end{bmatrix} \tag{6.196}$$

and $\mathbb{B}$ is the coefficient matrix given by

$$\mathbb{B} = \begin{bmatrix} b_{11} & \cdots & b_{1n} \\ \vdots & \ddots & \vdots \\ b_{m1} & \cdots & b_{mn} \end{bmatrix} \tag{6.197}$$

If an n-dimensional vector RV $\boldsymbol{X}$ has the mean vector and the covariance matrix defined by (6.164a) and (6.168), respectively, the mean vector and the covariance matrix of $\boldsymbol{W}$ are given by the following equations:

$$\boldsymbol{\mu}_W = \mathbb{B}\boldsymbol{\mu}_X \qquad \mathbb{C}_{WW} = \mathbb{B}\mathbb{C}_{XX}\mathbb{B}^{\mathrm{T}} \tag{6.198}$$

Proof To determine the mean vector of $\boldsymbol{W}$, consider its element W_i first. By matrix multiplication, we have

$$W_{\mathrm{i}} = [\, b_{\mathrm{i}1} \quad b_{\mathrm{i}2} \ldots \quad b_{in} \,]\boldsymbol{X} = \sum_{j=1}^{n} b_{ij}X_j$$

$$\mu_{W_i} = E\left\{\textstyle\sum_{j=1}^{n} b_{ij}X_j\right\}$$

By (5.13) and (5.22), we obtain the following equation:

$$\mu_{W_i} = \sum_{j=1}^{n} b_{ij}E(X_j) = \sum_{j=1}^{n} b_{ij}\mu_{X_j} = [\, b_{i1} \quad b_{i2} \ldots \quad b_{in} \,]\boldsymbol{\mu}_X$$

Collecting the above into a vector, we have

$$\boldsymbol{\mu}_W = \mathbb{B}\boldsymbol{\mu}_X \tag{6.199}$$

Now, to determine the covariance matrix of $\boldsymbol{W}$, using (6.195) and (6.199), we obtain the following equation:

$$\boldsymbol{W} - \boldsymbol{\mu}_W = \mathbb{B}\boldsymbol{X} - \mathbb{B}\boldsymbol{\mu}_X = \mathbb{B}(\boldsymbol{X} - \boldsymbol{\mu}_X) \tag{6.200}$$

Applying the definition given by (6.170) and substituting the above equation in the expression given below, we obtain the following equation:

$$\begin{aligned} \mathbb{C}_{WW} &= E\left\{(\boldsymbol{W} - \boldsymbol{\mu}_W)(\boldsymbol{W} - \boldsymbol{\mu}_W)^{\mathrm{T}}\right\} \\ &= E\left[\{\mathbb{B}(\boldsymbol{X} - \boldsymbol{\mu}_X)\}\{\mathbb{B}(\boldsymbol{X} - \boldsymbol{\mu}_X)\}^{\mathrm{T}}\right] \\ &= E\left[\left\{\mathbb{B}(\boldsymbol{X} - \boldsymbol{\mu}_X)(\boldsymbol{X} - \boldsymbol{\mu}_X)^{\mathrm{T}}\mathbb{B}^{\mathrm{T}}\right\}\right] \\ &= \mathbb{B}E\left\{(\boldsymbol{X} - \boldsymbol{\mu}_X)(\boldsymbol{X} - \boldsymbol{\mu}_X)^{\mathrm{T}}\right\}\mathbb{B}^{\mathrm{T}} \end{aligned} \tag{6.201}$$

By (6.170), we have

$$E\left\{(\boldsymbol{X} - \boldsymbol{\mu}_X)(\boldsymbol{X} - \boldsymbol{\mu}_X)^{\mathrm{T}}\right\} = \mathbb{C}_{XX}$$

Substituting the above into (6.201), we obtain (6.198).

Q.E.D.

Theorem 6.9.4 In a linear combination $\boldsymbol{W} = \mathbb{B}\boldsymbol{X}$, if $\mathbb{B}$ is chosen to diagonalize $\mathbb{C}_{XX}$ to produce a diagonal $\mathbb{C}_{XX}$ in (6.198), $\boldsymbol{W}$ is a vector RV of m uncorrelated RVs. This is a useful theorem that allows converting an arbitrary vector RV to uncorrelated RVs.

Proof If $\mathbb{B}$ diagonalizes $\mathbb{C}_{XX}$ so that $\mathbb{C}_{WW}$ becomes a diagonal matrix, the off-diagonal elements of $\mathbb{C}_{WW}$, which are Cov (W_i, W_j), are zero. Therefore, W_i and W_j are uncorrelated.

Chapter 7
Gaussian Distributions

The Gaussian distribution is one of the five important distributions discussed in this book. Because of its importance, a separate chapter is devoted here to this topic. Chapter 8 provides further discussion on the Gaussian distribution and the Gaussian stochastic process as applied to communications channel modeling.

This chapter starts with a brief discussion on the central limit theorem, which will show one of the fundamental reasons why the Gaussian distribution is important. This chapter discusses a Gaussian RV treated singly, two Gaussian RVs treated jointly, and multiple Gaussian RVs as vector RVs. This chapter derives the mean and the variance of the Gaussian RV and derives the characteristic function of the scalar Gaussian RV and that of the vector Gaussian RV. Finally, this chapter defines the Gaussian stochastic process.

7.1 Central Limit Theorem

It has been empirically observed that the statistical behaviors of many natural phenomena follow a Gaussian distribution. The Gaussian assumption on the thermal noise of a communications channel may also be viewed from this perspective. A theorem that theoretically supports the observation of these natural phenomena is the central limit theorem, which states that the distribution of the sum of n independent RVs with arbitrary distributions approaches the Gaussian distribution as n approaches the infinity. This is quite remarkable and is briefly introduced here.

The theorem is stated in several different forms depending upon the conditions under which it is applied. In a general form, the theorem is applicable for independent RVs with arbitrary individual distributions. In a more restricted form, the theorem is applicable for independent RVs with a common, albeit arbitrary, distribution. This latter form of the theorem is discussed below.

K.I. Park, *Fundamentals of Probability and Stochastic Processes with Applications to Communications*, https://doi.org/10.1007/978-3-319-68075-0_7

Let W be the sum of n mutually independent RVs with a common distribution as follows:

$$W = X_1 + X_2 + \cdots X_n = \sum_{i=1}^{n} X_i \tag{7.1}$$

$$E(X_i) = \mu_X \qquad \text{Var}\,(X_i) = \sigma_X^2,\ i = 1,\ .\ .\ .\ ,n \tag{7.2}$$

$$\begin{aligned} \mu_W &= E(W) = \sum_{i=1}^{n} \mu_X = n\mu_X \\ \sigma_X^2 &= \text{Var}(W) = \sum_{i=1}^{n} \sigma_X^2 = n\sigma_X^2 \quad \sigma_W = \text{S.D}(W) = \sigma_X\sqrt{n} \end{aligned} \tag{7.3}$$

The central limit theorem states that the CDF and the *pdf* of the RV W approaches the normal distribution with the mean and the variance given by (7.3). Referring to (7.11) and (7.13), we have

$$\lim_{n\to\infty} f_W(w) \ \rightarrow \ \frac{1}{\sigma_W\sqrt{2\pi}}\, e^{-\frac{1}{2}\left(\frac{w-\mu_W}{\sigma_W}\right)^2} \tag{7.4}$$

$$\lim_{n\to\infty} F_W(w) \ \rightarrow \ \int_{-\infty}^{\frac{w-\mu_W}{\sigma_W}} \frac{1}{\sqrt{2\pi}} e^{-\frac{1}{2}\lambda'^2} d\lambda' = \Phi\left(\frac{w-\mu_W}{\sigma_W}\right) \tag{7.5}$$

7.2 Single Gaussian RV

An RV X is said to be Gaussian, or normal, if its *pdf* is given by the following equation:

$$f_X(x) = \frac{1}{\beta\sqrt{2\pi}} e^{-\frac{1}{2}\left(\frac{x-\alpha}{\beta}\right)^2} \tag{7.6}$$

where α and β ($\beta > 0$) are arbitrary parameters. Later, we will show that the mean and the variance of a Gaussian-distributed RV are α and β^2, respectively. Since the mean and the variance of an RV is commonly denoted by μ and σ^2, respectively, a normal *pdf* is often given by the above equation with α and β replaced by μ and σ, respectively. When an RV X is normally distributed with parameters μ and σ we write

$$X \sim N\left(\mu, \sigma^2\right) \tag{7.7}$$

The standard normal *pdf*, denoted by $\varphi(x)$, is the normal distribution with $\mu = 0$ and $\sigma = 1$ as follows:

$$\varphi(x) = \frac{1}{\sqrt{2\pi}} e^{-\frac{1}{2}x^2} \tag{7.8}$$

The standard normal CDF, denoted by $\Phi(x)$, is given by the following integral of the above *pdf*:

$$\Phi(x) = \int_{-\infty}^{x} \frac{1}{\sqrt{2\pi}} e^{-\frac{1}{2}\lambda^2} \mathrm{d}\lambda \tag{7.9}$$

The above integral does not have a closed-form solution. The values of $\Phi(x)$ are tabulated in a mathematical table. For a given value of x, the value of $\Phi(x)$ can be found from a mathematical table.

The general normal *pdf* with μ and σ, denoted by $f_X(x)$, is given by the following equation:

$$f_X(x) = \frac{1}{\sigma\sqrt{2\pi}} e^{-\frac{1}{2}\left(\frac{x-\mu}{\sigma}\right)^2} \tag{7.10}$$

The general normal CDF is given by the integral of the normal *pdf* as follows:

$$F_X(x) = \int_{-\infty}^{x} \frac{1}{\sigma\sqrt{2\pi}} e^{-\frac{1}{2}\left(\frac{\lambda-\mu}{\sigma}\right)^2} \mathrm{d}\lambda \tag{7.11}$$

The above integral has no closed-form solution. The value of the CDF can be found from the mathematical table of the standard normal CDF by changing the variable of integration as follows:

$$\lambda' = \frac{\lambda - \mu}{\sigma} \qquad \mathrm{d}\lambda' = \frac{\mathrm{d}\lambda}{\sigma} \tag{7.12}$$

and the upper limit of integral from $\lambda = x$ to $\lambda' = \frac{x-\mu}{\sigma}$. The original integral becomes

$$F_X(x) = \int_{-\infty}^{\frac{x-\mu}{\sigma}} \frac{1}{\sqrt{2\pi}} e^{-\frac{1}{2}\lambda'^2} \mathrm{d}\lambda' = \Phi\left(\frac{x-\mu}{\sigma}\right) \tag{7.13}$$

The value of $\Phi\left(\frac{x-\mu}{\sigma}\right)$ is found from a mathematical table of the standard normal distribution function.

In Fig. 7.1, the left figure shows the standard normal *pdf* $\varphi(x)$. It is symmetric with respect to the origin. The shaded area shows the value of $\Phi(x)$. The right figure shows the general normal *pdf*s with $\mu \neq 0$ and $\sigma \neq 1$ overlaid on the standard normal *pdf* for comparison. The bigger the σ becomes, the broader the *pdf* curve becomes.

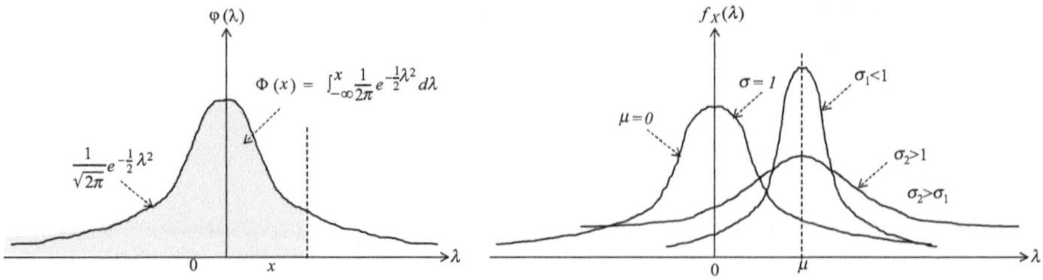

Fig. 7.1 Normal *pdf*s

Conversely, the smaller the σ becomes, the narrower the *pdf* curve becomes. The *pdf* curves with non-zero μ are symmetric with respect to $\lambda = \mu$.

One of the conditions that a *pdf* must satisfy is that the total area under the *pdf* curve must be equal to 1. Therefore, a normal *pdf* must also satisfy this condition.

Theorem 7.2.1 A *pdf* must satisfy the condition that the total area under the *pdf* curve be equal to 1. This theorem shows that the total area under a normal *pdf* curve is equal to 1 for arbitrary parameters μ and σ as follows:

$$F_X(+\infty) = \Phi(+\infty) = \int_{-\infty}^{+\infty} \frac{1}{\sigma\sqrt{2\pi}} e^{-\frac{1}{2}\left(\frac{x-\mu}{\sigma}\right)^2} \mathrm{d}x = 1 \tag{7.14}$$

Proof Although the integral of a normal *pdf* does not have a closed-form solution for a finite upper limit x, it can be shown that, for $x = +\infty$, the integral is equal to 1. To show this, start with the change of variable given by (7.12) as follows:

$$I = \int_{-\infty}^{+\infty} \frac{1}{\sigma\sqrt{2\pi}} e^{-\frac{1}{2}\left(\frac{x-\mu}{\sigma}\right)^2} \mathrm{d}x = \int_{-\infty}^{+\infty} \frac{1}{\sqrt{2\pi}} e^{-\frac{1}{2}\lambda^2} \mathrm{d}\lambda \tag{7.15}$$

Squaring both sides of (7.15), we have

$$\begin{aligned} I^2 &= \left\{\int_{-\infty}^{+\infty} \frac{1}{\sqrt{2\pi}} e^{-\frac{1}{2}\omega^2} \mathrm{d}\omega\right\}\left\{\int_{-\infty}^{+\infty} \frac{1}{\sqrt{2\pi}} e^{-\frac{1}{2}\gamma^2} \mathrm{d}\gamma\right\} \\ &= \frac{1}{2\pi}\int_{-\infty}^{+\infty}\int_{-\infty}^{+\infty} e^{-\frac{1}{2}(\omega^2+\gamma^2)} \mathrm{d}\gamma \mathrm{d}\omega \end{aligned} \tag{7.16}$$

Using the polar coordinate shown in Fig. 7.2 with the following change of variables and the limits of integral, we can rewrite (7.16) as follows:

$$\omega = r\cos\theta \quad \gamma = r\sin\theta \quad \mathrm{d}\omega\,\mathrm{d}\gamma = r\,\mathrm{d}r\,\mathrm{d}\theta \quad 0 \le \theta \le 2\pi \quad 0 \le r < \infty \tag{7.17}$$

Fig. 7.2 Polar coordinate

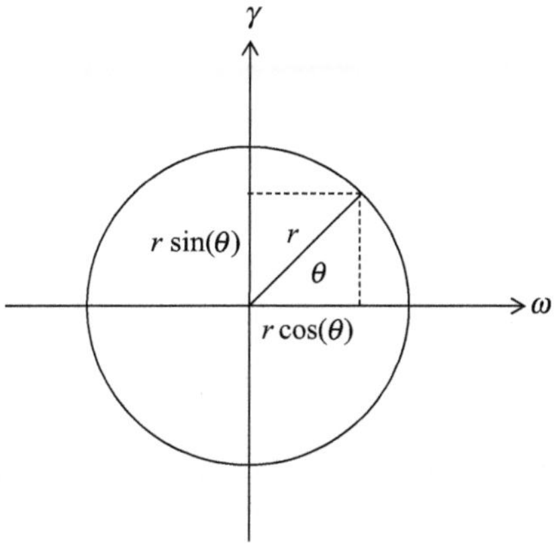

$$I^2 = \frac{1}{2\pi} \int_0^{2\pi} \left\{ \int_0^{+\infty} r e^{-\frac{1}{2}r^2} \mathrm{d}r \right\} \mathrm{d}\theta \tag{7.18}$$

The integral in braces in (7.18) can be evaluated by the following change of variable:

$$z = -\frac{1}{2}r^2 \qquad \mathrm{d}z = -r\mathrm{d}r$$

$$\int_0^{-\infty} r e^{-\frac{1}{2}r^2} \mathrm{d}r = -\int_0^{-\infty} e^z \mathrm{d}z = [-e^z]_0^{\infty} = \left[-e^{-\frac{1}{2}r^2}\right]_0^{-\infty} = 1 \tag{7.19}$$

Note that (7.19) will be used in proving other theorems later. Substituting (7.19) into (7.18), we obtain the following equation: $I^2 = \frac{1}{2\pi} \int_0^{2\pi} \mathrm{d}\theta = 1$ or $I = 1$ and

$$\begin{aligned} I &= \frac{1}{\sigma\sqrt{2\pi}} \int_{-\infty}^{+\infty} e^{-\frac{1}{2}\left(\frac{x-\mu}{\sigma}\right)^2} \mathrm{d}\lambda \\ &= 1 \end{aligned}$$

Q.E.D.

Theorem 7.2.2 If an RV X is normally distributed with parameters α and β, $X \sim N(\alpha, \beta^2)$, the mean of X is equal to α:

$$\mu_X = E(X) = \alpha \tag{7.20}$$

Proof

$$E(X) = \int_{-\infty}^{+\infty} x f_X(x)\mathrm{d}x = \frac{1}{\beta\sqrt{2\pi}} \int_{-\infty}^{+\infty} x e^{-\frac{1}{2}\left(\frac{x-\alpha}{\beta}\right)^2} \mathrm{d}x \tag{7.21}$$

Let

$$\lambda = \frac{x-\alpha}{\beta} \qquad x = \beta\lambda + \alpha \qquad \mathrm{d}x = \beta \mathrm{d}\lambda \tag{7.22}$$

Substituting (7.22) into (7.21), we have

$$\begin{aligned} E(X) &= \frac{1}{\beta\sqrt{2\pi}} \int_{-\infty}^{+\infty} (\beta\lambda + \alpha) e^{-\frac{1}{2}\lambda^2} \beta \mathrm{d}\lambda \\ &= \frac{\beta}{\beta\sqrt{2\pi}} \int_{-\infty}^{+\infty} \lambda e^{-\frac{1}{2}\lambda^2} \mathrm{d}\lambda + \alpha \left\{ \frac{1}{\sqrt{2\pi}} \int_{-\infty}^{+\infty} e^{-\frac{1}{2}\lambda^2} \mathrm{d}\lambda \right\} \end{aligned} \tag{7.23}$$

By (7.14), the term in braces in (7.23) is equal to 1. The first integral on the right-hand side of (7.23) can be evaluated as in (7.19):

$$\begin{aligned} \frac{\beta}{\beta\sqrt{2\pi}} \int_{-\infty}^{+\infty} \lambda e^{-\frac{1}{2}\lambda^2} \mathrm{d}\lambda &= -\frac{\beta}{\beta\sqrt{2\pi}} \left[e^{-\frac{1}{2}\lambda^2} \right]_{+\infty}^{-\infty} \\ &= 0 \end{aligned} \tag{7.24}$$

Substituting the above result into (7.23), we obtain the following result:

$$E(X) = 0 + \alpha = \alpha$$

Q.E.D.

Theorem 7.2.3 If an RV X is normally distributed with parameters α and β, $X \sim N(\alpha, \beta^2)$, the variance of X is equal to β^2:

$$\sigma_X^2 = \mathrm{Var}(X) = \beta^2 \tag{7.25}$$

Proof Using (5.26) and (7.20), we have

$$\mathrm{Var}\,(X) = E(X^2) - \{E(X)\}^2 = \frac{1}{\beta\sqrt{2\pi}} \int_{-\infty}^{+\infty} x^2 e^{-\frac{1}{2}\left(\frac{x-\alpha}{\beta}\right)^2} \mathrm{d}x - \alpha^2 \tag{7.26}$$

Using the change of variables given by (7.22), we evaluate the integral in the above equation as follows:

$$
\begin{aligned}
E(X^2) &= \frac{1}{\beta\sqrt{2\pi}} \int_{-\infty}^{+\infty} x^2 e^{-\frac{1}{2}\left(\frac{x-\alpha}{\beta}\right)^2} \mathrm{d}x = \frac{1}{\beta\sqrt{2\pi}} \int_{-\infty}^{+\infty} (\beta\lambda+\alpha)^2 e^{-\frac{1}{2}\lambda^2} \beta \mathrm{d}\lambda \\
&= \frac{1}{\sqrt{2\pi}} \int_{-\infty}^{+\infty} (\beta^2\lambda^2 + 2\alpha\beta\lambda + \alpha^2) e^{-\frac{1}{2}\lambda^2} \mathrm{d}\lambda \\
&= \frac{1}{\sqrt{2\pi}} \int_{-\infty}^{+\infty} \beta^2\lambda^2 e^{-\frac{1}{2}\lambda^2} \mathrm{d}\lambda + 2\alpha\beta \frac{1}{\sqrt{2\pi}} \int_{-\infty}^{+\infty} \lambda e^{-\frac{1}{2}\lambda^2} \mathrm{d}\lambda + \alpha^2 \frac{1}{\sqrt{2\pi}} \int_{-\infty}^{+\infty} e^{-\frac{1}{2}\lambda^2} \mathrm{d}\lambda \\
&= I_1 + I_2 + I_3
\end{aligned}
\tag{7.27a}
$$

where

$$
I_1 = \frac{1}{\sqrt{2\pi}} \int_{-\infty}^{+\infty} \beta^2\lambda^2 e^{-\frac{1}{2}\lambda^2} \mathrm{d}\lambda \tag{7.27b}
$$

$$
I_2 = 2\,\alpha\beta \frac{1}{\sqrt{2\pi}} \int_{-\infty}^{+\infty} \lambda e^{-\frac{1}{2}\lambda^2} \mathrm{d}\lambda \tag{7.27c}
$$

$$
I_3 = \alpha^2 \frac{1}{\sqrt{2\pi}} \int_{-\infty}^{+\infty} e^{-\frac{1}{2}\lambda^2} \mathrm{d}\lambda \tag{7.27d}
$$

The above three integrals are evaluated below.

To evaluate the first integral I_1, make the following change of variables in (7.27b):

$u = \lambda \qquad \mathrm{d}u = \mathrm{d}\lambda \qquad \mathrm{d}v = \lambda e^{-\frac{1}{2}\lambda^2} \mathrm{d}\,\lambda \qquad v = -e^{-\frac{1}{2}\lambda^2}$

Using the above change of variables, we can evaluate (7.27a) by the integration by parts as follows:

$$
I_1 = \frac{\beta^2}{\sqrt{2\pi}} \left\{ uv - \int_{-\infty}^{+\infty} v\mathrm{d}u \right\} = \frac{\beta^2}{\sqrt{2\pi}} \left[-\lambda e^{-\frac{1}{2}\lambda^2} \right]_{-\infty}^{+\infty} + \beta^2 \left\{ \frac{1}{\sqrt{2\pi}} \int_{-\infty}^{+\infty} e^{-\frac{1}{2}\lambda^2} \mathrm{d}\lambda \right\} \tag{7.28}
$$

In (7.28), the first term is zero by the l'Hopital's rule, and the term in braces is equal to 1 as shown by the same integral in (7.23) and (7.28) and thus becomes

$$I_1 = \beta^2 \tag{7.29}$$

Using the result of (7.19), we evaluate the second integral (7.27c) as follows:

$$I_2 = -\left[\frac{2\alpha\beta}{\sqrt{2\pi}} e^{-\frac{1}{2}\lambda^2}\right]_{-\infty}^{+\infty} = 0 \tag{7.30}$$

Using the result of (7.14), we evaluate the third integral (7.27d) as follows:

$$I_3 = \alpha^2 \tag{7.31}$$

By substituting (7.29), (7.30), and (7.31) into (7.27a), we obtain

$$E(X^2) = \beta^2 + \alpha^2 \tag{7.32}$$

Substituting (7.32) into (7.26), we obtain the following equation:

$$\mathrm{Var}(X) = E(X^2) - [E(X)]^2 = \beta^2 + \alpha^2 - \alpha^2 = \beta^2$$

Q.E.D.

For a normal distribution with parameters α and β, (7.20) and (7.25) show that $\alpha = \mu$ and $\beta = \sigma$, i.e.,

$$X \sim N(\alpha, \beta^2) \rightarrow \mu = \alpha; \quad \sigma^2 = \beta^2 \tag{7.33}$$

This is the reason why the two parameters of a normal distribution are commonly denoted by μ and σ.

Given a *pdf*, parameters such as the mean and the variance can be derived from the *pdf*, but the converse is not necessarily true. In the case of a normal distribution, however, the converse is true. If it is known that an RV X is normally distributed, it suffices to determine its mean and variance to determine its *pdf* completely. This is one of the important properties of a normal distribution.

Theorem 7.2.4 Suppose that RVs X and Y are independent and normally distributed with the means and the variances, μ_X and σ_X^2 and μ_Y and σ_Y^2, respectively:

$$X \sim N(\mu_X, \sigma_X^2) \qquad Y \sim N(\mu_Y, \sigma_Y^2)$$

then, the sum $W = X + Y$ is normally distributed with mean $(\mu_X + \mu_Y)$ and variance $(\sigma_X^2 + \sigma_Y^2)$:

$$W \sim N\{(\mu_X + \mu_X), (\sigma_X^2 + \sigma_X^2)\} \tag{7.34}$$

Proof By (4.99), we obtain the following equation:

$$
\begin{aligned}
f_W(w) &= f_X(x)*f_Y(x) = \int_{-\infty}^{+\infty} f_X(x)f_Y(w-x)\mathrm{d}x \\
&= \int_{-\infty}^{+\infty} \frac{1}{\sigma_X\sqrt{2\pi}} e^{-\frac{1}{2}\left(\frac{x-\mu_X}{\sigma_X}\right)^2} \frac{1}{\sigma_Y\sqrt{2\pi}} e^{-\frac{1}{2}\left(\frac{w-x-\mu_Y}{\sigma_Y}\right)^2} \mathrm{d}x \\
&= \int_{-\infty}^{+\infty} \frac{1}{\sigma_X\sigma_Y\sqrt{2\pi}\sqrt{2\pi}} e^{-\frac{1}{2}\times\frac{1}{\sigma_X^2\sigma_Y^2}\left\{\sigma_Y^2(x-\mu_X)^2+\sigma_X^2(w-x-\mu_Y)^2\right\}} \mathrm{d}x \\
&= \int_{-\infty}^{+\infty} I(w,x)\mathrm{d}x
\end{aligned}
\tag{7.35}
$$

where $I(w, x)$ denotes the integrand of the above integral as follows:

$$
\begin{aligned}
I(w,x) &= \frac{1}{\sigma_X\sigma_Y\sqrt{2\pi}\sqrt{2\pi}} e^{-\frac{1}{2}\times\frac{1}{\sigma_X^2\sigma_Y^2}\left\{\sigma_Y^2(x-\mu_X)^2+\sigma_X^2(w-x-\mu_Y)^2\right\}} \\
&= \frac{1}{\sigma_X\sigma_Y\sqrt{2\pi}\sqrt{2\pi}} e^{-\frac{1}{2}\frac{1}{\sigma_X^2\sigma_Y^2}\left\{(\sigma_X^2+\sigma_Y^2)x^2-2\{\mu_X\sigma_Y^2+\sigma_X^2(w-\mu_Y)\}x+\mu_X^2\sigma_Y^2+\sigma_X^2(w-\mu_Y)^2\right\}} \\
&= \frac{1}{\sigma_X\sigma_Y\sqrt{2\pi}\sqrt{2\pi}} e^{-\frac{1}{2}\frac{\sigma_X^2+\sigma_Y^2}{\sigma_X^2\sigma_Y^2}\left\{x^2-2\frac{\mu_X\sigma_Y^2+\sigma_X^2(w-\mu_Y)}{\sigma_X^2+\sigma_Y^2}x+\frac{\mu_X^2\sigma_Y^2+\sigma_X^2(w-\mu_Y)^2}{\sigma_X^2+\sigma_Y^2}\right\}} \\
&= \frac{1}{\sigma_X\sigma_Y\sqrt{2\pi}\sqrt{2\pi}} e^{M}
\end{aligned}
\tag{7.36}
$$

where M is the exponent of the above equation as follows:

$$
M = -\frac{1}{2}\frac{\sigma_X^2+\sigma_Y^2}{\sigma_X^2\sigma_Y^2}\left\{x^2-2\frac{\mu_X\sigma_Y^2+\sigma_X^2(w-\mu_Y)}{\sigma_X^2+\sigma_Y^2}x+\frac{\mu_X^2\sigma_Y^2+\sigma_X^2(w-\mu_Y)^2}{\sigma_X^2+\sigma_Y^2}\right\} \tag{7.37}
$$

In (7.37), let

$$
A = \frac{\mu_X\sigma_Y^2+\sigma_X^2(w-\mu_Y)}{\sigma_X^2+\sigma_Y^2} \tag{7.38}
$$

and, using (7.38), rewrite (7.37) as follows:

$$
\begin{aligned}
M &= -\frac{1}{2}\frac{\sigma_X^2+\sigma_Y^2}{\sigma_X^2\sigma_Y^2}\left\{x^2-2Ax+\frac{\mu_X^2\sigma_Y^2+\sigma_X^2(w-\mu_Y)^2}{\sigma_X^2+\sigma_Y^2}\right\}\\
&= -\frac{1}{2}\frac{\sigma_X^2+\sigma_Y^2}{\sigma_X^2\sigma_Y^2}\left\{x^2-2Ax+A^2-A^2+\frac{\mu_X^2\sigma_Y^2+\sigma_X^2(w-\mu_Y)^2}{\sigma_X^2+\sigma_Y^2}\right\}\\
&= -\frac{1}{2}\frac{\sigma_X^2+\sigma_Y^2}{\sigma_X^2\sigma_Y^2}\left\{(x-A)^2-A^2+\frac{\mu_X^2\sigma_Y^2+\sigma_X^2(w-\mu_Y)^2}{\sigma_X^2+\sigma_Y^2}\right\}\\
&= -\frac{1}{2}\frac{\sigma_X^2+\sigma_Y^2}{\sigma_X^2\sigma_Y^2}\left\{(x-A)^2+L\right\}
\end{aligned}
\tag{7.39}
$$

where

$$
L=-A^2+\frac{\mu_X^2\sigma_Y^2+\sigma_X^2(w-\mu_Y)^2}{\sigma_X^2+\sigma_Y^2} \tag{7.40}
$$

Substituting(7.38) into (7.40), we evaluate the above equation as follows:

$$
\begin{aligned}
L &= -\left\{\frac{\mu_X\sigma_Y^2+\sigma_X^2(w-\mu_Y)}{\sigma_X^2+\sigma_Y^2}\right\}^2+\frac{\mu_X^2\sigma_Y^2+\sigma_X^2(w-\mu_Y)^2}{\sigma_X^2+\sigma_Y^2}\\
&= \frac{-\left\{\mu_X\sigma_Y^2+\sigma_X^2(w-\mu_Y)\right\}^2+\left(\sigma_X^2+\sigma_Y^2\right)\left\{\mu_X^2\sigma_Y^2+\sigma_X^2(w-\mu_Y)^2\right\}}{\left(\sigma_X^2+\sigma_Y^2\right)^2}\\
&= \frac{-\left\{\mu_X^2\sigma_Y^4+2\mu_X\sigma_X^2\sigma_Y^2(w-\mu_Y)+\sigma_X^4(w-\mu_Y)^2\right\}+\mu_X^2\sigma_X^2\sigma_Y^2+\mu_X^2\sigma_Y^4+\sigma_X^2\left(\sigma_X^2+\sigma_Y^2\right)(w-\mu_Y)^2}{\left(\sigma_X^2+\sigma_Y^2\right)^2}\\
&= \frac{\left\{\sigma_X^2\left(\sigma_X^2+\sigma_Y^2\right)-\sigma_X^4\right\}(w-\mu_Y)^2-2\mu_X\sigma_X^2\sigma_Y^2(w-\mu_Y)+\mu_X^2\sigma_X^2\sigma_Y^2}{\left(\sigma_X^2+\sigma_Y^2\right)^2}\\
&= \frac{\sigma_X^2\sigma_Y^2(w-\mu_Y)^2-2\mu_X\sigma_X^2\sigma_Y^2(w-\mu_Y)+\mu_X^2\sigma_X^2\sigma_Y^2}{\left(\sigma_X^2+\sigma_Y^2\right)^2}\\
&= \frac{\sigma_X^2\sigma_Y^2}{\left(\sigma_X^2+\sigma_Y^2\right)^2}\left\{(w-\mu_Y)^2-2\mu_X(w-\mu_Y)+\mu_X^2\right\}=\frac{\sigma_X^2\sigma_Y^2}{\left(\sigma_X^2+\sigma_Y^2\right)^2}\left\{(w-\mu_X-\mu_Y)^2\right\}
\end{aligned}
\tag{7.41}
$$

Substituting L given by (7.41) into (7.39), we obtain the following equation:

$$
\begin{aligned}
M &= -\frac{1}{2}\frac{\left(\sigma_X^2+\sigma_Y^2\right)}{\sigma_X^2\sigma_Y^2}(x-A)^2-\frac{1}{2}\frac{\left(\sigma_X^2+\sigma_Y^2\right)}{\sigma_X^2\sigma_Y^2}\times\frac{\sigma_X^2\sigma_Y^2}{\left(\sigma_X^2+\sigma_Y^2\right)^2}\left\{(w-\mu_X-\mu_Y)^2\right\}\\
&= -\frac{1}{2}\frac{\left(\sigma_X^2+\sigma_Y^2\right)}{\sigma_X^2\sigma_Y^2}(x-A)^2-\frac{1}{2}\frac{1}{\left(\sigma_X^2+\sigma_Y^2\right)}\{w-(\mu_X+\mu_Y)\}^2\\
&= -\frac{1}{2}\left\{\frac{x-A}{\sigma_X\sigma_Y/\sqrt{\sigma_X^2+\sigma_Y^2}}\right\}^2-\frac{1}{2}\left\{\frac{w-(\mu_X+\mu_Y)}{\sqrt{\sigma_X^2+\sigma_Y^2}}\right\}^2
\end{aligned}
\tag{7.42}
$$

Substituting M given by (7.42) into (7.36), we have the following form for the integrand:

$$
\begin{aligned}
I(w,x) &= \frac{1}{\sigma_X\sigma_Y\sqrt{2\pi}\sqrt{2\pi}} e^{-\frac{1}{2}\left\{\frac{x-A}{\sigma_X\sigma_Y/\sqrt{\sigma_X^2+\sigma_Y^2}}\right\}^2 - \frac{1}{2}\left\{\frac{w-(\mu_X+\mu_Y)}{\sqrt{\sigma_X^2+\sigma_Y^2}}\right\}^2} \\
&= \frac{1}{\sqrt{\sigma_X^2+\sigma_Y^2}\sqrt{2\pi}} e^{-\frac{1}{2}\left\{\frac{w-(\mu_X+\mu_Y)}{\sqrt{\sigma_X^2+\sigma_Y^2}}\right\}^2} \times \frac{\sqrt{\sigma_X^2+\sigma_Y^2}}{\sigma_X\sigma_Y\sqrt{2\pi}} e^{-\frac{1}{2}\left\{\frac{x-A}{\sigma_X\sigma_Y/\sqrt{\sigma_X^2+\sigma_Y^2}}\right\}^2} \\
&= \frac{1}{\sqrt{\sigma_X^2+\sigma_Y^2}\sqrt{2\pi}} e^{-\frac{1}{2}\left\{\frac{w-(\mu_X+\mu_Y)}{\sqrt{\sigma_X^2+\sigma_Y^2}}\right\}^2} \times \frac{1}{\left(\sigma_X\sigma_Y/\sqrt{\sigma_X^2+\sigma_Y^2}\right)\sqrt{2\pi}} e^{-\frac{1}{2}\left\{\frac{x-A}{\sigma_X\sigma_Y/\sqrt{\sigma_X^2+\sigma_Y^2}}\right\}^2}
\end{aligned}
\tag{7.43}
$$

Substituting (7.43) into (7.35), we have

$$
\begin{aligned}
f_W(w) &= \int_{-\infty}^{+\infty} \frac{1}{\sqrt{\sigma_X^2+\sigma_Y^2}\sqrt{2\pi}} e^{-\frac{1}{2}\left\{\frac{w-(\mu_X+\mu_Y)}{\sqrt{\sigma_X^2+\sigma_Y^2}}\right\}^2} \times \frac{1}{\left(\sigma_X\sigma_Y/\sqrt{\sigma_X^2+\sigma_Y^2}\right)\sqrt{2\pi}} e^{-\frac{1}{2}\left\{\frac{x-A}{\sigma_X\sigma_Y/\sqrt{\sigma_X^2+\sigma_Y^2}}\right\}^2} dx \\
&= \frac{1}{\sqrt{\sigma_X^2+\sigma_Y^2}\sqrt{2\pi}} e^{-\frac{1}{2}\left\{\frac{w-(\mu_X+\mu_Y)}{\sqrt{\sigma_X^2+\sigma_Y^2}}\right\}^2} \int_{-\infty}^{+\infty} \frac{1}{\left(\sigma_X\sigma_Y/\sqrt{\sigma_X^2+\sigma_Y^2}\right)\sqrt{2\pi}} e^{-\frac{1}{2}\left\{\frac{x-A}{\sigma_X\sigma_Y/\sqrt{\sigma_X^2+\sigma_Y^2}}\right\}^2} dx
\end{aligned}
\tag{7.44}
$$

By (7.14), the integral in (7.44) is equal to 1 and we obtain the following results:

$$
f_W(w) = \frac{1}{\sqrt{\sigma_X^2+\sigma_Y^2}\sqrt{2\pi}} e^{-\frac{1}{2}\left\{\frac{w-(\mu_X+\mu_Y)}{\sqrt{\sigma_X^2+\sigma_Y^2}}\right\}^2}
$$

$$
W \sim N\{(\mu_X+\mu_Y), (\sigma_X^2+\sigma_Y^2)\}
$$

Q.E.D.

Theorem 7.2.5 Let

$$
X_i \sim N(\mu_i, \sigma_i^2), \quad i = 1, 2, \ldots, n \qquad X = \sum_{i=1}^{n} X_i \tag{7.45}
$$

If X_i's are independent RVs, the sum RV X is normally distributed with the mean and the variance given by

$$\mu_X = \sum_{i=1}^{n} \mu_i \qquad \sigma_X^2 = \sum_{i=1}^{n} \sigma_i^2 \tag{7.46}$$

$$X \sim N\left\{\sum_{i=1}^{n} \mu_i, \sum_{i=1}^{n} \sigma_i^2\right\} \tag{7.47}$$

Proof Consider the case of $n = 3$ as follows:

$$X = X_1 + X_2 + X_3 \tag{7.48}$$

Let

$$Y = X_1 + X_2 \tag{7.49}$$

By (7.34),

$$Y \sim N(\mu_1 + \mu_2, \sigma_1^2 + \sigma_2^2)$$

Substituting (7.49) into (7.48), we have

$$X = Y + X_3$$

By (7.34),

$$X \sim N(\mu_Y + \mu_3, \sigma_Y^2 + \sigma_3^2) \tag{7.50}$$

By (5.21),

$$\mu_Y = E(Y) = E(X_1 + X_2) = E(X_1) + E(X_2) = \mu_1 + \mu_2 \tag{7.51}$$

By (5.52), and, since Cov(X_1, X_2) = 0 because X_1 and X_2 are independent, we have

$$\sigma_Y^2 = \mathrm{Var}(Y) = \mathrm{Var}(X_1) + \mathrm{Var}(X_2) = \sigma_1^2 + \sigma_2^2 \tag{7.52}$$

Substituting (7.51) and (7.52) into (7.50), we have

$$X \sim N(\mu_1 + \mu_2 + \mu_3, \sigma_1^2 + \sigma_2^2 + \sigma_3^2)$$

This proves (7.47) for $n = 3$. By mathematical induction, (7.48) is proven for an arbitrary n.

Q.E.D.

7.3 Two Jointly Gaussian RVs

Two RVs X and Y are said to be jointly normal, if its joint *pdf* is given by the following equation:

$$f_{XY}(x,y) = \frac{1}{2\pi\sigma_X\sigma_Y\sqrt{1-\rho^2}} e^{-\frac{1}{2(1-\rho^2)}\left\{\left(\frac{x-\mu_X}{\sigma_X}\right)^2 - \frac{2\rho(x-\mu_X)(y-\mu_Y)}{\sigma_X\sigma_Y} + \left(\frac{y-\mu_Y}{\sigma_Y}\right)^2\right\}} \tag{7.53}$$

where $\mu_X, \mu_Y, \sigma_X^2, \sigma_Y^2$, and $-1 \le \rho \le 1$ are constants. When X and Y are jointly normal, we write

$$(X, Y) \sim N\left(\mu_X, \mu_Y, \sigma_X^2, \sigma_Y^2, \rho\right) \tag{7.54}$$

The above *pdf* is called the bivariate normal *pdf*. The bivariate normal *pdf* satisfies the following conditions.

Theorem 7.3.1 A bivariate normal *pdf* given by (7.53) satisfies the following condition:

$$\int_{-\infty}^{+\infty}\int_{-\infty}^{+\infty} f_{XY}(x,y)\mathrm{d}x\mathrm{d}y = 1 \tag{7.55}$$

This is a condition that any *pdf* must satisfy and is not a special condition for a bivariate normal *pdf*. In other words, by the definition of a *pdf*, the integral of a *pdf* over the entire real line(s) must be one.

Proof Rewrite (7.53) as follows:

$$f_{XY}(x,y) = \frac{1}{2\pi\sigma_X\sigma_Y\sqrt{1-\rho^2}} e^{E(x,y)} \tag{7.56}$$

where the exponent is given by the following equation:

$$\begin{aligned} E(x,y) &= -\frac{1}{2(1-\rho^2)}\left\{\left(\frac{x-\mu_X}{\sigma_X}\right)^2 - \frac{2\rho(x-\mu_X)(y-\mu_Y)}{\sigma_X\sigma_Y} + \left(\frac{y-\mu_Y}{\sigma_Y}\right)^2\right\} \\ &= -\frac{1}{2(1-\rho^2)}\left\{\left(\frac{x-\mu_X}{\sigma_X}\right)^2 - 2\frac{(x-\mu_X)}{\sigma_X}\frac{\rho(y-\mu_Y)}{\sigma_Y} + \rho^2\left(\frac{y-\mu_Y}{\sigma_Y}\right)^2 - \rho^2\left(\frac{y-\mu_Y}{\sigma_Y}\right)^2 + \left(\frac{y-\mu_Y}{\sigma_Y}\right)^2\right\} \\ &= -\frac{1}{2(1-\rho^2)}\left[\left\{\left(\frac{x-\mu_X}{\sigma_X}\right) - \rho\left(\frac{y-\mu_Y}{\sigma_Y}\right)\right\}^2 + (1-\rho^2)\left(\frac{y-\mu_Y}{\sigma_Y}\right)^2\right] \end{aligned} \tag{7.57}$$

Substituting (7.57) into (7.56), we obtain the following equation:

$$
\begin{aligned}
f_{XY}(x,y) &= \frac{1}{2\pi\sigma_X\sigma_Y\sqrt{1-\rho^2}} e^{-\frac{1}{2(1-\rho^2)}\left[\left\{\left(\frac{x-\mu_X}{\sigma_X}\right)-\rho\left(\frac{y-\mu_Y}{\sigma_Y}\right)\right\}^2+(1-\rho^2)\left(\frac{y-\mu_Y}{\sigma_Y}\right)^2\right]} \\
&= \frac{1}{\sqrt{2\pi}\sigma_X\sqrt{1-\rho^2}} e^{-\frac{1}{2(1-\rho^2)}\left\{\left(\frac{x-\mu_X}{\sigma_X}\right)-\rho\left(\frac{y-\mu_Y}{\sigma_Y}\right)\right\}^2} \frac{1}{\sigma_Y\sqrt{2\pi}} e^{-\frac{1}{2}\left(\frac{y-\mu_Y}{\sigma_Y}\right)^2} \\
&= \frac{1}{\sqrt{2\pi}\sigma_X\sqrt{1-\rho^2}} e^{-\frac{1}{2(1-\rho^2)\sigma_X^2}\left\{(x-\mu_X)-\rho\frac{\sigma_X}{\sigma_Y}(y-\mu_Y)\right\}^2} \frac{1}{\sigma_Y\sqrt{2\pi}} e^{-\frac{1}{2}\left(\frac{y-\mu_Y}{\sigma_Y}\right)^2} \\
&= \frac{1}{\sqrt{2\pi}\sigma_X\sqrt{1-\rho^2}} e^{-\frac{1}{2(1-\rho^2)\sigma_X^2}\left\{x-\left(\rho\frac{\sigma_X}{\sigma_Y}y+\mu_X-\rho\frac{\sigma_X}{\sigma_Y}\mu_Y\right)\right\}^2} \frac{1}{\sigma_Y\sqrt{2\pi}} e^{-\frac{1}{2}\left(\frac{y-\mu_Y}{\sigma_Y}\right)^2} \\
&= \frac{1}{\sqrt{2\pi}\sigma_X\sqrt{1-\rho^2}} e^{-\frac{1}{2(1-\rho^2)\sigma_X^2}\{x-A\}^2} \frac{1}{\sigma_Y\sqrt{2\pi}} e^{-\frac{1}{2}\left(\frac{y-\mu_Y}{\sigma_Y}\right)^2} \\
&= \frac{1}{\sqrt{2\pi}\sigma_X\sqrt{1-\rho^2}} e^{-\frac{1}{2}\left\{\frac{x-A}{\sigma_X\sqrt{1-\rho^2}}\right\}^2} \frac{1}{\sigma_Y\sqrt{2\pi}} e^{-\frac{1}{2}\left(\frac{y-\mu_Y}{\sigma_Y}\right)^2}
\end{aligned}
\tag{7.58}
$$

where

$$
A = \rho\frac{\sigma_X}{\sigma_Y}y + \mu_X - \rho\frac{\sigma_X}{\sigma_Y}\mu_Y \tag{7.59}
$$

Substituting (7.58) into (7.55), we obtain the following equation:

$$
\int_{-\infty}^{+\infty}\int_{-\infty}^{+\infty} f_{XY}(x,y)\mathrm{d}x\mathrm{d}y = \int_{-\infty}^{+\infty}\left\{\int_{-\infty}^{+\infty}\frac{1}{\sqrt{2\pi}\sigma_X\sqrt{1-\rho^2}} e^{-\frac{1}{2}\left\{\frac{x-A}{\sigma_X\sqrt{1-\rho^2}}\right\}^2}\mathrm{d}x\right\}\frac{e^{-\frac{1}{2}\left(\frac{y-\mu_Y}{\sigma_Y}\right)^2}}{\sigma_Y\sqrt{2\pi}}\mathrm{d}y \tag{7.60}
$$

By (7.14), the integral in braces with respect to x is equal to 1 and the outer integral with respect to y is also equal to 1 to yield the right-hand side of the above to be equal to 1.

Q.E.D.

Theorem 7.3.2 If RVs X and Y are jointly normal

$$
(X,Y) \sim N\left(\mu_X, \mu_Y, \sigma_X^2, \sigma_Y^2, \rho\right) \tag{7.61}
$$

The marginal distributions of X and Y are normal with means and variances μ_X, μ_Y, σ_X^2 and σ_Y^2, respectively, as follows:

$$
X \sim N\left(\mu_X, \sigma_X^2\right) \qquad f_X(x) = \frac{1}{\sigma_X\sqrt{2\pi}} e^{-\frac{1}{2}\left(\frac{x-\mu_X}{\sigma_X}\right)^2} \tag{7.62}
$$

$$Y \sim N\left(\mu_Y, \sigma_Y^2\right) \qquad f_Y(x) = \frac{1}{\sigma_Y\sqrt{2\pi}} e^{-\frac{1}{2}\left(\frac{y-\mu_Y}{\sigma_Y}\right)^2} \tag{7.63}$$

This theorem states that, if two RVs X and Y are jointly normal, X and Y are individually normal. However, the converse is not necessarily true. In other words, two normally distributed RVs X and Y are not necessarily jointly normal.

Proof By (4.60), the marginal *pdf* of X is given by

$$\begin{aligned} f_Y(y) &= \int_{-\infty}^{+\infty} f_{XY}(x,y)\mathrm{d}x \\ &= \int_{-\infty}^{+\infty} \frac{1}{2\pi\sigma_X\sigma_Y\sqrt{1-\rho^2}} e^{-\frac{1}{2(1-\rho^2)}\left\{\left(\frac{x-\mu_X}{\sigma_X}\right)^2 - \frac{2\rho(x-\mu_X)(y-\mu_Y)}{\sigma_X\sigma_Y} + \left(\frac{y-\mu_Y}{\sigma_Y}\right)^2\right\}}\mathrm{d}x \end{aligned}$$

Through the process of (7.58) through (7.60), the bivariate normal *pdf* in the above integral can be expressed by (7.58). Substituting (7.58) into the above equation, we obtain the following equation:

$$\begin{aligned} f_Y(y) &= \int_{-\infty}^{\infty} \frac{1}{\sqrt{2\pi}\sigma_X\sqrt{1-\rho^2}} e^{-\frac{1}{2}\left\{\frac{x-A}{\sigma_X\sqrt{1-\rho^2}}\right\}^2} \frac{1}{\sigma_Y\sqrt{2\pi}} e^{-\frac{1}{2}\left[\left(\frac{y-\mu_Y}{\sigma_Y}\right)^2\right]}\mathrm{d}x \\ &= \frac{1}{\sigma_Y\sqrt{2\pi}} e^{-\frac{1}{2}\left(\frac{y-\mu_Y}{\sigma_Y}\right)^2} \int_{-\infty}^{\infty} \frac{1}{\sqrt{2\pi}\sigma_X\sqrt{1-\rho^2}} e^{-\frac{1}{2}\left\{\frac{x-A}{\sigma_X\sqrt{1-\rho^2}}\right\}^2}\mathrm{d}x \end{aligned} \tag{7.64}$$

The integrand in the last integral is in the form of a normal *pdf* with $\mu = A$ and $\sigma = \sigma_X\sqrt{1-\rho^2}$. Therefore, as proven by (7.14), the integral is equal to 1 and (7.64) becomes

$$f_Y(y) = \frac{1}{\sigma_Y\sqrt{2\pi}} e^{-\frac{1}{2}\left(\frac{y-\mu_Y}{\sigma_Y}\right)^2}$$

This proves (7.63). Eq. (7.62) is proven similarly.

Theorem 7.3.3 If RVs X and Y are jointly normal

$$(X,Y) \sim N\left(\mu_X, \mu_Y, \sigma_X^2, \sigma_Y^2, \rho\right) \tag{7.65}$$

ρ is the correlation coefficient of X and Y, and the covariance of X and Y is given by

$$c_{XY} = \mathrm{Cov}(X,Y) = \rho\sqrt{\sigma_X^2\sigma_Y^2} = \rho\sigma_X\sigma_Y \tag{7.66}$$

$$\rho = \frac{c_{XY}}{\sigma_X \sigma_Y} \tag{7.67}$$

Proof By (5.33), we have the following equation:

$$c_{XY} = \mathrm{Cov}(X, Y) = E(XY) - \mu_X \mu_Y \tag{7.68}$$

$$E(XY) = \int_{-\infty}^{+\infty}\int_{-\infty}^{+\infty} xyf_{XY}(x, y)\mathrm{d}x\mathrm{d}y$$

$$= \int_{-\infty}^{+\infty}\int_{-\infty}^{+\infty} xy\frac{1}{2\pi\sigma_X\sigma_Y\sqrt{1-\rho^2}} e^{-\frac{1}{2(1-\rho^2)}\left\{\left(\frac{x-\mu_X}{\sigma_X}\right)^2 - \frac{2\rho(x-\mu_X)(y-\mu_Y)}{\sigma_X\sigma_Y} + \left(\frac{y-\mu_Y}{\sigma_Y}\right)^2\right\}}\mathrm{d}x\mathrm{d}y \tag{7.69}$$

Through the process of (7.58) through (7.60), the last integral can be expressed as the following where A is given by (7.59)

$$E(XY) = \int_{-\infty}^{+\infty}\left\{\int_{-\infty}^{+\infty}\frac{x}{\sqrt{2\pi}\sigma_X\sqrt{1-\rho^2}}e^{-\frac{1}{2}\left\{\frac{x-A}{\sigma_X\sqrt{1-\rho^2}}\right\}^2}\mathrm{d}x\right\}\frac{y}{\sigma_Y\sqrt{2\pi}}e^{-\frac{1}{2}\left(\frac{y-\mu_Y}{\sigma_Y}\right)^2}\mathrm{d}y \tag{7.70}$$

Through the process of (7.21) through (7.24), the integral in braces in the above integral is the expected value of X given Y fixed at y as follows:

$$\int_{-\infty}^{+\infty}\frac{x}{\sqrt{2\pi}\sigma_X\sqrt{1-\rho^2}}e^{-\frac{1}{2}\left\{\frac{x-A}{\sigma_X\sqrt{1-\rho^2}}\right\}^2}\mathrm{d}x = A$$

where A is given by (7.59)

$$A = \rho\frac{\sigma_X}{\sigma_Y}y + \mu_X - \rho\frac{\sigma_X}{\sigma_Y}\mu_Y \tag{7.71}$$

Substituting (7.71) into (7.70), we obtain

$$E(XY) = \int_{-\infty}^{+\infty}\left\{\rho\frac{\sigma_X}{\sigma_Y}y + \mu_X - \rho\frac{\sigma_X}{\sigma_Y}\mu_Y\right\}\frac{y}{\sigma_Y\sqrt{2\pi}}e^{-\frac{1}{2}\left(\frac{y-\mu_Y}{\sigma_Y}\right)^2}\mathrm{d}y$$

$$= \rho\left(\frac{\sigma_X}{\sigma_Y}\right)\int_{-\infty}^{+\infty}\frac{y^2}{\sigma_Y\sqrt{2\pi}}e^{-\frac{1}{2}\left(\frac{y-\mu_Y}{\sigma_Y}\right)^2}\mathrm{d}y + \left\{\mu_X - \rho\frac{\sigma_X}{\sigma_Y}\mu_Y\right\}\int_{-\infty}^{+\infty}\frac{y}{\sigma_Y\sqrt{2\pi}}e^{-\frac{1}{2}\left(\frac{y-\mu_Y}{\sigma_Y}\right)^2}\mathrm{d}y \tag{7.72}$$

By (7.32), the first and the second integral in the above equation are, respectively, given by the following:

$$\int_{-\infty}^{+\infty} \frac{y^2}{\sigma_Y\sqrt{2\pi}} e^{-\frac{1}{2}\left(\frac{y-\mu_Y}{\sigma_Y}\right)^2} dy = E(Y^2) = \sigma_Y^2 + \mu_Y^2 \tag{7.73}$$

$$\int_{-\infty}^{+\infty} \frac{y}{\sigma_Y\sqrt{2\pi}} e^{-\frac{1}{2}\left(\frac{y-\mu_Y}{\sigma_Y}\right)^2} dy = E(Y) = \mu_Y \tag{7.74}$$

Substituting (7.73) and (7.74) into (7.72), we have

$$E(XY) = \rho\frac{\sigma_X}{\sigma_Y}\left(\sigma_Y^2 + \mu_Y^2\right) + \left(\mu_X - \rho\frac{\sigma_X}{\sigma_Y}\mu_Y\right)\mu_Y = \rho\sigma_X\sigma_Y + \mu_X\,\mu_Y \tag{7.75}$$

Substituting (7.75) into (7.68), we have

$$c_{XY} = \mathrm{Cov}(X, Y) = E(XY) - \mu_X\mu_Y = \rho\sigma_X\sigma_Y + \mu_X\,\mu_Y - \mu_X\mu_Y = \rho\sigma_X\sigma_Y$$

The above proof also proves that the coefficient ρ is the correlation coefficient

$$\rho = \frac{c_{XY}}{\sigma_X\sigma_Y}$$

Hence, both (7.66) and (7.67) are proven.

Q.E.D.

7.4 Vector Gaussian RV

In the general discussion of RVs in earlier chapters, the RVs are not a function of time. However, in this section, we will consider the RVs of the specific time points of a stochastic process. The vector RVs discussed in this section follow the definitions given in Sect. 6.3 and, therefore, have the time points as the arguments.

The vector RV $\boldsymbol{X}(t_1, t_2, \ldots, t_n)$ is said to be jointly normal if its n-dimensional joint *pdf* is given by the following equation:

$$f_{\boldsymbol{X}(t_1,t_2,\ldots,t_n)}(x_1, x_2, \ldots, x_n; t_1, t_2, \ldots, t_n) = \frac{1}{\sqrt{(2\pi)^n \det\mathbb{C}_{XX}}} e^{-\frac{1}{2}(\boldsymbol{x}-\boldsymbol{\mu}_X)^{\mathrm{T}}\mathbb{C}_{XX}^{-1}(\boldsymbol{x}-\boldsymbol{\mu}_X)} \tag{7.76}$$

where the vector of specific values $\boldsymbol{x}$ and the vector RV $\boldsymbol{X}$ are as defined by (6.32a) and (6.35), respectively, and the mean vector $\boldsymbol{\mu}_X$ and the autocovariance matrix $\mathbb{C}_{XX}$

are given by (6.164a) and (6.168), respectively. By (2.33) and (2.41), we have the following equations for the terms occurring in (7.76):

$$\det \mathbb{C}_{XX} \triangleq \sum_{j} (-1)^{t(j)} C_{1j_1} C_{2j_2} \cdot \ . \ . C_{nj_n} \tag{7.77}$$

$$\mathbb{C}_{XX}^{-1} = \frac{1}{\det \mathbb{C}_{XX}} \mathrm{adj} \mathbb{C}_{XX} \tag{7.78}$$

where adj $\mathbb{C}_{XX}$ is defined by (2.40) and $t(j)$ is the total number of inversions of the permutation j defined by (2.34).

Consider the dimensions of the matrices in the following matrix multiplication term in the exponent in (7.76):

$$(\boldsymbol{x} - \boldsymbol{\mu}_X)^{\mathrm{T}} \mathbb{C}_{XX}^{-1} (\boldsymbol{x} - \boldsymbol{\mu}_X) \tag{7.79}$$

The following are the dimensions of the three matrices in the above multiplication:

$(\boldsymbol{x} - \boldsymbol{\mu}_X)^{\mathrm{T}}$ $1 \times n$ $\mathbb{C}_{XX}^{-1}$ $n \times n$ $(\boldsymbol{x} - \boldsymbol{\mu}_X)$ $n \times 1$

The multiplication of $(n \times n)$ matrix $\mathbb{C}_{XX}^{-1}$ and $(n \times 1)$ matrix $(\boldsymbol{x} - \boldsymbol{\mu}_X)$ yields a $n \times 1$ column vector. Therefore, multiplication of $(1 \times n)$ row vector $(\boldsymbol{x} - \boldsymbol{\mu}_X)^{\mathrm{T}}$ and $(n \times 1)$ column vector $\{\mathbb{C}_{XX}^{-1}(\boldsymbol{x} - \boldsymbol{\mu}_X)\}$ yields a scalar. The determinant $\det \mathbb{C}_{XX}$ is a scalar. Therefore, the multivariate joint normal *pdf* given by (7.76) is a scalar function.

By evaluating the matrix multiplication as discussed above, we may rewrite the multivariate joint *pdf* given by (7.76) as follows with the exponent in an expanded form:

$$\begin{aligned} & f_{X(t_1, t_2, \ldots, t_n)}(x_1, x_2, \ldots, x_n; t_1, t_2, \ldots, t_n) \\ & = \frac{1}{\sqrt{(2\pi)^n \det \mathbb{C}_{XX}}} e^{-\frac{1}{2} \times \frac{1}{\det \mathbb{C}_{XX}} \sum_{i=1}^{n} \sum_{j=1}^{n} (x_i - \mu_i) A_{ij} (x_j - \mu_j)} \end{aligned} \tag{7.80}$$

where $\det \mathbb{C}_{XX}$ and cofactor A_{ij} are given by (7.77) and (2.39), respectively.

Two vector RVs $\boldsymbol{X}(t_1, .., t_n)$ and $\boldsymbol{Y}(t'_1, .., t'_m)$ are jointly Gaussian, if the $(m + n)$-dimensional concatenated vector RV $\boldsymbol{Z}_{XY}(t_1, .., t_n, t'_1, .., t'_m)$ defined by (6.33) consisting of m and n RVs have the following $(m + n)$-dimensional multivariate normal *pdf*

$$\begin{aligned} & f_{Z_{XY}(t_1, .., t_n, t'_1, .., t'_m)}(x_1, .., x_n, y_1, .., y_m; t_1, .., t_n, t'_1, .., t'_m) \\ & = \frac{1}{\sqrt{(2\pi)^n \det \mathbb{C}_{Z_{XY} Z_{XY}}}} e^{-\frac{1}{2}(z_{XY} - \boldsymbol{\mu}_{Z_{XY}})^{\mathrm{T}} \mathbb{C}_{Z_{XY} Z_{XY}}^{-1} (z_{XY} - \boldsymbol{\mu}_{Z_{XY}})} \end{aligned} \tag{7.81}$$

where z_{XY} is the vector of specific values of the vector RV $\boldsymbol{Z}_{XY}$ defined by (6.35) and $\boldsymbol{\mu}_{Z_{XY}}$ and $\mathbb{C}_{Z_{XY} Z_{XY}}$ are defined by (6.165) and (6.191), respectively.

Example 7.4.1
Consider the case of $n = 3$.

$$\boldsymbol{X} = \begin{bmatrix} X_1 \\ X_2 \\ X_3 \end{bmatrix} \qquad \boldsymbol{\mu}_X = E(\boldsymbol{X}) = \begin{bmatrix} \mu_1 \\ \mu_2 \\ \mu_3 \end{bmatrix} \tag{7.82}$$

$$\mathbb{C}_{XX} = E\left\{(\boldsymbol{X} - \boldsymbol{\mu}_X)(\boldsymbol{X} - \boldsymbol{\mu}_X)^{\mathrm{T}}\right\} = E\left(\boldsymbol{X}\boldsymbol{X}^{\mathrm{T}}\right) - \boldsymbol{\mu}_X\boldsymbol{\mu}_X^{\mathrm{T}}$$

where

$$\boldsymbol{X}\boldsymbol{X}^{\mathrm{T}} = \begin{bmatrix} X_1 \\ X_2 \\ X_3 \end{bmatrix} [X_1 \ X_2 \ X_3] = \begin{bmatrix} X_1X_1 & X_1X_2 & X_1X_3 \\ X_2X_1 & X_2X_2 & X_2X_3 \\ X_3X_1 & X_3X_2 & X_3X_3 \end{bmatrix}$$

$$\boldsymbol{\mu}_X\boldsymbol{\mu}_X^{\mathrm{T}} = \begin{bmatrix} \mu_1 \\ \mu_2 \\ \mu_3 \end{bmatrix} [\mu_1 \quad \mu_2 \quad \mu_3] = \begin{bmatrix} \mu_1\mu_1 & \mu_1\mu_2 & \mu_1\mu_3 \\ \mu_2\mu_1 & \mu_2\mu_2 & \mu_2\mu_3 \\ \mu_3\mu_1 & \mu_3\mu_2 & \mu_3\mu_3 \end{bmatrix}$$

$$\begin{aligned} \mathbb{C}_{XX} &= E\left\{ \begin{bmatrix} (X_1 - \mu_1)(X_1 - \mu_1) & (X_1 - \mu_1)(X_2 - \mu_2) & (X_1 - \mu_1)(X_3 - \mu_3) \\ (X_2 - \mu_2)(X_1 - \mu_1) & (X_2 - \mu_2)(X_2 - \mu_2) & (X_2 - \mu_2)(X_3 - \mu_3) \\ (X_3 - \mu_3)(X_1 - \mu_1) & (X_3 - \mu_3)(X_3 - \mu_3) & (X_3 - \mu_3)(X_1 - \mu_1) \end{bmatrix} \right\} \\ &= \begin{bmatrix} c_{11} & c_{12} & c_{13} \\ c_{21} & c_{22} & c_{23} \\ c_{31} & c_{32} & c_{33} \end{bmatrix} \end{aligned} \tag{7.83}$$

7.5 Characteristic Function of a Gaussian RV

This section derives the characteristic function of a Gaussian RV first for a scalar RV and then for a vector RV.

7.5.1 Characteristic Function of a Scalar Gaussian RV

We first consider a scalar RV X. The characteristic function of the normal distribution is, by definition, given by

$$\Psi_X(\omega) = E\{e^{j\omega X}\} = \int_{-\infty}^{+\infty} e^{j\omega x} \frac{1}{\sigma\sqrt{2\pi}} e^{-\frac{1}{2}\left(\frac{x-\mu}{\sigma}\right)^2} dx = \int_{-\infty}^{+\infty} \frac{1}{\sigma\sqrt{2\pi}} e^{j\omega x-\frac{1}{2}\left(\frac{x-\mu}{\sigma}\right)^2} dx \tag{7.84}$$

To evaluate the above integral, complete the square in the exponent in the integral as follows:

$$\begin{aligned}
& j\omega x - \frac{1}{2}\left(\frac{x-\mu}{\sigma}\right)^2 = j\omega x - \frac{1}{2\sigma^2}(x^2 - 2\mu x + \mu^2) \\
&= -\frac{1}{2\sigma^2}(x^2 - j\omega 2\sigma^2 x - 2\mu x + \mu^2) = -\frac{1}{2\sigma^2}\{x^2 - 2(j\omega\sigma^2 + \mu)x + \mu^2\} \\
&= -\frac{1}{2\sigma^2}\left\{x^2 - 2(j\omega\sigma^2 + \mu)x + (j\omega\sigma^2 + \mu)^2 - (j\omega\sigma^2 + \mu)^2 + \mu^2\right\} \\
&= -\frac{1}{2\sigma^2}\left\{\left[x - (j\omega\sigma^2 + \mu)\right]^2 - (j\omega\sigma^2 + \mu)^2 + \mu^2\right\} \\
&= -\frac{1}{2\sigma^2}\left\{\left[x - (j\omega\sigma^2 + \mu)\right]^2 + \omega^2\sigma^4 - 2j\omega\mu\sigma^2 - \ \mu^2 + \mu^2 \ (+\mu)^2\right\} \\
&= -\frac{1}{2\sigma^2}\left\{\left[x - (j\omega\sigma^2 + \mu)\right]^2 + \omega^2\sigma^4 - 2j\omega\mu\sigma^2 - \ \mu^2 + \ \mu^2\right\} \\
&= -\frac{1}{2\sigma^2}\left[x - (j\omega\sigma^2 + \mu)\right]^2 - \frac{1}{2}\omega^2\sigma^2 + j\omega\mu \\
&= -\frac{1}{2}\left[\frac{x - (j\omega\sigma^2 + \mu)}{\sigma}\right]^2 + j\omega\mu - \frac{1}{2}\omega^2\sigma^2
\end{aligned}$$

Substituting the last expression in the exponent in (7.84), we obtain the following equation:

$$\begin{aligned}
\Psi_X(\omega) &= \int_{-\infty}^{+\infty} \frac{1}{\sigma\sqrt{2\pi}} e^{-\frac{1}{2}\left[\frac{x-(j\omega\sigma^2+\mu)}{\sigma}\right]^2 + j\omega\mu - \frac{1}{2}\omega^2\sigma^2} dx \\
&= e^{j\omega\mu - \frac{\omega^2\sigma^2}{2}} \int_{-\infty}^{+\infty} \frac{1}{\sigma\sqrt{2\pi}} e^{-\frac{1}{2}\left[\frac{x-(j\omega\sigma^2+\mu)}{\sigma}\right]^2} dx
\end{aligned} \tag{7.85}$$

By (7.15) through (7.19), we have shown that the last integral, which is the Gaussian integral, is equal to 1. Therefore, we have the following characteristic function of the normal scalar RV X:

$$\psi_X(\omega) = e^{j\omega\mu - \frac{\omega^2\sigma^2}{2}} \tag{7.86}$$

7.5.2 Characteristic Function of a Gaussian Vector RV

In this section, the time points as the arguments of the RV $\boldsymbol{X}(t_1, .., t_n)$ are not shown for simplicity. By the definition given by (6.62), the characteristic function of a Gaussian vector RV $\boldsymbol{X}$ is as follows:

$$\Psi_X(\omega_1, \omega_2, \ldots, \omega_n) \triangleq E\{e^{j\omega^T X}\} = E\{e^{j(\omega_1 X_1 + \omega_2 X_2 + \cdots + \omega_n X_n)}\} \tag{7.87}$$

$$= \int_{-\infty}^{+\infty} \cdots \int_{-\infty}^{+\infty} e^{j\omega^T x} \frac{1}{\sqrt{(2\pi)^n \det \mathbb{C}_{XX}}} e^{-\frac{1}{2}(x-\mu_X)^T \mathbb{C}_{XX}^{-1}(x-\mu_X)} dx_1 dx_2 \ldots dx_n \tag{7.88}$$

To evaluate the above integral, let us first evaluate the following simpler integral and then return to complete the above integral:

$$\Psi_X(\omega_1, \omega_2, \ldots, \omega_n) = \int_{-\infty}^{+\infty} \cdots \int_{-\infty}^{+\infty} e^{j\omega^T x} e^{-\frac{1}{2}x^T \mathbb{A} x} dx_1 dx_2 \ldots dx_n \tag{7.89}$$

Consider the second exponent without -½ in the above integral, $x^T \mathbb{A} x$, in which $\mathbb{A}$ is assumed to be an $(n \times n)$ symmetric matrix. Since $\mathbb{P}\mathbb{P}^{-1} = \mathbb{I}$, by pre-multiplying and post-multiplying $\mathbb{A}$ by $\mathbb{P}\mathbb{P}^{-1}$, we have the following expression:

$$x^T \mathbb{A} x = x^T (\mathbb{P}\mathbb{P}^{-1}) \mathbb{A} (\mathbb{P}\mathbb{P}^{-1}) x = x^T \mathbb{P} (\mathbb{P}^{-1} \mathbb{A} \mathbb{P}) \mathbb{P}^{-1} x$$

In the above equation, we choose $\mathbb{P}$ to diagonalize $\mathbb{A}$ so that $(\mathbb{P}^{-1}\mathbb{A}\mathbb{P}) = \mathbb{D}$, yielding the following expression:

$$x^T \mathbb{A} x = x^T \mathbb{P} \mathbb{D} \mathbb{P}^{-1} x \tag{7.90}$$

where

$$\mathbb{D} = \mathbb{P}^{-1} \mathbb{A} \mathbb{P}, \quad \det \mathbb{P} \neq 0 \tag{7.91}$$

$\mathbb{D}$ is a diagonal matrix with the eigenvalues of $\mathbb{A}$ as the diagonal elements as follows:

$$\mathbb{D} = \begin{bmatrix} \lambda_1 & \cdots & 0 \\ \vdots & \ddots & \vdots \\ 0 & \cdots & \lambda_n \end{bmatrix}, \lambda_i = \text{eigenvalue}, i = 1, \ldots, n. \tag{7.92}$$

From (7.91), we have

$$\mathbb{A} = \mathbb{P} \mathbb{D} \mathbb{P}^{-1} \tag{7.93}$$

Taking the transposition of both sides of the above and using the matrix identity (2.20d), we have

$$\mathbb{A}^T = (\mathbb{P}\mathbb{D}\mathbb{P}^{-1})^T = (\mathbb{P}^{-1})^T \mathbb{D}^T \mathbb{P}^T = (\mathbb{P}^{-1})^T \mathbb{D} \mathbb{P}^T \tag{7.94}$$

Since $\mathbb{A}$ is symmetric, $\mathbb{A}^T = \mathbb{A}$, and, so, by equating (7.93) with (7.94), we obtain the following equation:

$$\mathbb{P}\mathbb{D}\mathbb{P}^{-1} = (\mathbb{P}^{-1})^T \mathbb{D} \mathbb{P}^T \tag{7.95}$$

or

$$\mathbb{D} = \mathbb{P}^{-1}\left(\mathbb{P}^{-1}\right)^{\mathrm{T}}\mathbb{D}\mathbb{P}^{\mathrm{T}}\mathbb{P} \tag{7.96}$$

From the above, we derive the following relationship that holds true with the diagonalizing matrix $\mathbb{P}$:

$$\mathbb{P}^{\mathrm{T}}\mathbb{P} = \mathbb{I}$$

By left multiplying both sides, we obtain

$$\mathbb{P}^{-1} = \mathbb{P}^{\mathrm{T}} \tag{7.97}$$

The above result shows that the matrix $\mathbb{P}$ diagonalizing $\mathbb{A}$ is an orthogonal matrix if $\mathbb{A}$ is a symmetric matrix. In (7.90), let

$$\boldsymbol{z} = \begin{bmatrix} z_1 \\ z_2 \\ \cdot \\ z_i \\ \cdot \\ z_n \end{bmatrix} = \mathbb{P}^{-1}\boldsymbol{x} \tag{7.98}$$

Then, we have

$$\boldsymbol{z}^{\mathrm{T}} = \left(\mathbb{P}^{-1}\boldsymbol{x}\right)^{\mathrm{T}} = \boldsymbol{x}^{\mathrm{T}}\left(\mathbb{P}^{-1}\right)^{\mathrm{T}} \tag{7.99}$$

Substituting (7.97) into the above equation, we have

$$\boldsymbol{z}^{\mathrm{T}} = [z_1 \quad z_2 \quad \cdot \quad z_i \quad \cdot \quad z_n] = \boldsymbol{x}^{\mathrm{T}}\left(\mathbb{P}^{\mathrm{T}}\right)^{\mathrm{T}} = \boldsymbol{x}^{\mathrm{T}}\mathbb{P} \tag{7.100}$$

Substituting (7.98) and (7.100) into (7.90), we obtain the following equation:

$$\boldsymbol{x}^{\mathrm{T}}\mathbb{A}\boldsymbol{x} = \boldsymbol{z}^{\mathrm{T}}\mathbb{D}\boldsymbol{z} = \sum_{i=1}^{n}\lambda_i z_i^2 \tag{7.101}$$

Following the method described in Sect. 2.2.6, we can find the diagonalizing matrix $\mathbb{P}$ as follows. First, find n eigenvalues of $\mathbb{A}$ by solving the following equation:

$$\det\left(\mathbb{A} - \lambda\mathbb{I}\right) = 0 \tag{7.102}$$

Then, obtain the eigenvectors corresponding to the eigenvalues from the following n linearly independent relationships:

$$\mathbb{A}\boldsymbol{b}_i = \lambda_i\boldsymbol{b}_i, i = 1, \ldots, n \tag{7.103}$$

where the components of the eigenvectors are denoted as follows:

$$\boldsymbol{b}_i = \begin{bmatrix} b_{1i} \\ b_{2i} \\ \cdot \\ b_{ki} \\ \cdot \\ b_{ni} \end{bmatrix}, i = 1, \ldots, n \tag{7.104}$$

For the symmetric matrix $\mathbb{A}$, its diagonalizing matrix $\mathbb{P}$ is orthonormal. The eigenvectors are orthogonal to each other and their norms are unity. Therefore, the inner products, $\langle .,. \rangle$, of the eigenvectors are given by

$$\langle \boldsymbol{b}_i, \boldsymbol{b}_j \rangle = \sum_{k=1}^{n} b_{ki} b_{kj} = \delta_{ij} \tag{7.105}$$

The diagonalizing matrix $\mathbb{P}$ is given by

$$\mathbb{P} = [\boldsymbol{b}_1 \ \ \boldsymbol{b}_2 \ \ldots \ \boldsymbol{b}_n] = \begin{bmatrix} b_{11} & \cdot & b_{1i} & \cdot & b_{1n} \\ \cdot & \cdot & \cdot & \cdot & \cdot \\ b_{k1} & \cdot & b_{ki} & \cdot & b_{kn} \\ \cdot & \cdot & \cdot & \cdot & \cdot \\ b_{n1} & \cdot & b_{ni} & \cdot & b_{nn} \end{bmatrix} \tag{7.106}$$

$$\mathbb{P}^{\mathrm{T}} = \mathbb{P}^{-1} = \begin{bmatrix} b_{11} & \cdot & b_{k1} & \cdot & b_{n1} \\ \cdot & \cdot & \cdot & \cdot & \cdot \\ b_{1i} & \cdot & b_{ki} & \cdot & b_{ni} \\ \cdot & \cdot & \cdot & \cdot & \cdot \\ b_{1n} & \cdot & b_{kn} & \cdot & b_{nn} \end{bmatrix} \tag{7.107}$$

Substituting (7.107) into (7.98), the transformation of $\boldsymbol{x}$ by $\mathbb{P}^{\mathrm{T}}$ to $\boldsymbol{z}$ is given by

$$\boldsymbol{z} = \begin{bmatrix} z_1 \\ \cdot \\ z_i \\ \cdot \\ z_n \end{bmatrix} = \mathbb{P}^{\mathrm{T}} \boldsymbol{x} = \begin{bmatrix} b_{11} & \cdot & b_{j1} & \cdot & b_{n1} \\ \cdot & \cdot & \cdot & \cdot & \cdot \\ b_{1i} & \cdot & b_{ji} & \cdot & b_{ni} \\ \cdot & \cdot & \cdot & \cdot & \cdot \\ b_{1n} & \cdot & b_{jn} & \cdot & b_{nn} \end{bmatrix} \begin{bmatrix} x_1 \\ \cdot \\ x_j \\ \cdot \\ x_n \end{bmatrix}$$

$$= \begin{bmatrix} b_{11}x_1 + .. + b_{j1}x_j + .. + b_{n1}x_n \\ \cdot \\ b_{1i}x_1 + .. + b_{ji}x_j + .. + b_{ni}x_n \\ \cdot \\ b_{1n}x_1 + .. + b_{jn}x_j + .. + b_{nn}x_n \end{bmatrix} = \begin{bmatrix} \sum_{j=1}^{n} b_{j1}x_j \\ \cdot \\ \cdot \\ \sum_{j=1}^{n} b_{ji}x_j \\ \cdot \\ \cdot \\ \sum_{j=1}^{n} b_{jn}x_j \end{bmatrix} \tag{7.108}$$

Now, consider the first exponent in (7.89)

$$j\boldsymbol{\omega}^{\mathrm{T}}\boldsymbol{x} \tag{7.109}$$

From (7.98), we have

$$\boldsymbol{x} = \mathbb{P}\boldsymbol{z} \tag{7.110}$$

Substituting (7.110) into (7.109), we have

$$j\boldsymbol{\omega}^{\mathrm{T}}\boldsymbol{x} = j\boldsymbol{\omega}^{\mathrm{T}}\mathbb{P}\boldsymbol{z} \tag{7.111}$$

In the above equation, let

$$\boldsymbol{\omega}^{\mathrm{T}}\mathbb{P} = \left(\mathbb{P}^{\mathrm{T}}\boldsymbol{\omega}\right)^{\mathrm{T}} = \left(\boldsymbol{\omega}^{*}\right)^{\mathrm{T}} \tag{7.112}$$

where

$$\boldsymbol{\omega}^{*} = \begin{bmatrix} \omega_1^* \\ \omega_2^* \\ \cdot \\ \omega_i^* \\ \cdot \\ \omega_n^* \end{bmatrix} = \mathbb{P}^{\mathrm{T}}\boldsymbol{\omega} = \mathbb{P}^{-1}\boldsymbol{\omega} = \begin{bmatrix} b_{11} & \cdot & b_{j1} & \cdot & b_{n1} \\ \cdot & \cdot & \cdot & \cdot & \cdot \\ b_{1i} & \cdot & b_{ji} & \cdot & b_{ni} \\ \cdot & \cdot & \cdot & \cdot & \cdot \\ b_{1n} & \cdot & b_{jn} & \cdot & b_{nn} \end{bmatrix} \begin{bmatrix} \omega_1 \\ \cdot \\ \omega_j \\ \cdot \\ \omega_n \end{bmatrix} = \begin{bmatrix} \sum_{j=1}^{n} b_{j1}\omega_j \\ \cdot \\ \cdot \\ \sum_{j=1}^{n} b_{ji}\omega_j \\ \cdot \\ \cdot \\ \sum_{j=1}^{n} b_{jn}\omega_j \end{bmatrix} \tag{7.113}$$

where the identity given by (7.97) is used. Substituting (7.112) into (7.111), we have

$$j\boldsymbol{\omega}^{\mathrm{T}}\boldsymbol{x} = j\boldsymbol{\omega}^{*\mathrm{T}}\boldsymbol{z} = \sum_{i=1}^{n} j\omega_i^* z_i \tag{7.114}$$

Substituting (7.101) and (7.114) into (7.89), we have

$$\begin{aligned} \Psi_X(\omega_1, \omega_2, \ldots, \omega_n) &= \int_{-\infty}^{+\infty} \cdots \int_{-\infty}^{+\infty} e^{\sum_{i=1}^{n} j\omega_i^* z_i}\, e^{-\frac{1}{2}\sum_{i=1}^{n}\lambda_i z_i^2}\, \mathrm{d}x_1\mathrm{d}x_2 \ldots \mathrm{d}x_n \\ &= \int_{-\infty}^{+\infty} \cdots \int_{-\infty}^{+\infty} e^{\sum_{i=1}^{n}\left(j\omega_i^* z_i - \frac{1}{2}\lambda_i z_i^2\right)}\, \mathrm{d}x_1\mathrm{d}x_2 \ldots \mathrm{d}x_n \end{aligned} \tag{7.115}$$

Since $\boldsymbol{z}$ is a transformation of $\boldsymbol{x}$ by (7.98),

$$dz_1 dz_2 \ldots dz_n = \mid \mathbb{J} \mid \mathrm{d}x_1 \mathrm{d}x_2 \ldots \mathrm{d}x_n \tag{7.116}$$

where $\mathbb{J}$ is the Jacobian matrix defined below and $\mid \mathbb{J} \mid$ is the determinant of $\mathbb{J}$

$$\mathbb{J} = \begin{bmatrix} \frac{\partial z_1}{\partial x_1} & \cdot & \cdot & \cdot & \frac{\partial z_1}{\partial x_n} \\ \cdot & \cdot & \cdot & \cdot & \cdot \\ \cdot & \cdot & \frac{\partial z_i}{\partial x_j} & \cdot & \cdot \\ \cdot & \cdot & \cdot & \cdot & \cdot \\ \frac{\partial z_n}{\partial x_1} & \cdot & \cdot & \cdot & \frac{\partial z_n}{\partial x_n} \end{bmatrix} \tag{7.117}$$

The diagonalizing matrix $\mathbb{P}$ provides the orthogonal transformation, and the Jacobian of the orthogonal transformation is unity. Therefore, (7.116) becomes

$$\mathrm{d}z_1 \mathrm{d}z_2 \ldots \mathrm{d}z_n = \mathrm{d}x_1 \mathrm{d}x_2 \ldots \mathrm{d}x_n \tag{7.118}$$

Substituting the above into (7.115), we have

$$\begin{aligned} \Psi_X(\omega_1, \omega_2, \ldots, \omega_n) &= \int_{-\infty}^{+\infty} \cdots \int_{-\infty}^{+\infty} e^{\sum_{i=1}^{n} \left(j\omega_i^* z_i - \frac{1}{2}\lambda_i z_i^2\right)} \mathrm{d}z_1 \mathrm{d}z_2 \ldots \mathrm{d}z_n \\ &= \left\{ \int_{-\infty}^{+\infty} e^{\left(j\omega_i^* z_i - \frac{1}{2}\lambda_i z_i^2\right)} \mathrm{d}z_1 \right\} \cdots \left\{ \int_{-\infty}^{+\infty} e^{\left(j\omega_i^* z_i - \frac{1}{2}\lambda_i z_i^2\right)} \mathrm{d}z_i \right\} \\ &\quad \cdots \left\{ \int_{-\infty}^{+\infty} e^{\left(j\omega_i^* z_i - \frac{1}{2}\lambda_i z_i^2\right)} \mathrm{d}z_n \right\} \\ &= \prod_{i=1}^{n} \int_{-\infty}^{+\infty} e^{\left(j\omega_i^* z_i - \frac{1}{2}\lambda_i z_i^2\right)} \mathrm{d}z_i \end{aligned} \tag{7.119}$$

The above integral is a multiplication of n integrals of the following form. The integral is rewritten in the form of (7.84) as follows:

$$\int_{-\infty}^{+\infty} e^{\left(j\omega_i^* z_i - \frac{1}{2}\lambda_i z_i^2\right)} \mathrm{d}z_i = \frac{\sqrt{2\pi}}{\sqrt{\lambda_i}} \int_{-\infty}^{+\infty} \frac{1}{\frac{1}{\sqrt{\lambda_i}}\sqrt{2\pi}} e^{j\omega_i^* z_i} e^{-\frac{1}{2}\left(\frac{z_i}{\frac{1}{\sqrt{\lambda_i}}}\right)^2} \mathrm{d}z_i \tag{7.120}$$

Comparing the above integral with (7.84), we see that it is the characteristic function of a Gaussian RV with $\mu = 0$, $\sigma = \frac{1}{\sqrt{\lambda_i}}$ and its solution is given by (7.86) as follows

$$\int_{-\infty}^{+\infty} e^{\left(j\omega_i^* z_i - \frac{1}{2}\lambda_i z_i^2\right)} dz_i = \frac{\sqrt{2\pi}}{\sqrt{\lambda_i}} e^{j\omega_i^* \mu - \frac{\omega_i^{*2}\sigma^2}{2}} \tag{7.121}$$

in which, we make the following substitutions $\mu = 0, \ \sigma = \frac{1}{\sqrt{\lambda_i}}$ to obtain

$$\int_{-\infty}^{+\infty} e^{\left(j\omega_i^* z_i - \frac{1}{2}\lambda_i z_i^2\right) dz_i = \frac{\sqrt{2\pi}}{\sqrt{\lambda_i}} e^{-\frac{1}{2\lambda_i}\omega_i^{*2}}} \tag{7.122}$$

Substituting (7.122) into (7.119), we have

$$\begin{aligned}\psi_X(\omega_1, \omega_2, \ldots, \omega_n) &= \prod_{i=1}^{n} \int_{-\infty}^{+\infty} e^{\left(j\omega_i^* z_i - \frac{1}{2}\lambda_i z_i^2\right)} dz_i \prod_{i=1}^{n} \frac{\sqrt{2\pi}}{\sqrt{\lambda_i}} e^{-\frac{1}{2\lambda_i}\omega_i^{*2}} \\ &= \frac{(2\pi)^{\frac{n}{2}}}{\left\{\prod_{i=1}^{n} \lambda_i\right\}^{\frac{1}{2}}} e^{-\frac{1}{2}\sum_{i=1}^{n}\frac{\omega_i^{*2}}{\lambda_i}}\end{aligned} \tag{7.123}$$

Since the diagonal elements of $\mathbb{D}$ are λ_i's, the diagonal elements of $\mathbb{D}^{-1}$ are $1/\lambda_i$'s. Therefore, the summation in the exponent of the last term in the above equation can be expressed in terms of the following matrix multiplication:

$$\sum_{i=1}^{n} \frac{\omega_i^{*2}}{\lambda_i} = \boldsymbol{\omega}^{*\mathrm{T}} \mathbb{D}^{-1} \boldsymbol{\omega}^* \tag{7.124}$$

The product of n eigenvalues in the denominator of the last term of (7.123) is equal to the determinant of $\mathbb{D}$ as follows:

$$\det\mathbb{D} = \prod_{i=1}^{n} \lambda_i \tag{7.125}$$

Substituting (7.124) and (7.125) into (7.123), we obtain the following equation:

$$\psi_X(\omega_1, \omega_2, \ldots, \omega_n) = \frac{(2\pi)^{\frac{n}{2}}}{\{\det\mathbb{D}\}^{\frac{1}{2}}} e^{-\frac{1}{2}\boldsymbol{\omega}^{*\mathrm{T}}\mathbb{D}^{-1}\boldsymbol{\omega}^*} \tag{7.126}$$

By (7.91), we have $\mathbb{D} = \mathbb{P}^{-1}\mathbb{A}\mathbb{P}$, and, taking the inverse of both sides and using the matrix identity (2.68a) and (7.97), we obtain the following equation:

$$\begin{aligned}\mathbb{D}^{-1} &= \left(\mathbb{P}^{-1}\mathbb{A}\mathbb{P}\right)^{-1} = (\mathbb{A}\mathbb{P})^{-1}\left(\mathbb{P}^{-1}\right)^{-1} \\ &= \mathbb{P}^{-1}\mathbb{A}^{-1}\left(\mathbb{P}^{-1}\right)^{-1} = \mathbb{P}^{-1}\mathbb{A}^{-1}\mathbb{P} = \mathbb{P}^{\mathrm{T}}\mathbb{A}^{-1}\mathbb{P}\end{aligned} \tag{7.127}$$

Substituting the above equation into the exponent of (7.126) and using (7.113) for $\boldsymbol{\omega}^*$, we obtain the following equation:

$$\begin{aligned}\boldsymbol{\omega}^{*\mathrm{T}}\mathbb{D}^{-1}\boldsymbol{\omega}^* &= \boldsymbol{\omega}^{*\mathrm{T}}\mathbb{P}^{\mathrm{T}}\mathbb{A}^{-1}\mathbb{P}\boldsymbol{\omega}^* = \left(\mathbb{P}^{-1}\boldsymbol{\omega}\right)^{\mathrm{T}}\mathbb{P}^{\mathrm{T}}\mathbb{A}^{-1}\mathbb{P}\mathbb{P}^{-1}\boldsymbol{\omega}\\ &= \boldsymbol{\omega}^{\mathrm{T}}\left(\mathbb{P}^{-1}\right)^{\mathrm{T}}\mathbb{P}^{\mathrm{T}}\mathbb{A}^{-1}\mathbb{P}\mathbb{P}^{-1}\boldsymbol{\omega}\end{aligned} \tag{7.128}$$

In the above equation,

$$\mathbb{P}\mathbb{P}^{-1} = \mathbb{I} \qquad \left(\mathbb{P}^{-1}\right)^{\mathrm{T}}\mathbb{P}^{\mathrm{T}} = \left(\mathbb{P}^{\mathrm{T}}\right)^{\mathrm{T}}\mathbb{P}^{-1} = \mathbb{P}\mathbb{P}^{-1} = \mathbb{I}$$

Substituting the above relationships into (7.128), we obtain the following equation:

$$\boldsymbol{\omega}^{*\mathrm{T}}\mathbb{D}^{-1}\boldsymbol{\omega}^* = \boldsymbol{\omega}^{\mathrm{T}}\mathbb{A}^{-1}\boldsymbol{\omega} \tag{7.129}$$

Finally,

$$\det\mathbb{D} = \det\mathbb{A} \tag{7.130}$$

Substituting (7.129) and (7.130) into (7.126), we obtain the following solution for the "simpler integral," ψ_X, as follows:

$$\psi_X(\omega_1, \omega_2, \ldots, \omega_n) = \frac{(2\pi)^{\frac{n}{2}}}{(\det\mathbb{A})^{\frac{1}{2}}}e^{-\frac{1}{2}\boldsymbol{\omega}^{\mathrm{T}}\mathbb{A}^{-1}\boldsymbol{\omega}} \tag{7.131}$$

Now, to evaluate the characteristic function of the normal distribution with the mean vector $\boldsymbol{\mu}_X$, recognize that the autocovariance matrix $\mathbb{C}_{XX}$ is symmetric. Now, return to the original integral (7.88). To change $\Psi_X(\omega_1, \omega_2, \ldots, \omega_n)$ to the form of $\psi_X(\omega_1, \omega_2, \ldots, \omega_n)$, substitute $\boldsymbol{z} = \boldsymbol{x} - \boldsymbol{\mu}_X, \boldsymbol{x} = \boldsymbol{z} + \boldsymbol{\mu}_X, \mathbb{A} = \mathbb{C}_{XX}{}^{-1}$ and $\mathbb{C}_{XX} = \mathbb{A}^{-1}$ in (7.88) to obtain the following equation:

$$\begin{aligned}\Psi_X(\omega_1, \omega_2, \ldots, \omega_n) &= \int_{-\infty}^{+\infty}\cdots\int_{-\infty}^{+\infty} e^{j\boldsymbol{\omega}^{\mathrm{T}}(\boldsymbol{z}+\boldsymbol{\mu}_X)}\frac{1}{(2\pi)^{\frac{n}{2}}(\det\mathbb{A}^{-1})^{\frac{1}{2}}}e^{-\frac{1}{2}\boldsymbol{z}^{\mathrm{T}}\mathbb{A}\boldsymbol{z}}\mathrm{d}z_1\mathrm{d}z_2\ldots\mathrm{d}z_n\\ &= \frac{1}{(2\pi)^{\frac{n}{2}}(\det\mathbb{A}^{-1})^{\frac{1}{2}}}e^{j\boldsymbol{\omega}^{\mathrm{T}}\boldsymbol{\mu}_X}\int_{-\infty}^{+\infty}..\int_{-\infty}^{+\infty} e^{j\boldsymbol{\omega}^{\mathrm{T}}\boldsymbol{z}}e^{-\frac{1}{2}\boldsymbol{z}^{\mathrm{T}}\mathbb{A}\boldsymbol{z}}\mathrm{d}z_1..\mathrm{d}z_n\\ &= \frac{1}{(2\pi)^{\frac{n}{2}}(\det\mathbb{A}^{-1})^{\frac{1}{2}}}e^{j\boldsymbol{\omega}^{\mathrm{T}}\boldsymbol{\mu}_X}\Psi_X(\omega_1, .., \omega_n)\end{aligned} \tag{7.132}$$

Substituting (7.131) into the above equation, we obtain the following equation:

$$\Psi_X(\omega_1, \omega_2, \ldots, \omega_n) = \frac{1}{(2\pi)^{\frac{n}{2}}\left(\det\mathbb{A}^{-1}\right)^{\frac{1}{2}}} e^{j\boldsymbol{\omega}^{\mathrm{T}}\boldsymbol{\mu}_X} \frac{(2\pi)^{\frac{n}{2}}}{(\det\mathbb{A})^{\frac{1}{2}}} e^{-\frac{1}{2}\boldsymbol{\omega}^{\mathrm{T}}\mathbb{A}^{-1}\boldsymbol{\omega}} \tag{7.133}$$

By (2.32), substituting $\det\mathbb{A}^{-1} = (\det\mathbb{A})^{-1}$ into the above equation and canceling the terms, we obtain the following characteristic function of a Gaussian vector RV $\boldsymbol{X}$:

$$\Psi_X(\omega_1, \omega_2, \ldots, \omega_n) = e^{j\boldsymbol{\omega}^{\mathrm{T}}\boldsymbol{\mu}_X - \frac{1}{2}\boldsymbol{\omega}^{\mathrm{T}}\mathbb{C}_{XX}\boldsymbol{\omega}} \tag{7.134}$$

Example 7.5.1

Find the characteristic function of a bivariate normal distribution, that is, $n = 2$. Let $C_{ij} = C_{X_iX_j}$.

$$\begin{aligned}
\boldsymbol{\omega}^{\mathrm{T}}\mathbb{C}_{XX}\boldsymbol{\omega} &= [\omega_1 \quad \omega_2]\begin{bmatrix} C_{11} & C_{12} \\ C_{21} & C_{22} \end{bmatrix}\begin{bmatrix} \omega_1 \\ \omega_2 \end{bmatrix} = [\omega_1 \quad \omega_2]\begin{bmatrix} C_{11}\omega_1 + C_{12}\omega_2 \\ C_{21}\omega_1 + C_{22}\omega_2 \end{bmatrix} \\
&= \omega_1(C_{11}\omega_1 + C_{12}\omega_2) + \omega_2(C_{21}\omega_1 + C_{22}\omega_2) \\
&= C_{11}\omega_1^2 + C_{12}\omega_1\omega_2 + C_{21}\omega_1\omega_2 + C_{22}\omega_2^2 \\
&= C_{11}\omega_1^2 + 2C_{12}\omega_1\omega_2 + C_{22}\omega_2^2
\end{aligned}$$

$$j\boldsymbol{\omega}^{\mathrm{T}}\boldsymbol{\mu}_X = j[\omega_1 \quad \omega_2]\begin{bmatrix} \mu_1 \\ \mu_2 \end{bmatrix} = j(\omega_1\mu_1 + \omega_2\mu_2)$$

$$j\boldsymbol{\omega}^{\mathrm{T}}\boldsymbol{\mu}_X - \frac{1}{2}\boldsymbol{\omega}^{\mathrm{T}}\mathbb{C}_{XX}\boldsymbol{\omega} = j(\omega_1\mu_1 + \omega_2\mu_2) - \frac{1}{2}\left(C_{11}\omega_1^2 + 2C_{12}\omega_1\omega_2 + C_{22}\omega_2^2\right)$$

$$\Psi_X(\omega_1, \omega_2) = e^{j(\omega_1\mu_1 + \omega_2\mu_2) - \frac{1}{2}\left(C_{11}\omega_1^2 + 2C_{12}\omega_1\omega_2 + C_{22}\omega_2^2\right)}$$

7.6 Gaussian Stochastic Process

A stochastic process $X(t)$ is Gaussian, if the vector RV $\boldsymbol{X}(t_1, t_2, \ldots, t_n)$ defined by (6.32a) has the n-dimensional multivariate normal *pdf* given by (7.76) for all values of n and for all possible values of the n time points.

Two stochastic processes $X(t)$ and $Y(t)$ are jointly Gaussian, if the $(m + n)$-dimensional concatenated vector RV $\boldsymbol{Z}_{XY}\left(t_1, .., t_n, t_1', .., t_m'\right)$ defined by (6.33) consisting of n and m RVs at n and m arbitrarily selected time points, $X(t_i)$, $i = 1, 2, \ldots, n$, and $Y\left(t_i'\right), i = 1, 2, \ldots, m$, have the $(m + n)$-dimensional multivariate normal *pdf* given by (7.81) for all values of n and m and all values of t_i and t_i'.

Chapter 8
Applications

In this chapter, we discuss three examples of applications—communications channel modeling, estimation theory, and queueing analysis.

8.1 Analysis of Communications System

A communications system is used to convey a message or information from a source to a destination. Figure 8.1 shows a simple primitive communications system in which light is used to communicate a message from a source to a destination.

It is ironic that, lately, an optical signal, which was the signal used in primitive communications, is coming into vogue as the signal of choice for information transmission. Switching a signal optically, however, is a much harder technical challenge and, even if an optical signal is used for transmission, it needs to be converted to an electrical signal for switching at signal switching points along a communications path. All in all, the primary signal of choice for communications systems is the electrical signal. In this section, we review the basic elements of the communications system.

Figure 8.2 shows the major components of a modern communications system. The transducer is a device that converts one form of signal to another.

In Fig. 8.3, the acoustic signal is converted into the electrical signal, which is transmitted toward the destination for communications of the source information.

8.1.1 Linear Time-Invariant (LTI) System

In this chapter, we apply the statistical characterization of RVs and stochastic processes to the analysis of communications channels. We will assume that the

K.I. Park, *Fundamentals of Probability and Stochastic Processes with Applications to Communications*, https://doi.org/10.1007/978-3-319-68075-0_8

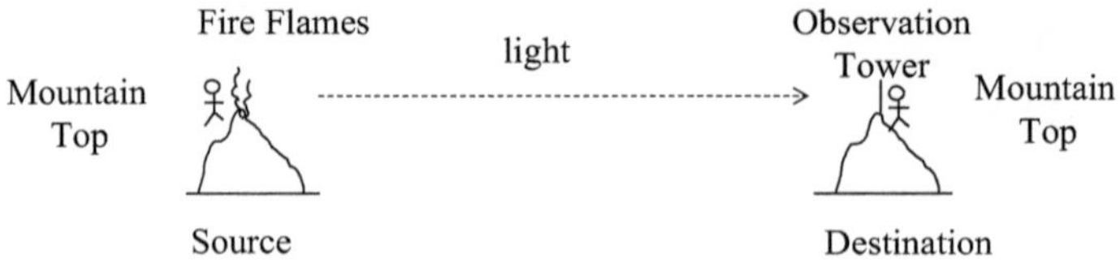

Fig. 8.1 A simple primitive communications system

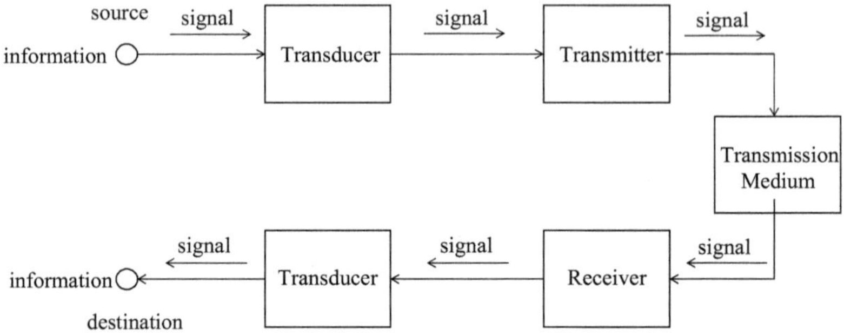

Fig. 8.2 Major components of a modern communications system

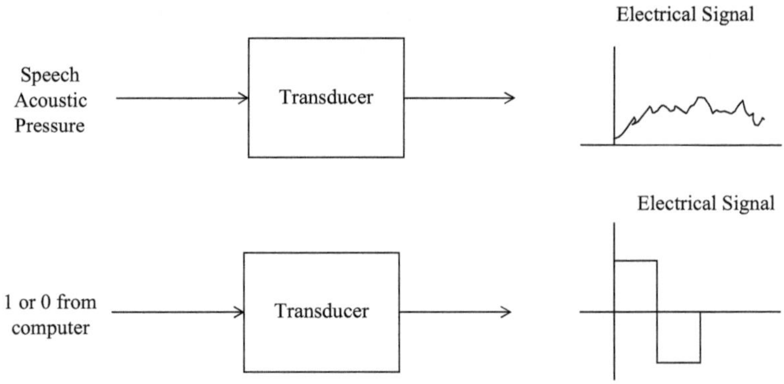

Fig. 8.3 Transducer functions

communications channels are mathematically modeled as a linear time-invariant (LTI) system.

In a linear system, the superposition theorem holds true, which states that the system response to the sum of individual inputs is equal to the sum of the responses to the individual inputs. To illustrate the concept, suppose that two voltages x_1 and x_2 are applied across an impedance individually at two separate instants and the corresponding currents are measured across the impedance as the outputs y_1 and y_2, respectively. Denote the sum of two responses $y_1 + y_2$ by y. Next, the sum of the two

Fig. 8.4 (a) Impulse response, (b) Output $y(t)$ in response to input $x(t)$

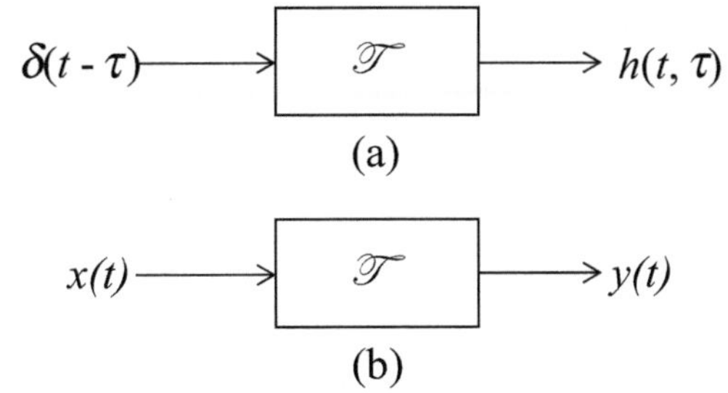

inputs $x_1 + x_2$ is applied across the same impedance, and the response is measured, which is denoted by y^*. In a linear electrical system, the two results are equal, $y = y^*$.

In linear algebra, linear transformation $\mathcal{T}$ is defined by the following equations, where $\boldsymbol{x}_1$ and $\boldsymbol{x}_2$ are vectors and α is a scalar:

$$\mathcal{T}(\boldsymbol{x}_1 + \boldsymbol{x}_2) = \mathcal{T}(\boldsymbol{x}_1) + \mathcal{T}(\boldsymbol{x}_2) \tag{8.1}$$

$$\mathcal{T}(\alpha\boldsymbol{x}_1) = \alpha\mathcal{T}(\boldsymbol{x}_1) \tag{8.2}$$

The following equation shows the linear transformation at time t denoted by $\mathcal{T}_t$ operated on the scalar input time function $x(t)$ to produce the scalar output time function $y(t)$:

$$y(t) = \mathcal{T}_t\{x(t)\} \tag{8.3}$$

The subscript t of the operator $\mathcal{T}_t$ shows that the operation is performed at t so that the above transformation applies for both the time-invariant and time-varying systems. For the time-invariant system, $\mathcal{T}_t$ is independent of t and the subscript t may be dropped (Fig. 8.4).

A linear system is completely characterized by determining the impulse response function $h(t, \tau)$ for all values of t and τ. The impulse response function $h(t, \tau)$ is the system response measured at time t due to the impulse applied at time τ. In the linear system, $h(t, \tau)$ is the linear transformation of $\delta(t-\tau)$ as follows:

$$h(t, \tau) = \mathcal{T}_t\{\delta(t - \tau)\} \tag{8.4}$$

Using the Dirac delta function defined by (4.26) and (4.27), the input function $x(t)$ can be expressed as follows (Fig. 8.5):

$$x(t) = \int_{-\infty}^{+\infty} x(\tau)\delta(t - \tau)\mathrm{d}\tau \tag{8.5}$$

Fig. 8.5 Impulse function at $t = \tau$

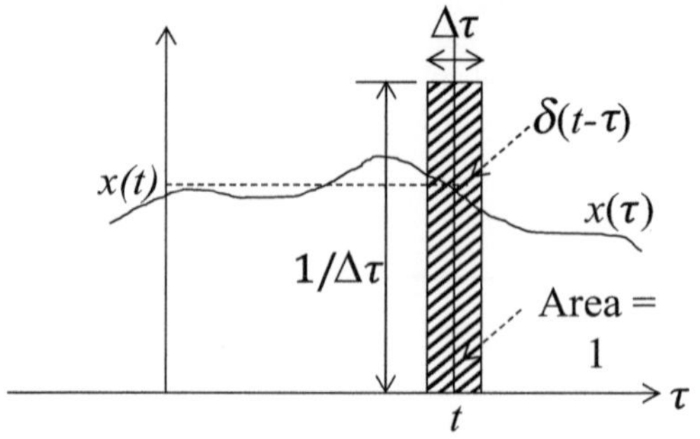

Substituting (8.5) into (8.3) and using (8.4), we obtain the following equation:

$$y(t) = \mathcal{T}_t\{x(t)\} = \mathcal{T}_t\left\{\int_{-\infty}^{+\infty} x(\tau)\delta(t-\tau)\mathrm{d}\tau\right\}$$
$$= \int_{-\infty}^{+\infty} x(\tau)\mathcal{T}_t\{\delta(t-\tau)\}\mathrm{d}\tau = \int_{-\infty}^{+\infty} x(\tau)h(t,\tau)\mathrm{d}\tau \tag{8.6}$$

We now derive the above integral by convolution of the input and the impulse response, which will provide further insight into the transformation. Let t be the current time. For $\tau_i < t$, the product $x(\tau_i)h(t,\tau_i)$ is the contribution to the output as measured at the current time t of the past input atτ_i. For a noncausal system, $x(\tau_i)h(t,\tau_i)$ for $\tau_i > t$ gives the contribution of the future input as measured at the current time t. $x(\tau_i)h(t,\tau_i)$ for $\tau_i = t$ gives the contribution of the current value of the input. The following expression gives the approximate contribution to the output at time t due to the input values at various time points in the past, present, and, for a noncausal system, the future:

$$y(t,\tau_i) \approx x(\tau_i)h(t,\tau_i)\Delta\tau \tag{8.7}$$

The sum of all contributions to the output measured at time t is given by

$$y(t) \approx \sum_{i=-\infty}^{+\infty} y(t,\tau_i) = \sum_{i=-\infty}^{+\infty} x(\tau_i)h(t,\tau_i)\Delta\tau \tag{8.8}$$

In the limit as $\Delta\tau \rightarrow 0$, the above summation becomes the following integral:

$$y(t) = \int_{-\infty}^{+\infty} x(\tau)h(t,\tau)\mathrm{d}\tau \tag{8.9}$$

which coincides with the result given by (8.6).

For the LTI system, the impulse response is a function of the interval length $t-\tau$ only and does not depend on t and τ independently. This means that the response of the LTI system to the impulse remains unchanged regardless of the time t at which it is measured if the interval length $t-\tau$ remains the same. For the LTI system, therefore, by setting $\tau=0$, we have the following relationship:

$$h(t,\tau) = h(t-\tau) = h(t) \tag{8.10}$$

The general equation given by (8.9), which is applicable for both the LTI and the linear time-varying (LTV) systems, can now be expressed by the following convolution:

$$y(t) = x(t)*h(t) = \int_{-\infty}^{+\infty} x(\tau)h(t-\tau)\mathrm{d}\tau \tag{8.11}$$

The convolution given by (8.11) is illustrated by Fig. 8.6.

8.1.2 Fourier Transform

The Fourier transform converts a time function to a frequency function so that the analysis of the function can be done in the frequency domain. The results of the frequency domain analysis can then be converted back to the time domain by the inverse Fourier transform. The time function and the corresponding frequency function obtained by the Fourier transform are called a Fourier pair.

Since, in this book, capital letters such as X and Y are used to denote RVs and the boldface letters **X** and **Y**, to denote vectors, we will use a special font to distinguish the Fourier transforms. The letter case will be maintained so that the variables can be kept track of. For example, the Fourier transform of $X(t)$ and $x(t)$ will be denoted by $\mathrm{X}(\omega)$ and $\mathrm{x}(\omega)$, respectively. The Fourier transform and the inverse Fourier transform operators will be denoted by $\mathscr{F}[.]$ and $\mathscr{F}^{-1}[.]$, respectively.

For a given time function $f(t)$, its Fourier transform and the corresponding inverse Fourier transform are defined by the following equations, respectively:

$$f(\omega) = \mathscr{F}[f(t)] = \int_{-\infty}^{+\infty} f(t)e^{-j\omega t}\mathrm{d}t \tag{8.12}$$

$$f(t) = \mathscr{F}^{-1}[f(\omega)] = \frac{1}{2\pi}\int_{-\infty}^{+\infty} f(\omega)e^{j\omega t}\mathrm{d}\omega \tag{8.13}$$

To analyze the transient response of the system, the Laplace transform is used, where the parameter of transformation is $s=\sigma+j\omega$. For steady-state analysis, the Fourier transform is used, where the parameter of transformation is $j\omega$, which is

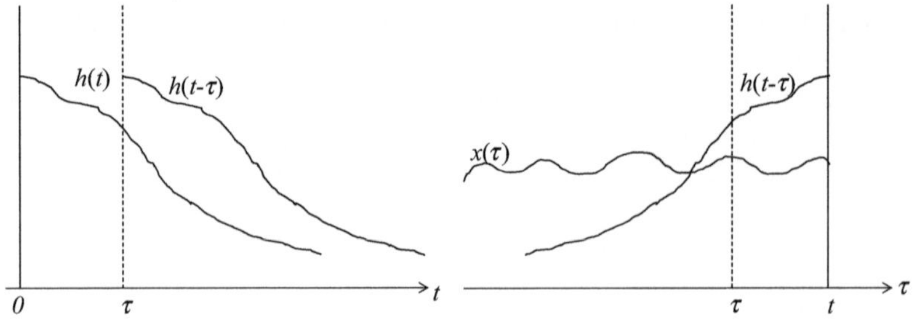

Fig. 8.6 Illustration of convolution

obtained by setting the real component $\sigma = 0$ in the Laplace transformation parameter s.

Transfer Function

The Fourier transforms of two specific time functions are particularly important. One is called the transfer function and, the other, the power spectrum. These two functions are defined below.

The transfer function of a linear system, denoted by $h(\omega)$, is as follows:

$$h(\omega) = \mathcal{F}\left[h(t)\right] = \int_{-\infty}^{+\infty} h(t)e^{-j\omega t}\mathrm{d}t \tag{8.14}$$

Given, the transfer function, the impulse response is obtained by the inverse Fourier transform of the transfer function as follows:

$$h(t) = \mathcal{F}^{-1}[h(\omega)] = \frac{1}{2\pi}\int_{-\infty}^{+\infty} h(\omega)e^{j\omega t}\mathrm{d}\omega \tag{8.15}$$

Power Spectrum

The power spectrum of a wide sense stationary stochastic process $X(t)$ is the Fourier transform of the autocorrelation function of $X(t)$

$$R(\omega) = \mathcal{F}\left[R(t)\right] = \int_{-\infty}^{+\infty} R(t)e^{-j\omega t}\mathrm{d}t \tag{8.16}$$

$$R(t) = \mathcal{F}^{-1}[R(\omega)] = \frac{1}{2\pi}\int_{-\infty}^{+\infty} R(\omega)e^{j\omega t}\mathrm{d}\omega \tag{8.17}$$

8.1.3 Input-Output Relationship

By taking the Fourier transform of both sides of (8.11), we have

$$\begin{aligned}\mathcal{F}[y(t)] &= \mathcal{F}[x(t)*h(t)] \\ &= \mathcal{F}\left[\int_{-\infty}^{+\infty} x(\tau)h(t-\tau)\mathrm{d}\tau\right] \\ &= \int_{-\infty}^{+\infty}\left[\int_{-\infty}^{+\infty} x(\tau)h(t-\tau)\mathrm{d}\tau\right]e^{-j\omega t}\mathrm{d}t \end{aligned} \tag{8.18}$$

Letting $-\tau = \lambda$ or $t = \tau + \lambda$ and $\mathrm{d}t = \mathrm{d}\lambda$, we have

$$\begin{aligned} y(\omega) &= \mathcal{F}[y(t)] \\ &= \int_{-\infty}^{+\infty}\int_{-\infty}^{+\infty} x(\tau)h(\lambda)\mathrm{d}\tau e^{-j\omega(\lambda+\tau)}\mathrm{d}t \\ &= \int_{-\infty}^{+\infty}\int_{-\infty}^{+\infty} x(\tau)h(\lambda)\mathrm{d}\tau e^{-j\omega}e^{-j\omega\tau}\mathrm{d}\lambda \\ &= \int_{-\infty}^{+\infty} x(\tau)e^{-j\omega\tau}\mathrm{d}\tau \int_{-\infty}^{+\infty} h(\lambda)e^{-j\omega\lambda}\mathrm{d}\lambda = x(\omega)h(\omega) \end{aligned} \tag{8.19}$$

8.1.4 White Noise Process

A process $X(t)$ is called a white noise process if the two RVs $X(t_i)$ and $X(t_j)$ are uncorrelated for all possible pairs of t_i and t_j. For a zero-mean process, this condition is met if the autocorrelation is given by the following equation:

$$R(t_1, t_2) = a(t_1)\delta(t_1 - t_2) = a\delta(\tau) \tag{8.20}$$

The power spectrum of the white noise process is given by

$$R(\omega) = \mathcal{F}[R(t)] = \int_{-\infty}^{+\infty} a\delta(t)e^{-j\omega t}\mathrm{d}t = a\, e^{-j\omega 0} = a$$

The white noise process has a constant power spectrum at all frequencies.

Using (8.13) and (6.60), we obtain the inverse Fourier transform of the power spectrum of the white noise process as follows:

$$R(t) = \mathscr{F}^{-1}[R(\omega)] = \frac{1}{2\pi}\int_{-\infty}^{+\infty} ae^{j\omega t}\mathrm{d}\,\omega = a\delta(t)$$

8.1.5 Properties of Gaussian RVs and Gaussian Processes

Property 1
If n RVs have the nth-order multivariate normal distribution, any subset of m RVs of them has the mth-order multivariate normal distribution for $m < n$. This property is shown for $n = 2$ by (7.62) and (7.63). The general proof can be given by a similar method.

Property 2
Let $\boldsymbol{X}$ be an n-dimensional vector RV and $\boldsymbol{W}$, an m-dimensional vector RV formed by m linear combinations of the components of $\boldsymbol{X}$ as follows:

$$\boldsymbol{W} = \mathbb{B}\boldsymbol{X} \tag{8.21}$$

where

$$\boldsymbol{W} = \begin{bmatrix} W_1 \\ \cdot \\ \cdot \\ W_m \end{bmatrix} \quad \mathbb{B} = \begin{bmatrix} b_{11} & \cdots & b_{1n} \\ \vdots & \ddots & \vdots \\ b_{m1} & \cdots & b_{mn} \end{bmatrix} \quad \boldsymbol{X} = \begin{bmatrix} X_1 \\ \cdot \\ \cdot \\ X_n \end{bmatrix} \tag{8.22}$$

If $\boldsymbol{X}$ is jointly Gaussian with an nth-order multivariate Gaussian distribution with the mean vector and covariance matrix, $\boldsymbol{\mu}_{\boldsymbol{X}}$ and $\mathbb{C}_{\boldsymbol{WW}}$, the linear combination $\boldsymbol{W}$ is also jointly Gaussian with an mth-order multivariate Gaussian distribution with the mean vector and the covariance matrix, $\boldsymbol{\mu}_{\boldsymbol{W}}$ and $\mathbb{C}_{\boldsymbol{XX}}$, given below

$$\boldsymbol{\mu}_{W} = \mathbb{B}\boldsymbol{\mu}_{X} \qquad \mathbb{C}_{WW} = \mathbb{B}\mathbb{C}_{XX}\mathbb{B}^{\mathrm{T}} \tag{8.23}$$

Since $\boldsymbol{W}$ is jointly Gaussian, the individual linear combinations, W_i, are normally distributed with the mean and variance

$$\mu_{W_i} = [b_{i1} \;\; b_{i2} \ldots b_{in}]\boldsymbol{\mu}_X = \sum_{j=1}^{n} b_{ij}\mu_{X_j}$$

$$\sigma_{W_i}^2 = [b_{i1} \;\; b_{i2} \ldots \;\; b_{in}]\mathbb{C}_{XX}\begin{bmatrix} b_{i1} \\ b_{i2} \\ \cdot \\ \cdot \\ b_{in} \end{bmatrix} \tag{8.24}$$

Proof By the definition given by (6.62), the characteristic function of $\boldsymbol{W}$ is the following expected value:

$$\Psi_W(\omega_1, \omega_2, \ldots, \omega_n) = E\left(e^{j\boldsymbol{\omega}^{\mathrm{T}}\boldsymbol{W}}\right) \tag{8.25}$$

Substituting (8.21) into the above equation, we have

$$\Psi_W(\omega_1, \omega_2, \ldots, \omega_n) = E(e^{j\boldsymbol{\omega}^{\mathrm{T}}\mathbb{B}\boldsymbol{X}}) = E\left(e^{j\tilde{\boldsymbol{\omega}}^{\mathrm{T}}\boldsymbol{X}}\right) \tag{8.26}$$

where

$$\tilde{\boldsymbol{\omega}}^{\mathrm{T}} = \boldsymbol{\omega}^{\mathrm{T}}\mathbb{B} \tag{8.27}$$

Since, in the definition of a characteristic function, $\boldsymbol{\omega}^{\mathrm{T}}$ is a parameter of the integral of the expectation operation and is not a variable of integration, the integral with $\tilde{\boldsymbol{\omega}}^{\mathrm{T}}$ gives the same value given by (7.133) with $\boldsymbol{\omega}^{\mathrm{T}}$ replaced with $\tilde{\boldsymbol{\omega}}^{\mathrm{T}}$ as follows:

$$\Psi_W(\omega_1, \omega_2, \ldots, \omega_n) = E\left(e^{j\tilde{\boldsymbol{\omega}}^{\mathrm{T}}\boldsymbol{X}}\right) = e^{j\tilde{\boldsymbol{\omega}}^{\mathrm{T}}\boldsymbol{\mu}_X - \frac{1}{2}\tilde{\boldsymbol{\omega}}^{\mathrm{T}}\mathbb{C}_{XX}\tilde{\boldsymbol{\omega}}} \tag{8.28}$$

Taking the transposition of (8.27), we have

$$\tilde{\boldsymbol{\omega}} = (\tilde{\boldsymbol{\omega}}^{\mathrm{T}})^{\mathrm{T}} = (\boldsymbol{\omega}^{\mathrm{T}}\mathbb{B})^{\mathrm{T}} = \mathbb{B}^{\mathrm{T}}(\boldsymbol{\omega}^{\mathrm{T}})^{\mathrm{T}} = \mathbb{B}^{\mathrm{T}}\boldsymbol{\omega} \tag{8.29}$$

Now, substituting (8.27) and (8.29) into the exponent of (8.28), we obtain the following equation:

$$\Psi_W(\omega_1, \omega_2, \ldots, \omega_n) = e^{j\boldsymbol{\omega}^{\mathrm{T}}\mathbb{B}\boldsymbol{\mu}_X - \frac{1}{2}\boldsymbol{\omega}^{\mathrm{T}}\mathbb{B}\mathbb{C}_{XX}\mathbb{B}^{\mathrm{T}}\boldsymbol{\omega}} \tag{8.30}$$

In the above equation, let

$$\boldsymbol{\mu} = \mathbb{B}\boldsymbol{\mu}_X \qquad \mathbb{C} = \mathbb{B}\mathbb{C}_{XX}\mathbb{B}^{\mathrm{T}} \tag{8.31}$$

to obtain

$$\Psi_W(\omega_1, \omega_2, \ldots, \omega_n) = E\left(e^{j\tilde{\omega}^{\mathrm{T}} X}\right) = e^{j\omega^{\mathrm{T}}\mu - \frac{1}{2}\omega^{\mathrm{T}}\mathbb{C}\,\omega} \tag{8.32}$$

The last equation is the characteristic function of a multivariate normal distribution with the mean vector $\boldsymbol{\mu}$ and the autocovariance matrix $\mathbb{C}$. The remainder of the proof is to show that $\boldsymbol{\mu}$ and $\mathbb{C}$ are indeed the mean vector and the autocovariance matrix of $\boldsymbol{W}$.

By (8.21), we have the following mean vector of $\boldsymbol{W}$:

$$\boldsymbol{\mu}_W = E(\boldsymbol{W}) = E(\mathbb{B}\boldsymbol{X}) = \mathbb{B}E(\boldsymbol{X}) = \mathbb{B}\boldsymbol{\mu}_X \tag{8.33}$$

The covariance matrix of $\boldsymbol{W}$ is given by

$$\mathbb{C}_{WW} = E(\boldsymbol{W}\boldsymbol{W}^{\mathrm{T}}) - \boldsymbol{\mu}_W\boldsymbol{\mu}_W^{\mathrm{T}} \tag{8.34}$$

Substituting (8.21) into the above equation and using the matrix identity (2.20d), we obtain the following autocovariance matrix of $\boldsymbol{W}$:

$$\begin{aligned}
\mathbb{C}_{WW} &= E\{\mathbb{B}\boldsymbol{X}(\mathbb{B}\boldsymbol{X})^{\mathrm{T}}\} - \mathbb{B}\boldsymbol{\mu}_X(\mathbb{B}\boldsymbol{\mu}_X)^{\mathrm{T}} \\
&= E(\mathbb{B}\boldsymbol{X}\boldsymbol{X}^{\mathrm{T}}\mathbb{B}^{\mathrm{T}}) - \mathbb{B}\boldsymbol{\mu}_X{\boldsymbol{\mu}_X}^{\mathrm{T}}\mathbb{B}^{\mathrm{T}} \\
&= \mathbb{B}E(\boldsymbol{X}\boldsymbol{X}^{\mathrm{T}})\mathbb{B}^{\mathrm{T}} - \mathbb{B}\boldsymbol{\mu}_X{\boldsymbol{\mu}_X}^{\mathrm{T}}\mathbb{B}^{\mathrm{T}} \\
&= \mathbb{B}\{E(\boldsymbol{X}\boldsymbol{X}^{\mathrm{T}}) - \boldsymbol{\mu}_X{\boldsymbol{\mu}_X}^{\mathrm{T}}\}\mathbb{B}^{\mathrm{T}} \\
&= \mathbb{B}\mathbb{C}_{XX}\mathbb{B}^{\mathrm{T}}
\end{aligned} \tag{8.35}$$

By comparing (8.33) and (8.35) with (8.31), we establish that

$$\boldsymbol{\mu} = \boldsymbol{\mu}_W; \qquad \mathbb{C} = \mathbb{C}_{WW}$$

Q.E.D.

Property 3

If $X_1, X_2, \ldots, X_n$ are mutually independent, the off-diagonal elements of the covariance matrix, $\mathbb{C}_{XX}$, are zero, and, so, $\mathbb{C}_{XX}$ becomes the following:

$$\mathbb{C}_{WW} = \mathbb{B}\mathbb{C}_{XX}\mathbb{B}^{\mathrm{T}}$$

$$= \begin{bmatrix} b_{11} & \cdots & b_{1n} \\ \vdots & \ddots & \vdots \\ b_{m1} & \cdots & b_{mn} \end{bmatrix} \begin{bmatrix} \sigma_{11}^2 & \cdots & 0 \\ \vdots & \ddots & \vdots \\ 0 & \cdots & \sigma_{nn}^2 \end{bmatrix} \begin{bmatrix} b_{11}b_{21}\ldots b_{m1} \\ b_{12}b_{22}\ldots b_{m2} \\ . \\ . \\ b_{1n}b_{2n}\ldots b_{mn} \end{bmatrix}$$

$$= \begin{bmatrix} b_{11} & \cdots & b_{1n} \\ \vdots & \ddots & \vdots \\ b_{m1} & \cdots & b_{mn} \end{bmatrix} \begin{bmatrix} \sigma_{11}^2 b_{11}\sigma_{11}^2 b_{21}\ldots\sigma_{11}^2 b_{m1} \\ \sigma_{22}^2 b_{12}\sigma_{22}^2 b_{22}\ldots\sigma_{22}^2 b_{m2} \\ . \\ . \\ \sigma_{nn}^2 b_{1n}\sigma_{nn}^2 b_{2n}\ldots\sigma_{nn}^2 b_{mn} \end{bmatrix} \tag{8.36}$$

$$= \begin{bmatrix} \sigma_{11}^2\sum_{k=1}^{n} b_{1k}b_{1k} & \sigma_{11}^2\sum_{k=1}^{n} b_{1k}b_{2k} & . & . & \sigma_{11}^2\sum_{k=1}^{n} b_{1k}b_{mk} \\ \sigma_{22}^2\sum_{k=1}^{n} b_{2k}b_{1k} & \sigma_{22}^2\sum_{k=1}^{n} b_{2k}b_{2k} & . & . & \sigma_{22}^2\sum_{k=1}^{n} b_{2k}b_{mk} \\ . & . & . & . & . \\ . & . & . & . & . \\ \sigma_{nn}^2\sum_{k=1}^{n} b_{mk}b_{1k} & \sigma_{nn}^2\sum_{k=1}^{n} b_{mk}b_{2k} & . & . & \sigma_{nn}^2\sum_{k=1}^{n} b_{mk}b_{mk} \end{bmatrix}$$

$$c_{W_{ij}} = \mathrm{Cov}\left(W_i, W_j\right) = \sigma_{ii}^2 \sum_{k=1}^{n} b_{ik}b_{jk} \tag{8.37}$$

where

$$\sigma_{ii}^2 = \mathrm{Var}\left(X_i\right)$$

The vector RV $\boldsymbol{W}$ given by (8.21) has an mth-order multivariate normal distribution with the mean vector $\boldsymbol{\mu_W}$ given by (8.33) and the covariance matrix, $\mathbb{C}_{\boldsymbol{WW}}$, given by (8.36).

The individual RV W_i is normal. The mean μ_{W_i} is given by (8.24). By setting $j = i$ in (8.37), we obtain the following equation for the variance of individual RVs:

$$\sigma_{W_{ii}}^2 = \mathbb{C}_{W_{ii}} = \sigma_{ii}^2 \sum_{k=1}^{n} b_{ik}b_{ik} = \sigma_{ii}^2 \sum_{k=1}^{n} b_{ik}^2 \tag{8.38}$$

Finally, if $W = X_1 + X_2 + \cdots + X_n$, by setting $b_{ij} = 1, j = 1, \ldots, n$, we obtain the following equation:

$$\sigma_{W_i}^2 = \sum_{i=1}^{n} \sigma_{ii}^2 \tag{8.39}$$

Property 4

In conjunction with (5.53), we have discussed that the independence of two RVs implies the uncorrelatedness of the RVs, whereas the converse is not necessarily true. In the case of the normal RVs, however, the converse is also true, that is, if two normal RVs are uncorrelated, they are also mutually independent.

By the definition of the uncorrelatedness and the independence given by (5.53) and (4.82b), we have the following inference:

$$\text{If } E(XY) = E(X)E(Y), \text{ then } \quad f_{XY}(x,y) = f_X(x)f_Y(y) \tag{8.40}$$

Proof By (5.44) and (7.67), if X and Y are uncorrelated, $\mathbb{C}_{XY}=0$, and $\rho=0$. Substituting $\rho = 0$ into the bivariate normal *pdf* given by (7.53), we obtain the following equation:

$$f_{XY}(x,y) = \frac{1}{2\pi\sigma_X\sigma_Y} e^{-\frac{1}{2}\left\{\left(\frac{x-\mu_X}{\sigma_X}\right)^2 + \left(\frac{y-\mu_Y}{\sigma_Y}\right)^2\right\}}$$

$$= \frac{1}{\sigma_X\sqrt{2\pi}} e^{-\frac{1}{2}\left(\frac{x-\mu_X}{\sigma_X}\right)^2} \frac{1}{\sigma_Y\sqrt{2\pi}} e^{-\frac{1}{2}\left(\frac{y-\mu_Y}{\sigma_Y}\right)^2} = f_X(x)f_Y(y)$$

Q.E.D.

Property 5

As can be seen in (7.76), the nth-order multivariate normal distribution is completely specified by the n-dimensional mean vector $\boldsymbol{\mu_X}$ and $n \times n$ covariance matrix $\mathbb{C}_{XX}$. The covariance of a stochastic process $X(t)$ is given by (6.125) in terms of the autocorrelation $r_{XX}(t_1, t_2)$ *and* the mean $\mu_X(t)$. Therefore, a Gaussian process is completely specified by the mean $\mu_X(t)$ and the autocorrelation $r_{XX}(t_1, t_2)$.

Property 6

An important consequence of Property 5 is that, if a process is Gaussian and is wide sense stationary, the process is also strict sense stationary.

Proof A Gaussian process is completely characterized by the nth-order multivariate normal distribution given by (7.76). Consider the nth-order multivariate normal distribution for the n new time points obtained by shifting the original n time points by the same amount τ. The new distribution is completely characterized by the new covariance matrix $\mathbb{C}_{XX}(t_1+\tau, \ldots, t_n+\tau)$ and $\mu_X(t_1+\tau, \ldots, t_n+\tau)$. The covariance matrix is specified by the following elements:

$$c_{X_iX_j}(t_i+\tau, t_j+\tau) = r_{X_iX_j}(t_i+\tau, t_j+\tau) - \mu_{X_i}(t_i+\tau)\mu_{X_j}(t_j+\tau) \tag{8.41}$$

If the process is wide sense stationary, we have the following equations:

$$r_{X_iX_j}(t_i+\tau, t_j+\tau) = r_{X_iX_j}(t_i, t_j) = r_{X_iX_j}(t_j - t_i) = r_{X_iX_j}(\delta_{ij}) \tag{8.42}$$

$$\mu_{X_i}(t_i+\tau) = \mu_{X_i}(t_i) = \mu_{X_i} \tag{8.43}$$

$$c_{X_iX_j}(t_i+\tau, t_j+\tau) = c_{X_iX_j}(t_i, t_j) \tag{8.44}$$

The wide sense stationarity states that the autocorrelation is invariant under a shift of time and the mean is a constant. Therefore, the covariance matrix is invariant with the shift of time. Since the mean is a constant and the covariance matrix is invariant with the shift of time, the complete statistical properties of the process is invariant with the shift of time, which is the condition for the strict sense stationarity.

Q.E.D.

Property 7
Under Property 1, if $X(t)$ is a Gaussian process, the RVs at any two instants of the process, $X(t_i)$ and $X(t_j)$, are jointly normal, and the RVs at any instants, $X(t_i)$,are individually normal.

Property 8
If two processes $X(t)$ and $Y(t)$ are jointly Gaussian and uncorrelated, the two processes are mutually independent. The proof of this property can be given in a similar way to that given for (8.40).

8.1.6 Input-Output Relations of a Stochastic Process

Theorem 8.1.1 The mean of the output process $W(t)$ of the LTI system is equal to the output of the mean of the input $X(t)$

$$\mu_W(t) = \mu_X(t) * h(t) \tag{8.45}$$

Proof The output of the LTI is a linear combination of the inputs given by (8.8). The mean of the output $W(t)$ can be approximately expressed as follows:

$$\mu_W(t) = E\{W(t)\} \approx E\left\{\sum_{i=-\infty}^{+\infty} W(t, \tau_i)\right\} = E\left\{\sum_{i=-\infty}^{+\infty} X(\tau_i)h(t, \tau_i)\Delta\tau\right\}$$

By (5.22), the expectation operation is moved inside the summation operation as follows:

$$\mu_W(t) \approx \sum_{i=-\infty}^{+\infty} E\{X(\tau_i)\}h(t, \tau_i)\Delta\tau$$

In the limit as $\Delta\tau \to 0$, the above approximation becomes the following convolution:

$$\mu_W(t) = \int_{-\infty}^{+\infty} E\{X(\tau)\}h(t-\tau)\mathrm{d}\tau = \int_{-\infty}^{+\infty} \mu_X(\tau)h(t-\tau)\mathrm{d}\tau = \mu_X(t) * h(t)$$

Q.E.D.

Theorem 8.1.2 Denote the input and output processes by $X(t)$ and $W(t)$ and the impulse response and its complex conjugate by $h(t)$ and $h^*(t)$. Then, the following relationships hold true with the autocorrelations of the input and output processes $r_{WW}(t_1, t_2)$ and $r_{XX}(t_1, t_2)$ and the cross-correlation between the input and output processes $r_{XW}(t_1, t_2)$.

$$\text{(a) } r_{XW}(t_1, t_2) = r_{XX}(t_1, t_2) * h^*(t_2) \tag{8.46}$$

$$\text{(b) } r_{WW}(t_1, t_2) = r_{XW}(t_1, t_2) * h(t_1) \tag{8.47}$$

$$\text{(c) } r_{WW}(t_1, t_2) = r_{XW}(t_1, t_2) * h^*(t_2) * h(t_1) \tag{8.48}$$

Proof

(a) By the definition given by (6.109), we have the following expression:

$$\begin{aligned} r_{XW}(t_1, t_2) = E\{X(t_1)W^*(t_2)\} &= E\left\{X(t_1)\left[\int_{-\infty}^{+\infty} X(\tau)h(t_2 - \tau)d\tau\right]^*\right\} \\ &= E\left\{X(t_1)\int_{-\infty}^{+\infty} X^*(\tau)h^*(t_2 - \tau)d\tau\right\} \\ &= \int_{-\infty}^{+\infty} E\{X(t_1)X^*(\tau)\}h^*(t_2 - \tau)d\tau \\ &= \int_{-\infty}^{+\infty} r_{XX}(t_1, \tau)h^*(t_2 - \tau)d\tau = r_{XX}(t_1, t_2) * h^*(t_2) \end{aligned}$$

Q.E.D.

(b)

$$\begin{aligned} r_{WW}(t_1, t_2) &= E\{W(t_1)W^*(t_2)\} \\ &= E\left\{\left[\int_{-\infty}^{+\infty} X(\tau)h(t_1 - \tau)d\tau\right]W^*(t_2)\right\} \\ &= \int_{-\infty}^{+\infty} E\{X(\tau)W^*(t_2)\}h(t_1 - \tau)d\tau \\ &= \int_{-\infty}^{+\infty} r_{XW}(\tau, t_2)h(t_1 - \tau)d\tau = r_{XW}(t_1, t_2) * h(t_1) \end{aligned}$$

Q.E.D.

(c) Substituting (8.46) into (8.47), we obtain the desired result.

Q.E.D.

Theorem 8.1.3 If the input *X*(*t*) is WSS, the output process *W*(*t*) is WSS, and the following relations hold true with $\delta = t_1 - t_2$:

$$\text{(a)} \quad r_{XW}(t_1, t_2) = r_{XX}(\delta) * h^*(-\delta) = r_{XW}(t_1 - t_2) = r_{XW}(\delta) \tag{8.49}$$

$$\text{(b)} \quad r_{WW}(t_1, t_2) = r_{XW}(\delta) * h(\delta) \tag{8.50}$$

$$\text{(c)} \quad r_{WW}(t_1, t_2) = r_{XX}(\delta) * h^*(-\delta) * h(\delta) \tag{8.51}$$

Proof

(a) If *X*(t) is WSS,

$$r_{XX}(t_1, t_2) = r_{XX}(t_1 - t_2)$$

Substituting the above equation into (8.46), we obtain the following equation:

$$r_{XW}(t_1, t_2) = r_{XX}(t_1 - t_2) * h^*(t_2) = \int_{-\infty}^{+\infty} r_{XX}\{t_1 - (t_2 - \tau)\} h^*(\tau) d\tau$$

$$= \int_{-\infty}^{+\infty} r_{XX}(t_1 - t_2 + \tau) h^*(\tau) d\tau = \int_{-\infty}^{+\infty} r_{XX}(\delta + \tau) h^*(\tau) d\tau$$

$$= -\int_{+\infty}^{-\infty} r_{XX}(\delta - \tau') h^*(-\tau') d\tau'$$

$$= \int_{-\infty}^{+\infty} r_{XX}(\delta - \tau') h^*(-\tau') d\tau' = r_{XX}(\delta) * h^*(-\delta)$$

Q.E.D.

(b) Substituting (8.49) into (8.47), we have

$$r_{WW}(t_1, t_2) = r_{XW}(t_1 - t_2) * h(t_1)$$

$$= \int_{-\infty}^{+\infty} r_{XW}(t_1 - \tau - t_2) h(\tau) d\tau$$

$$= \int_{-\infty}^{+\infty} r_{XW}(\delta - \tau) h(\tau) d\tau$$

$$= r_{XW}(\delta) * h(\delta)$$

Q.E.D.

(c) Substituting (8.49) into (8.50), we obtain the desired result.

Q.E.D.

Finally, if $X(t)$ is WSS, its mean is constant, and, using (8.45), we obtain the following equation:

$$\mu_W(t) = \mu_X(t) * h(t) = \mu_X * h(t) = \mu_X \int_{-\infty}^{+\infty} h(\tau) d\tau = \text{constant}$$

Since the mean is constant and, by (8.51), the autocorrelation depends only on δ, $W(t)$ is WSS.

Theorem 8.1.4 Denote the input and output processes by $X(t)$ and $W(t)$ and the impulse response and its complex conjugate by $h(t)$ and $h^*(t)$. Then, the following relationships hold true with the autocovariance of the input and output processes $c_{WW}(t_1, t_2)$ and $c_{XX}(t_1, t_2)$ and the cross-covariance between the input and output processes $c_{XW}(t_1, t_2)$.

$$\text{(a)} \quad c_{XW}(t_1, t_2) = c_{XX}(t_1, t_2) * h^*(t_2) \tag{8.52}$$

$$\text{(b)} \quad c_{WW}(t_1, t_2) = c_{XW}(t_1, t_2) * h(t_1) \tag{8.53}$$

$$\text{(c)} \quad c_{WW}(t_1, t_2) = r_{XW}(t_1, t_2) * h^*(t_2) * h(t_1) \tag{8.54}$$

Proof The proof is similar to the proof of (8.45).

Theorem 8.1.5 If the input process to a linear system is Gaussian, the output process is also Gaussian. This property holds true for both the LTI and the LTV systems.

Proof The proof is based on (8.8) and the Gaussian Property 2 stated in conjunction with (8.23). Eq. (8.8) shows an approximation to the output of a linear system as a linear combination of various time points of the input, in which the impulse response is used as weighting coefficients of linear combination. Consider m RVs at m time points on the output process given by (8.8) as follows:

$$W(t_j) \approx \sum_{i=-\infty}^{+\infty} W(t_j, \tau_i) = \sum_{i=-\infty}^{+\infty} X(\tau_i) h(t_j, \tau_i) \Delta\tau, \quad \text{for } j = 1, \ldots, m$$

The m RVs are linear combinations of the RVs of an input process. In the limit, the approximating summation becomes the convolution of the input and the impulse response given by (8.9). For a Gaussian input, the input signal at the various time points in the approximating summation and, in the limit, in the convolution is Gaussian RVs. Property 2 proves that a linear combination of Gaussian RVs is Gaussian. Therefore, the output signal given by the convolution of the input signal and the impulse response is Gaussian. Both (8.8) and (8.9) are general equations applicable for both the LTI and LTV systems. Therefore, the above property holds true for both the LTI and the LTV system.

8.2 Estimation Theory

This section discusses another example of application of probability and stochastic processes, the estimation theory. The matrix operation is used extensively throughout this section. The matrix operations needed in the discussion are reviewed in Sect 2.2.

8.2.1 Estimation Problem Statement

Let $\boldsymbol{X}$ be an n-dimensional vector RV consisting of n component RVs, X_i, $i = 1, \ldots, n$, which are called the state variables. The vector RV $\boldsymbol{X}$, which is also called the "true" vector, is to be estimated based on the observations or measurements on $\boldsymbol{X}$.

$$\boldsymbol{X} = \begin{bmatrix} X_1 \\ \cdot \\ \cdot \\ X_n \end{bmatrix} \quad n - \text{dimensional true vector to be estimated} \tag{8.55}$$

The vector RV $\boldsymbol{X}$ may be measured directly as in the example of temperature measurements discussed in Sect 4.1. On the other hand, the vector RV $\boldsymbol{X}$ may not be accessible directly and can only be observed by measuring other variables related to $\boldsymbol{X}$. Let $\boldsymbol{W}$ be an m-dimensional observation vector consisting of m component RVs, $W_j, j = 1, \ldots, m$, which is the data obtained as a function of the state vector $\boldsymbol{X}$.

$$\boldsymbol{W} = \begin{bmatrix} W_1 \\ \cdot \\ \cdot \\ W_m \end{bmatrix} \quad m - \text{dimensional obscvation (or data) vector} \tag{8.56}$$

The general problem of estimation is stated as follows. Given the observation vector $\boldsymbol{W}$ as a function of the state vector $\boldsymbol{X}$, determine an estimator $\hat{\boldsymbol{X}}$ as a function of the observation vector $\boldsymbol{W}$ to satisfy a certain criterion.

$$\hat{\boldsymbol{X}} = f(\boldsymbol{W}) \tag{8.57}$$

$$\boldsymbol{W} = g(\boldsymbol{X}) \tag{8.58}$$

8.2.2 Linear Minimum Mean Square Error (MMSE) Estimation

This is a sub-class of the general estimation problem stated by (8.57) and (8.58). In this class of estimation, the estimator function *f*(.) defined by (8.57) is postulated as a linear combination of the observation vector **W** as follows:

$$\hat{\boldsymbol{X}} = \mathbb{A}\,\boldsymbol{W} \tag{8.59}$$

where

$$\hat{\boldsymbol{X}} = \begin{bmatrix} \hat{X}_1 \\ \cdot \\ \cdot \\ \hat{X}_n \end{bmatrix} \qquad n - \text{dimensional estimation vector} \tag{8.60}$$

$$\mathbb{A} = \begin{bmatrix} a_{11} & \cdots & a_{1m} \\ \vdots & \ddots & \vdots \\ a_{n1} & \cdots & a_{nm} \end{bmatrix} \qquad (n \times m) \text{ coeficient matrix} \tag{8.61}$$

The covariance matrix of the vector RV is defined by (6.168). For simplicity, the RVs $\boldsymbol{X}$ and $\boldsymbol{W}$ are assumed to be zero mean, $\mu_X = \mu_W = 0$. Then, we have the following covariance matrices for the estimation problem:

$(n \times n)$ state covariance matrix

$$\mathbb{C}_{XX} = E\{\boldsymbol{X}\boldsymbol{X}^{\mathrm{T}}\} = \begin{bmatrix} c_{X_1X_1} & \cdots & c_{X_1X_n} \\ \vdots & \ddots & \vdots \\ c_{X_1X_1} & \cdots & c_{XX_n} \end{bmatrix} \tag{8.62}$$

$(m \times m)$ observation covariance matrix

$$\mathbb{C}_{WW} = E\{\boldsymbol{W}\boldsymbol{W}^{\mathrm{T}}\} = \begin{bmatrix} c_{W_1W_1} & \cdots & c_{W_1W_n} \\ \vdots & \ddots & \vdots \\ c_{W_mW_1} & \cdots & c_{W_mW_m} \end{bmatrix} \tag{8.63}$$

$(n \times m)$ covariance matrix of state and observation

$$\mathbb{C}_{XW} = E\{\boldsymbol{X}\boldsymbol{W}^{\mathrm{T}}\} = \begin{bmatrix} c_{X_1W_1} & \cdots & c_{X_1W_m} \\ \vdots & \ddots & \vdots \\ c_{X_nW_1} & \cdots & c_{X_nW_m} \end{bmatrix} \tag{8.64}$$

with the elements defined by (5.33) as follows:

$$c_{X_iX_j} = E(X_iX_j) \qquad c_{W_iW_j} = E(W_iW_j) \qquad c_{X_iW_j} = E(X_iW_j) \tag{8.65}$$

Error Covariance Matrix

An n-dimensional estimation error vector $\boldsymbol{E}$ is defined as the difference between the estimate and the true state as follows:

$$\boldsymbol{E} = \hat{\boldsymbol{X}} - \boldsymbol{X} \tag{8.66}$$

where

$$\boldsymbol{E} = \begin{bmatrix} E_1 \\ \cdot \\ \cdot \\ E_n \end{bmatrix} \qquad n - \text{dimensional estimation error vector} \tag{8.67}$$

Assuming that the error vector is zero mean, the estimation error, or simply error, covariance matrix is defined as follows:

$$\begin{aligned} \mathbb{C}_{\boldsymbol{EE}} &= E\{\boldsymbol{EE}^{\mathrm{T}}\} \\ &= E\left\{(\hat{\boldsymbol{X}} - \boldsymbol{X})(\hat{\boldsymbol{X}} - \boldsymbol{X})^{\mathrm{T}}\right\} \\ &= \begin{bmatrix} c_{E_1E_1} & \cdots & c_{E_1E_n} \\ \vdots & \ddots & \vdots \\ c_{E_nE_1} & \cdots & c_{E_nE_n} \end{bmatrix} \\ c_{E_iE_j} &= E(E_iE_j) \end{aligned} \tag{8.68}$$

Note that the diagonal elements of the error covariance matrix are the variances of the errors of the individual state variable components

$$\sigma_{ii}^2 = E(E_{ii}^2)$$

Linear MMSE General Formulation

Problem

The linear MMSE problem can be stated as follows:

The observation data $\boldsymbol{W}$ is given by (8.58) as a function of the state variable $\boldsymbol{X}$. Postulate a linear estimator model given by (8.59). Determine the coefficient matrix $\mathbb{A}$ in (8.59) that would minimize the diagonal elements of the error covariance matrix $\mathbb{C}_{\boldsymbol{EE}}$ defined by (8.68).

Solution

Substitute (8.59) into (8.68) and expand the terms as follows:

$$\begin{aligned}
\mathbb{C}_{EE} &= E\{\boldsymbol{E}\boldsymbol{E}^{\mathrm{T}}\} \\
&= E\left\{(\hat{\boldsymbol{X}} - \boldsymbol{X})(\hat{\boldsymbol{X}} - \boldsymbol{X})^{\mathrm{T}}\right\} \\
&= E\left\{(\mathbb{A}\boldsymbol{W} - \boldsymbol{X})(\mathbb{A}\boldsymbol{W} - \boldsymbol{X})^{\mathrm{T}}\right\} \\
&= E\{(\mathbb{A}\boldsymbol{W} - \boldsymbol{X})(\boldsymbol{W}^{\mathrm{T}}\mathbb{A}^{\mathrm{T}} - \boldsymbol{X}^{\mathrm{T}})\} \\
&= E(\mathbb{A}\boldsymbol{W}\boldsymbol{W}^{\mathrm{T}}\mathbb{A}^{\mathrm{T}} - \boldsymbol{X}\boldsymbol{W}^{\mathrm{T}}\mathbb{A}^{\mathrm{T}} - \mathbb{A}\boldsymbol{W}\boldsymbol{X}^{\mathrm{T}} + \boldsymbol{X}\boldsymbol{X}^{\mathrm{T}}) \\
&= \mathbb{A}E(\boldsymbol{W}\boldsymbol{W}^{\mathrm{T}})\mathbb{A}^{\mathrm{T}} - E(\boldsymbol{X}\boldsymbol{W}^{\mathrm{T}})\mathbb{A}^{\mathrm{T}} - \mathbb{A}E(\boldsymbol{W}\boldsymbol{X}^{\mathrm{T}}) + E(\boldsymbol{X}\boldsymbol{X}^{\mathrm{T}}) \\
&= \mathbb{A}\mathbb{C}_{WW}\mathbb{A}^{\mathrm{T}} - \mathbb{C}_{XW}\mathbb{A}^{\mathrm{T}} - \mathbb{A}\mathbb{C}_{WX} + \mathbb{C}_{XX}
\end{aligned} \tag{8.69}$$

To determine the coefficient matrix $\mathbb{A}$ satisfying the criterion defined in the problem statement, we start by factoring the first two terms of the right-hand side of the above equation as follows. By inserting $\mathbb{C}_{WW}^{-1}\mathbb{C}_{WW} = \mathbb{I}$ in front of the second term and factoring the terms, we obtain the following expression:

$$\begin{aligned}
\mathbb{A}\mathbb{C}_{WW}\mathbb{A}^{\mathrm{T}} - \mathbb{C}_{XW}\mathbb{A}^{\mathrm{T}} &= \mathbb{A}\mathbb{C}_{WW}\mathbb{A}^{\mathrm{T}} - \mathbb{C}_{XW}\mathbb{C}_{WW}^{-1}\mathbb{C}_{WW}\mathbb{A}^{\mathrm{T}} = (\mathbb{A} - \mathbb{C}_{XW}\mathbb{C}_{WW}^{-1})\mathbb{C}_{WW}\mathbb{A}^{\mathrm{T}} \\
&= (\mathbb{A} - \mathbb{C}_{XW}\mathbb{C}_{WW}^{-1})\mathbb{C}_{WW}\{\mathbb{A}^{\mathrm{T}} - (\mathbb{C}_{XW}\mathbb{C}_{WW}^{-1})^{\mathrm{T}} + (\mathbb{C}_{XW}\mathbb{C}_{WW}^{-1})^{\mathrm{T}}\} \\
&= (\mathbb{A} - \mathbb{C}_{XW}\mathbb{C}_{WW}^{-1})\mathbb{C}_{WW}\{(\mathbb{A} - \mathbb{C}_{XW}\mathbb{C}_{WW}^{-1})^{\mathrm{T}} + (\mathbb{C}_{XW}\mathbb{C}_{WW}^{-1})^{\mathrm{T}}\} \\
&= (\mathbb{A} - \mathbb{C}_{XW}\mathbb{C}_{WW}^{-1})\mathbb{C}_{WW}(\mathbb{A} - \mathbb{C}_{XW}\mathbb{C}_{WW}^{-1})^{\mathrm{T}} \\
&\quad + (\mathbb{A} - \mathbb{C}_{XW}\mathbb{C}_{WW}^{-1})\mathbb{C}_{WW}(\mathbb{C}_{XW}\mathbb{C}_{WW}^{-1})^{\mathrm{T}} \\
&= (\mathbb{A} - \mathbb{C}_{XW}\mathbb{C}_{WW}^{-1})\mathbb{C}_{WW}(\mathbb{A} - \mathbb{C}_{XW}\mathbb{C}_{WW}^{-1})^{\mathrm{T}} \\
&\quad + (\mathbb{A} - \mathbb{C}_{XW}\mathbb{C}_{WW}^{-1})\mathbb{C}_{WW}(\mathbb{C}_{WW}^{-1})^{\mathrm{T}}\mathbb{C}_{XW}^{\mathrm{T}}
\end{aligned}$$

Since $\mathbb{C}_{WW}$ is symmetric $\mathbb{C}_{WW}^{\mathrm{T}} = \mathbb{C}_{WW}$ and, by (2.68c), we have $(\mathbb{C}_{WW}^{-1})^{\mathrm{T}} = (\mathbb{C}_{WW}^{\mathrm{T}})^{-1} = \mathbb{C}_{WW}^{-1}$, we can rewrite the above equation as follows:

$$\begin{aligned}
&\mathbb{A}\mathbb{C}_{WW}\mathbb{A}^{\mathrm{T}} - \mathbb{C}_{XW}\mathbb{A}^{\mathrm{T}} = (\mathbb{A} - \mathbb{C}_{XW}\mathbb{C}_{WW}^{-1})\mathbb{C}_{WW}(\mathbb{A} - \mathbb{C}_{XW}\mathbb{C}_{WW}^{-1})^{\mathrm{T}} \\
&\quad + (\mathbb{A} - \mathbb{C}_{XW}\mathbb{C}_{WW}^{-1})\mathbb{C}_{XW}^{\mathrm{T}} = (\mathbb{A} - \mathbb{C}_{XW}\mathbb{C}_{WW}^{-1})\mathbb{C}_{WW}(\mathbb{A} - \mathbb{C}_{XW}\mathbb{C}_{WW}^{-1})^{\mathrm{T}} \\
&\quad + \mathbb{A}\mathbb{C}_{XW}^{\mathrm{T}} - \mathbb{C}_{XW}\mathbb{C}_{WW}^{-1}\mathbb{C}_{XW}^{\mathrm{T}}
\end{aligned}$$

Substituting the above into the right-hand side of (8.69), we obtain the following equation:

$$\begin{aligned}
\mathbb{C}_{EE} = {} & (\mathbb{A} - \mathbb{C}_{XW}{\mathbb{C}_{WW}}^{-1})\mathbb{C}_{WW}(\mathbb{A} - \mathbb{C}_{XW}{\mathbb{C}_{WW}}^{-1})^{\mathrm{T}} + \mathbb{A}{\mathbb{C}_{XW}}^{\mathrm{T}} \\
& - \mathbb{C}_{XW}{\mathbb{C}_{WW}}^{-1}{\mathbb{C}_{XW}}^{\mathrm{T}} - \mathbb{A}\mathbb{C}_{WX} + \mathbb{C}_{XX}
\end{aligned}$$

Since, for real RVs $\boldsymbol{X}$ and $\boldsymbol{W}$, $\mathbb{C}_{XW}^{T} = \mathbb{C}_{WX}$, the second and the fourth terms on the right-hand side cancel each other, and the above expression becomes the following expression:

$$\begin{aligned}\mathbb{C}_{EE} &= (\mathbb{A} - \mathbb{C}_{XW}\mathbb{C}_{WW}^{-1})\mathbb{C}_{WW}(\mathbb{A} - \mathbb{C}_{XW}\mathbb{C}_{WW}^{-1})^{T} - \mathbb{C}_{XW}\mathbb{C}_{WW}^{-1}\mathbb{C}_{XW}^{T} + \mathbb{C}_{XX} \\ &= \mathbb{G} - \mathbb{H} + \mathbb{C}_{XX}\end{aligned} \tag{8.70}$$

where

$$\mathbb{G} = (\mathbb{A} - \mathbb{C}_{XW}\mathbb{C}_{WW}^{-1})\mathbb{C}_{WW}(\mathbb{A} - \mathbb{C}_{XW}\mathbb{C}_{WW}^{-1})^{T} \tag{8.71}$$

$$\mathbb{H} = \mathbb{C}_{XW}\mathbb{C}_{WW}^{-1}\mathbb{C}_{XW}^{T} \tag{8.72}$$

In the last three equations, note that $\mathbb{G}$ is a function of $\mathbb{A}$, which should be selected in such a way that the diagonal elements of $\mathbb{C}_{EE}$ be minimized. $\mathbb{H}$ and $\mathbb{C}_{XX}$, on the other hand, are independent of $\mathbb{A}$. To determine the desired $\mathbb{A}$, we need to show next that the observation matrix $\mathbb{C}_{WW}$ and its inverse $\mathbb{C}_{WW}^{-1}$ appearing in (8.70) through (8.72) are nonnegative definite. To show this, form an arbitrary quadratic form of $\boldsymbol{W}$ as follows:

$$y = \boldsymbol{a}^{T}\boldsymbol{W} \qquad\qquad y = y^{T} = \boldsymbol{W}^{T}\boldsymbol{a} \tag{8.73}$$

where

$$\boldsymbol{a}^{T} = [a_1 \ . \ . \ a_m]$$

The quadratic form y given above is a scalar and, by squaring it, we obtain the following positive quantity:

$$y^2 = yy^{T} = \boldsymbol{a}^{T}\boldsymbol{W}\boldsymbol{W}^{T}\boldsymbol{a}$$

The expected value of the above quantity is nonnegative because the left-hand side is a squared quantity, and we have the following relation:

$$E(y^2) = E(\boldsymbol{a}^{T}\boldsymbol{W}\boldsymbol{W}^{T}\boldsymbol{a}) = \boldsymbol{a}^{T}E(\boldsymbol{W}\boldsymbol{W}^{T})\boldsymbol{a} = \boldsymbol{a}^{T}\mathbb{C}_{WW}\,\boldsymbol{a} \geq 0 \tag{8.74}$$

By the definition of the nonnegative definiteness of a matrix given by (2.62), the above result shows that $\mathbb{C}_{XX}$ is nonnegative definite. By Theorem 2.2.4, since $\mathbb{C}_{XX}$ is nonnegative definite, $\mathbb{C}_{WW}^{-1}$ is also nonnegative definite.

Both $\mathbb{C}_{WW}$ and $\mathbb{C}_{WW}^{-1}$ are symmetric matrices. $\mathbb{G}$ and $\mathbb{H}$ of (8.71) and (8.72) are transformations of $\mathbb{C}_{WW}$ and $\mathbb{C}_{WW}^{-1}$, respectively, as in Theorem 2.2.5, and the same theorem shows that the diagonal elements of $\mathbb{G}$ and $\mathbb{H}$ are nonnegative and therefore, the minimum value of their diagonal elements is zero. The diagonal elements of $\mathbb{G}$ can be made zero by selecting $\mathbb{A}$ to satisfy the following equation:

$$\mathbb{G} = (\mathbb{A} - \mathbb{C}_{XW}\mathbb{C}_{WW}^{-1})\mathbb{C}_{WW}(\mathbb{A} - \mathbb{C}_{XW}\mathbb{C}_{WW}^{-1})^{T} = 0$$

From the above equation, we derive the following solution:

$$\mathbb{A} - \mathbb{C}_{XW}\mathbb{C}_{WW}^{-1} = 0 \tag{8.75}$$

From the above equation, we obtain the coefficient matrix $\mathbb{A}$ that minimizes the diagonal elements of the error covariance matrix $\mathbb{C}_{EE}$ as follows:

$$\mathbb{A} = \mathbb{C}_{XW}\mathbb{C}_{WW}^{-1} \tag{8.76}$$

Substituting the above equation into (8.59), we obtain the following minimum mean square error estimator:

$$\hat{X} = \mathbb{C}_{XW}\mathbb{C}_{WW}^{-1}W \tag{8.77}$$

Substituting (8.76) into (8.70), we obtain the following minimum error covariance matrix:

$$\mathbb{C}_{EE} = \mathbb{C}_{XX} - \mathbb{H} = \mathbb{C}_{XX} - \mathbb{C}_{XW}\mathbb{C}_{WW}^{-1}\mathbb{C}_{XW}^{\mathrm{T}} \tag{8.78}$$

where $\mathbb{H}$ is given by (8.72).

Linear Observations

The optimum coefficient matrix $\mathbb{A}$ given by (8.76) is a general solution without postulating any particular model of observations. Now, postulate that the observation vector W is given by a linear combination of the state variables and that the observations contain an additive observation noise or error N as follows:

$$W = \mathbb{B}X + N \tag{8.79}$$

$$N = \begin{bmatrix} N_1 \\ \cdot \\ \cdot \\ N_m \end{bmatrix} \quad m - \text{dimensional observation error vector} \tag{8.80}$$

$$\mathbb{B} = \begin{bmatrix} b_{11} & \cdots & b_{1n} \\ \vdots & \ddots & \vdots \\ b_{m1} & \cdots & b_{mn} \end{bmatrix} \quad (m \times n) \text{ observation matrix} \tag{8.81}$$

Assuming that N is zero mean, the observation error covariance matrix $\mathbb{C}_{NN}$ is defined as follows:

$$\begin{aligned} \mathbb{C}_{NN} &= E\{NN^{\mathrm{T}}\} \\ &= \begin{bmatrix} E(N_1^2) & \cdots & N_1N_m \\ \vdots & \ddots & \vdots \\ N_mN_1 & \cdots & E(N_m^2) \end{bmatrix} \end{aligned} \tag{8.82}$$

Substituting (8.79) into (8.63) and (8.64), we obtain the covariance matrices in (8.76) as follows:

$$
\begin{aligned}
\mathbb{C}_{XW} &= E(XW^{\mathrm{T}}) \\
&= E\left\{X(\mathbb{B}X+N)^{\mathrm{T}}\right\} \\
&= E\{X(X^{\mathrm{T}}\mathbb{B}^{\mathrm{T}}+N^{\mathrm{T}})\} \\
&= E(XX^{\mathrm{T}})\mathbb{B}^{\mathrm{T}}+E(XN^{\mathrm{T}}) \\
&= \mathbb{C}_{XX}\mathbb{B}^{\mathrm{T}}+\mathbb{C}_{XN}
\end{aligned}
\tag{8.83}
$$

$$
\begin{aligned}
\mathbb{C}_{WW} &= E(WW^{\mathrm{T}}) \\
&= E\left\{(\mathbb{B}X+N)(\mathbb{B}X+N)^{\mathrm{T}}\right\} \\
&= E\{(\mathbb{B}X+N)(X^{\mathrm{T}}\mathbb{B}^{\mathrm{T}}+N^{\mathrm{T}})\} \\
&= E(\mathbb{B}XX^{\mathrm{T}}\mathbb{B}^{\mathrm{T}}+NX^{\mathrm{T}}\mathbb{B}^{\mathrm{T}}+\mathbb{B}XN^{\mathrm{T}}+NN^{\mathrm{T}}) \\
&= \mathbb{B}E(XX^{\mathrm{T}})\mathbb{B}^{\mathrm{T}}+E(NX^{\mathrm{T}})\mathbb{B}^{\mathrm{T}}+\mathbb{B}E(XN^{\mathrm{T}})+E(NN^{\mathrm{T}}) \\
&= \mathbb{B}\mathbb{C}_{XX}\mathbb{B}^{\mathrm{T}}+\mathbb{C}_{NX}\mathbb{B}^{\mathrm{T}}+\mathbb{B}\mathbb{C}_{XN}+\mathbb{C}_{NN} \\
&= \mathbb{B}\mathbb{C}_{XX}\mathbb{B}^{\mathrm{T}}+\mathbb{C}_{XN}^{\mathrm{T}}\mathbb{B}^{\mathrm{T}}+\mathbb{B}\mathbb{C}_{XN}+\mathbb{C}_{NN} \\
&= \mathbb{B}\mathbb{C}_{XX}\mathbb{B}^{\mathrm{T}}+\mathbb{C}_{NX}\mathbb{B}^{\mathrm{T}}+\mathbb{B}\mathbb{C}_{XN}+\mathbb{C}_{NN} \\
&= \mathbb{B}\mathbb{C}_{XX}\mathbb{B}^{\mathrm{T}}+(\mathbb{B}\mathbb{C}_{XN})^{\mathrm{T}}+\mathbb{B}\mathbb{C}_{XN}+\mathbb{C}_{NN}
\end{aligned}
\tag{8.84}
$$

Substituting (8.83) and (8.84) into the general solution (8.76) and (8.77), we obtain the following solution for linear observations:

$$
\mathbb{A} = (\mathbb{C}_{XX}\mathbb{B}^{\mathrm{T}}+\mathbb{C}_{XN})\left\{\mathbb{B}\mathbb{C}_{XX}\mathbb{B}^{\mathrm{T}}+(\mathbb{B}\mathbb{C}_{XN})^{\mathrm{T}}+\mathbb{B}\mathbb{C}_{XN}+\mathbb{C}_{NN}\right\}^{-1} \tag{8.85}
$$

$$
\hat{X} = (\mathbb{C}_{XX}\mathbb{B}^{\mathrm{T}}+\mathbb{C}_{XN})\left\{\mathbb{B}\mathbb{C}_{XX}\mathbb{B}^{\mathrm{T}}+(\mathbb{B}\mathbb{C}_{XN})^{\mathrm{T}}+\mathbb{B}\mathbb{C}_{XN}+\mathbb{C}_{NN}\right\}^{-1}W \tag{8.86}
$$

$$
\mathbb{C}_{EE} = \mathbb{C}_{XX}-(\mathbb{C}_{XX}\mathbb{B}^{\mathrm{T}}+\mathbb{C}_{XN})\left\{\mathbb{B}\mathbb{C}_{XX}\mathbb{B}^{\mathrm{T}}+(\mathbb{B}\mathbb{C}_{XN})^{\mathrm{T}}+\mathbb{B}\mathbb{C}_{XN}+\mathbb{C}_{NN}\right\}^{-1}(\mathbb{C}_{XX}\mathbb{B}^{\mathrm{T}}+\mathbb{C}_{XN})^{\mathrm{T}} \tag{8.87}
$$

Simplifying Conditions

Assume that the observation error N and the state variable X are uncorrelated. Then, by (5.45), the covariance matrix $\mathbb{C}_{XN}=0$, and (8.85), (8.86), and (8.87) can be simplified as follows:

$$
\mathbb{A} = \mathbb{C}_{XX}\mathbb{B}^{\mathrm{T}}(\mathbb{C}_{NN}+\mathbb{B}\mathbb{C}_{XX}\mathbb{B}^{\mathrm{T}})^{-1} \tag{8.88}
$$

$$
\hat{X} = \mathbb{C}_{XX}\mathbb{B}^{\mathrm{T}}(\mathbb{C}_{NN}+\mathbb{B}\mathbb{C}_{XX}\mathbb{B}^{\mathrm{T}})^{-1}W \tag{8.89}
$$

$$
\mathbb{C}_{EE} = \mathbb{C}_{XX}-\mathbb{C}_{XX}\mathbb{B}^{\mathrm{T}}(\mathbb{C}_{NN}+\mathbb{B}\mathbb{C}_{XX}\mathbb{B}^{\mathrm{T}})^{-1}(\mathbb{C}_{XX}\mathbb{B}^{\mathrm{T}})^{\mathrm{T}} \tag{8.90}
$$

The above three equations constitute the solution for the linear MMSE estimation problem with a linear observation model with the observation error uncorrelated with the state variable.

8.3 Kalman Filter

We first illustrate the concept of the Kalman filter using a scalar RV and then derive the general form of the Kalman filter for the vector RV.

8.3.1 Kalman Filter: Scalar Case

Suppose that the state variable to be estimated is a scalar X. In (8.88), the covariance matrix is a scalar $\mathbb{C}_{XX}=c$. The observation vector $\boldsymbol{W}$ is given by the following equation:

$$\boldsymbol{W} = \mathbb{B}X + \boldsymbol{N}$$

For simplicity, assume that

$$\mathbb{B} = \begin{bmatrix} 1 \\ \vdots \\ 1 \end{bmatrix}$$

Then, we have

$$\begin{aligned} \boldsymbol{W} &= \mathbb{B}X + \boldsymbol{N} \\ &= \begin{bmatrix} W_1 \\ \cdot \\ \cdot \\ W_m \end{bmatrix} \\ &= X \begin{bmatrix} 1 \\ \vdots \\ 1 \end{bmatrix} + \begin{bmatrix} N_1 \\ \cdot \\ \cdot \\ N_m \end{bmatrix} \end{aligned}$$

Further assume that

$$\sigma^2_{N_iN_i} = 1, i = 1, \ldots m \qquad \mathbb{C}^{-1}_{NN} = \left(\sigma^2_{N_iN_i}\right)^{-1}\mathbb{I} = \mathbb{I}$$

With the above simplifying assumptions, the optimum coefficient matrix is given by the following equation:

$$\mathbb{A} = \frac{1}{m}\mathbb{B}^{\mathrm{T}} \tag{8.91}$$

The corresponding MMSE estimator is given by the following arithmetic average of the measurements:

$$\hat{X} = \mathbb{A}W = \frac{1}{m}\mathbb{B}^{\mathrm{T}}\ W = \frac{1}{m}[1\ \ \ldots\ldots\ \ 1]\begin{bmatrix} W_1 \\ \cdot \\ \cdot \\ W_m \end{bmatrix} = \frac{1}{m}(W_1 + \ldots.. + W_m) \tag{8.92}$$

Consider the following problem.

Suppose that X is estimated by taking the arithmetic mean of m existing measurements using (8.92). Suppose that a set of s new measurements becomes available, and the current estimate is to be enhanced by using these additional measurements. Find the algorithm of incorporating the new measurements to enhance the current estimate.

The existing estimate and the enhanced estimate are distinguished by different notations. X is estimated by (8.92) based on m existing measurements as follows, where the superscript "old" stands for the current estimate to be updated

$$\hat{X}^{\mathrm{old}} = \frac{1}{m}\left(W_1^{\mathrm{old}} + \ldots + W_m^{\mathrm{old}}\right) \tag{8.93}$$

Call this the a priori estimate of X. A set of s new measurements becomes available. Derive a method of incorporating the new measurements to enhance the a priori estimate. The new estimate is called the a posteriori estimate of X.

For simplicity, assume that the linear observation model, the observation error variance, etc. stay the same with the new measurements so that the optimum estimator with the new measurements is still the same arithmetic mean. To make the illustration simple, let the number of new data $s = 1$.

The enhanced estimate is given by the same Eq. (8.92) with the number of data points $(m + 1)$ as follows:

$$\begin{aligned} \hat{X}\mathrm{total} &= \frac{1}{m+1}\left(W_1^{\mathrm{old}} + \cdots + W_m^{\mathrm{old}} + W_1^{\mathrm{new}}\right) \\ &= \frac{1}{m+1}\left(W_1^{\mathrm{old}} + \cdots + W_m^{\mathrm{old}}\right) + \frac{1}{m+1}W_1^{\mathrm{new}} \end{aligned} \tag{8.94}$$

The superscripts "total," "old," and "new" are used to distinguish the estimate and the data associated with the old, new, and total data.

Another way of "updating" the estimate instead of (8.94) is to use the knowledge of the existing estimate, which has been obtained by m data points. Multiplying the existing estimate given by (8.93) by m, we obtain the following equation:

$$W_1^{\mathrm{old}} + \ldots + W_m^{\mathrm{old}} = m\hat{X}^{\mathrm{old}} \tag{8.95}$$

Substituting the above equation into (8.94), we obtain the following equation:

$$\begin{aligned}
\hat{X}^{\text{tot}} &= \frac{1}{m+1} m\hat{X}^{\text{old}} + \frac{1}{m+1} W_1^{\text{new}} \\
&= \frac{m}{m+1}\hat{X}^{\text{old}} - \hat{X}^{\text{old}} + \hat{X}^{\text{old}} + \frac{1}{m+1} W_1^{\text{new}} \\
&= \hat{X}^{\text{old}} - \frac{1}{m+1}\hat{X}^{\text{old}} + \frac{1}{m+1} W_1^{\text{new}} = \hat{X}^{\text{old}} + \frac{1}{m+1}\left(W_1^{\text{new}} - \hat{X}^{\text{old}}\right) \\
&= \hat{X}^{\text{old}} + K\left(W_1^{\text{new}} - \hat{X}^{\text{old}}\right)
\end{aligned} \tag{8.96}$$

where

$$K = \frac{1}{m+1} \tag{8.97}$$

The MMSE estimator given by (8.96) and (8.97) is called the Kalman form of the estimator. The equation shows that the new estimate, the a posteriori estimate, is given by modifying the existing estimate by adding a correction term, which is the difference between the new data and the existing estimate weighted by a factor K. The weighting factor K is called the Kalman gain.

Computationally speaking, the first method of updating the estimate by (8.94) would involve m addition and one multiplication operations. The Kalman filter method by (8.96) would involve one addition, one subtraction, and one division operations. Depending on the problem at hand, the Kalman filter approach provides computational advantage over the first method.

8.3.2 *Kalman Filter: Vector Case*

The concept of the Kalman filter illustrated above by a simple scalar case becomes complicated for the general case of vector RVs. One of the challenges of the general Kalman filter formulation is dealing with the extensive matrix operations involved in the derivation. Therefore, before proceeding to derive the Kalman filter equations, we will first convert the MSSE solution given by (8.88), (8.89), and (8.90) into the following forms, respectively, which are more convenient for the Kalman filter analysis

$$\mathbb{A} = \left(\mathbb{C}_{XX}^{-1} + \mathbb{B}^{\text{T}}\mathbb{C}_{NN}^{-1}\mathbb{B}\right)^{-1}\mathbb{B}^{\text{T}}\mathbb{C}_{NN}^{-1} \tag{8.98}$$

$$\hat{\boldsymbol{X}} = \left(\mathbb{C}_{XX}^{-1} + \mathbb{B}^{\text{T}}\mathbb{C}_{NN}^{-1}\mathbb{B}\right)^{-1}\mathbb{B}^{\text{T}}\mathbb{C}_{NN}^{-1}\boldsymbol{W} \tag{8.99}$$

$$\mathbb{C}_{EE} = \left(\mathbb{C}_{XX}^{-1} + \mathbb{B}^{\text{T}}\mathbb{C}_{NN}^{-1}\mathbb{B}\right)^{-1} \tag{8.100}$$

The two equivalent forms of matrix expressions are identified below as the matrix identities so that the reverse direction of matrix transformation, that is, changing the equation in the form of the right-hand side to that in the form of the left-hand side, can be performed using the same matrix identities

Matrix Identities

$$\mathbb{C}_{XX}\mathbb{B}^{\mathrm{T}}(\mathbb{C}_{NN}+\mathbb{B}\mathbb{C}_{XX}\mathbb{B}^{\mathrm{T}})^{-1}=(\mathbb{C}_{XX}^{-1}+\mathbb{B}^{\mathrm{T}}\mathbb{C}_{NN}^{-1}\mathbb{B})^{-1}\mathbb{B}^{\mathrm{T}}\mathbb{C}_{NN}^{-1} \tag{8.101}$$

$$\mathbb{C}_{XX}-\mathbb{C}_{XX}\mathbb{B}^{\mathrm{T}}(\mathbb{C}_{NN}+\mathbb{B}\mathbb{C}_{XX}\mathbb{B}^{\mathrm{T}})^{-1}(\mathbb{C}_{XX}\mathbb{B}^{\mathrm{T}})^{\mathrm{T}}=(\mathbb{C}_{XX}^{-1}+\mathbb{B}^{\mathrm{T}}\mathbb{C}_{NN}^{-1}\mathbb{B})^{-1} \tag{8.102}$$

The conversions of the above forms can be obtained by the matrix operations using, among other matrix identities, the following two simple matrix identities given by the next two theorems.

Theorem 8.3.1

$$(\mathbb{I}+\mathbb{K})^{-1}=\mathbb{I}-(\mathbb{I}+\mathbb{K})^{-1}\mathbb{K} \tag{8.103}$$

Proof Let

$$(\mathbb{I}+\mathbb{K})^{-1}=\mathbb{A} \tag{8.104}$$

Left-multiplying both sides of the above equation by $(\mathbb{I}+\mathbb{K})$, we obtain the following sequence of expressions:

$$\begin{aligned}&(\mathbb{I}+\mathbb{K})^{-1}(\mathbb{I}+\mathbb{K})=\mathbb{A}(\mathbb{I}+\mathbb{K})\\&\mathbb{I}=\mathbb{A}+\mathbb{A}\mathbb{K}\qquad \mathbb{I}-\mathbb{A}\mathbb{K}=\mathbb{A}\end{aligned}$$

Substituting (8.104) into the last equation, we obtain the following equation:

$$\mathbb{I}-(\mathbb{I}+\mathbb{K})^{-1}\mathbb{K}=\mathbb{A}$$

which, by (8.104), is equal to $(\mathbb{I}+\mathbb{K})^{-1}$.

Q.E.D.

Theorem 8.3.2

$$(\mathbb{I}+\mathbb{M}\mathbb{N})^{-1}\mathbb{M}=\mathbb{M}(\mathbb{I}+\mathbb{N}\mathbb{M})^{-1} \tag{8.105}$$

This theorem will be used multiple times in the conversion process discussed in this section.

Proof We can change the matrix product as follows, proceeding from the left-hand side of the above equation and using the matrix identity (2.68a) twice in the process

$$
\begin{aligned}
(\mathbb{I}+\mathbb{M}\mathbb{N})^{-1}\mathbb{M} &= (\mathbb{I}+\mathbb{M}\mathbb{N})^{-1}\left(\mathbb{M}^{-1}\right)^{-1} \\
&= \left\{\left(\mathbb{M}^{-1}\right)(\mathbb{I}+\mathbb{M}\mathbb{N})\right\}^{-1} \\
&= \left(\mathbb{M}^{-1}+\mathbb{M}^{-1}\mathbb{M}\mathbb{N}\right)^{-1} \\
&= \left(\mathbb{M}^{-1}+\mathbb{N}\right)^{-1} \\
&= \left\{(\mathbb{I}+\mathbb{N}\mathbb{M})\mathbb{M}^{-1}\right\}^{-1} \\
&= \left(\mathbb{M}^{-1}\right)^{-1}(\mathbb{I}+\mathbb{N}\mathbb{M})^{-1} \\
&= \mathbb{M}(\mathbb{I}+\mathbb{N}\mathbb{M})^{-1}
\end{aligned}
$$

Q.E.D.

Conversion of Coefficient Matrix $\mathbb{A}$ and Estimator Equation $\hat{X}$

We will first convert the coefficient matrix (8.88) to (8.98) as follows. By taking out the first term inside the braces in (8.88), we obtain the following equation:

$$\mathbb{A} = \mathbb{C}_{XX}\mathbb{B}^{\mathrm{T}}\left\{\mathbb{C}_{NN}(\mathbb{I}+\mathbb{C}_{NN}^{-1}\mathbb{B}\mathbb{C}_{XX}\mathbb{B}^{\mathrm{T}})\right\}^{-1}$$

Applying the matrix identity (2.68a), $(\mathbb{A}\mathbb{B})^{-1} = \mathbb{B}^{-1}\mathbb{A}^{-1}$, to the inverse operation inside the braces, we obtain the following equation:

$$\mathbb{A} = \mathbb{C}_{XX}\mathbb{B}^{\mathrm{T}}\left(\mathbb{I}+\mathbb{C}_{NN}^{-1}\mathbb{B}\mathbb{C}_{XX}\mathbb{B}^{\mathrm{T}}\right)^{-1}\ \mathbb{C}_{NN}^{-1}$$

Apply the matrix identity given by (8.105), $\mathbb{M}(\mathbb{I}+\mathbb{N}\mathbb{M})^{-1} = (\mathbb{I}+\mathbb{M}\mathbb{N})^{-1}\mathbb{M}$, to the above equation with the following correspondences:

$$\mathbb{M} \to \mathbb{C}_{XX}\mathbb{B}^{\mathrm{T}} \qquad \mathbb{N} \to \mathbb{C}_{NN}^{-1}\mathbb{B}$$

The result is the following equation:

$$\mathbb{A} = (\mathbb{I}+\mathbb{C}_{XX}\mathbb{B}^{\mathrm{T}}\mathbb{C}_{NN}^{-1}\mathbb{B})^{-1}\mathbb{C}_{XX}\mathbb{B}^{\mathrm{T}}\mathbb{C}_{NN}^{-1}$$

Left multiplying the terms in the parentheses by $\mathbb{C}_{XX}\mathbb{C}_{XX}^{-1} = \mathbb{I}$ in the above equation before taking the inversion operation, we obtain the following expression:

$$
\begin{aligned}
\mathbb{A} &= \left\{\mathbb{C}_{XX}\mathbb{C}_{XX}^{-1}(\mathbb{I}+\mathbb{C}_{XX}\mathbb{B}^{\mathrm{T}}\mathbb{C}_{NN}^{-1}\mathbb{B})\right\}^{-1}\mathbb{C}_{XX}\mathbb{B}^{\mathrm{T}}\mathbb{C}_{NN}^{-1} \\
&= \left\{\mathbb{C}_{XX}(\mathbb{C}_{XX}^{-1}+\mathbb{C}_{XX}^{-1}\mathbb{C}_{XX}\mathbb{B}^{\mathrm{T}}\mathbb{C}_{NN}^{-1}\mathbb{B})\right\}^{-1}\mathbb{C}_{XX}\mathbb{B}^{\mathrm{T}}\mathbb{C}_{NN}^{-1}
\end{aligned}
$$

Applying the matrix identity (2.68a), $(\mathbb{A}\mathbb{B})^{-1} = \mathbb{B}^{-1}\mathbb{A}^{-1}$, to the inverse operation of the terms in the braces and replacing $\mathbb{C}_{XX}^{-1}\mathbb{C}_{XX}$ inside the parentheses with $\mathbb{I}$, we obtain the following equation, which is the desired Eq. (8.98):

$$\begin{aligned}\mathbb{A} &= (\mathbb{C}_{XX}^{-1} + \mathbb{B}^{\mathrm{T}}\mathbb{C}_{NN}^{-1}\mathbb{B})^{-1}\mathbb{C}_{XX}^{-1}\mathbb{C}_{XX}\mathbb{B}^{\mathrm{T}}\mathbb{C}_{NN}^{-1} \\ &= (\mathbb{C}_{XX}^{-1} + \mathbb{C}_{XX}^{-1}\mathbb{C}_{XX}\mathbb{B}^{\mathrm{T}}\mathbb{C}_{NN}^{-1}\mathbb{B})^{-1}\mathbb{C}_{XX}\mathbb{B}^{\mathrm{T}}\mathbb{C}_{NN}^{-1}\end{aligned}$$

The above equation completes the proof of the matrix identity (8.101).

Substituting the above matrix identity into (8.89), we transform the estimator equation from (8.89) to (8.99).

Conversion of Error Covariance Matrix $\mathbb{C}_{EE}$

By inserting $\mathbb{C}_{NN}\mathbb{C}_{NN}{}^{-1} = \mathbb{I}$ before the last term in (8.90), we obtain the following expression:

$$\begin{aligned}\mathbb{C}_{EE} &= \mathbb{C}_{XX} - \mathbb{C}_{XX}\mathbb{B}^{\mathrm{T}}(\mathbb{C}_{NN} + \mathbb{B}\mathbb{C}_{XX}\mathbb{B}^{\mathrm{T}})^{-1}\mathbb{C}_{NN}\mathbb{C}_{NN}^{-1}(\mathbb{C}_{XX}\mathbb{B}^{\mathrm{T}})^{\mathrm{T}} \\ &= \mathbb{C}_{XX} - \mathbb{C}_{XX}\mathbb{B}^{\mathrm{T}}(\mathbb{C}_{NN} + \mathbb{B}\mathbb{C}_{XX}\mathbb{B}^{\mathrm{T}})^{-1}\mathbb{C}_{NN}^{-1^{-1}}\mathbb{C}_{NN}^{-1}(\mathbb{C}_{XX}\mathbb{B}^{\mathrm{T}})^{\mathrm{T}}\end{aligned}$$

Applying the matrix identity (2.68a), $\mathbb{B}^{-1}\mathbb{A}^{-1} = (\mathbb{A}\mathbb{B})^{-1}$, to the first two inverse operations in the above equation, we obtain the following expression:

$$\begin{aligned}\mathbb{C}_{EE} &= \mathbb{C}_{XX} - \mathbb{C}_{XX}\mathbb{B}^{\mathrm{T}}\{\mathbb{C}_{NN}^{-1}(\mathbb{C}_{NN} + \mathbb{B}\mathbb{C}_{XX}\mathbb{B}^{\mathrm{T}})\}^{-1}\mathbb{C}_{NN}^{-1}(\mathbb{C}_{XX}\mathbb{B}^{\mathrm{T}})^{\mathrm{T}} \\ &= \mathbb{C}_{XX} - \mathbb{C}_{XX}\mathbb{B}^{\mathrm{T}}(\mathbb{I} + \mathbb{C}_{NN}^{-1}\mathbb{B}\mathbb{C}_{XX}\mathbb{B}^{\mathrm{T}})^{-1}\mathbb{C}_{NN}^{-1}(\mathbb{C}_{XX}\mathbb{B}^{\mathrm{T}})^{\mathrm{T}}\end{aligned}$$

Apply the matrix identity given by (8.105), $\mathbb{M}(\mathbb{I} + \mathbb{N}\mathbb{M})^{-1} = (\mathbb{I} + \mathbb{M}\mathbb{N})^{-1}\mathbb{M}$, to the above equation with the following correspondences:

$$\mathbb{M} \to \mathbb{B}^{\mathrm{T}} \qquad \mathbb{N} \to \mathbb{C}_{NN}^{-1}\mathbb{B}\mathbb{C}_{XX}$$

The result is the following equation:

$$\mathbb{C}_{EE} = \mathbb{C}_{XX} - \mathbb{C}_{XX}(\mathbb{I} + \mathbb{B}^{\mathrm{T}}\mathbb{C}_{NN}^{-1}\mathbb{B}\mathbb{C}_{XX})^{-1}\mathbb{B}^{\mathrm{T}}\mathbb{C}_{NN}^{-1}(\mathbb{C}_{XX}\mathbb{B}^{\mathrm{T}})^{\mathrm{T}}$$

Apply the same matrix identity again to the above equation with the following correspondences:

$$\mathbb{M} \to \mathbb{C}_{XX} \qquad \mathbb{N} \to \mathbb{B}^{\mathrm{T}}\mathbb{C}_{NN}^{-1}\mathbb{B}$$

The result is the following equation:

$$\begin{aligned}\mathbb{C}_{EE} &= \mathbb{C}_{XX} - (\mathbb{I} + \mathbb{C}_{XX}\mathbb{B}^{\mathrm{T}}\mathbb{C}_{NN}^{-1}\mathbb{B})^{-1}\mathbb{C}_{XX}\mathbb{B}^{\mathrm{T}}\mathbb{C}_{NN}^{-1}(\mathbb{C}_{XX}\mathbb{B}^{\mathrm{T}})^{\mathrm{T}} \\ &= \mathbb{C}_{XX} - (\mathbb{I} + \mathbb{C}_{XX}\mathbb{B}^{\mathrm{T}}\mathbb{C}_{NN}^{-1}\mathbb{B})^{-1}\mathbb{C}_{XX}\mathbb{B}^{\mathrm{T}}\mathbb{C}_{NN}^{-1}\mathbb{B}^{\mathrm{T}^{\mathrm{T}}}\mathbb{C}_{XX}^{\mathrm{T}} \\ &= \mathbb{C}_{XX} - (\mathbb{I} + \mathbb{C}_{XX}\mathbb{B}^{\mathrm{T}}\mathbb{C}_{NN}^{-1}\mathbb{B})^{-1}\mathbb{C}_{XX}\mathbb{B}^{\mathrm{T}}\mathbb{C}_{NN}^{-1}\mathbb{B}\mathbb{C}_{XX}^{\mathrm{T}}\end{aligned}$$

Since $\mathbb{C}_{XX} = \mathbb{C}_{XX}{}^{\mathrm{T}}$, we can factor it out to obtain the following equation:

$$\mathbb{C}_{EE} = \{\mathbb{I} - (\mathbb{I} + \mathbb{C}_{XX}\mathbb{B}^{\mathrm{T}}\mathbb{C}_{NN}^{-1}\mathbb{B})^{-1}\mathbb{C}_{XX}\mathbb{B}^{\mathrm{T}}\mathbb{C}_{NN}^{-1}\mathbb{B}\}\mathbb{C}_{XX}$$

Apply the matrix identity (8.103), $\mathbb{I} - (\mathbb{I} + \mathbb{K})^{-1}\mathbb{K} = (\mathbb{I} + \mathbb{K})^{-1}$, to the terms inside the braces in the above equation with the following correspondence:

$$\mathbb{K} \rightarrow \mathbb{C}_{XX}\mathbb{B}^{\mathrm{T}}\mathbb{C}_{NN}^{-1}\mathbb{B}$$

The result is the following equation:

$$\mathbb{C}_{EE} = (\mathbb{I} + \mathbb{C}_{XX}\mathbb{B}^{\mathrm{T}}\mathbb{C}_{NN}^{-1}\mathbb{B})^{-1}\mathbb{C}_{XX} = (\mathbb{I} + \mathbb{C}_{XX}\mathbb{B}^{\mathrm{T}}\mathbb{C}_{NN}^{-1}\mathbb{B})^{-1}\mathbb{C}_{XX}^{-1^{-1}}$$

Applying the matrix identity (2.68a), $\mathbb{B}^{-1}\mathbb{A}^{-1} = (\mathbb{A}\mathbb{B})^{-1}$, to the above equation, we obtain the following equation, which is the desired Eq. (8.100):

$$\mathbb{C}_{EE} = \{\mathbb{C}_{XX}^{-1}(\mathbb{I} + \mathbb{C}_{XX}\mathbb{B}^{\mathrm{T}}\mathbb{C}_{NN}^{-1}\mathbb{B})\}^{-1} = (\mathbb{C}_{XX}^{-1} + \mathbb{C}_{XX}^{-1}\mathbb{C}_{XX}\mathbb{B}^{\mathrm{T}}\mathbb{C}_{NN}^{-1}\mathbb{B})^{-1} = (\mathbb{C}_{XX}^{-1} + \mathbb{B}^{\mathrm{T}}\mathbb{C}_{NN}^{-1}\mathbb{B})^{-1}$$

The above equation completes the proof of the matrix identity (8.102).

Simplifying Assumption
Another simplifying assumption is that the variability of the states to be estimated is very large such that the inverse of its covariance matrix is zero

$$\mathbb{C}_{XX}^{-1} = 0 \tag{8.106}$$

The above assumption is also the worst-case assumption in the absence of prior information about the state variable. Substituting the above into (8.98), (8.99), and (8.100), we obtain the following simplified equations for the MSSE solution, respectively:

$$\mathbb{A} = (\mathbb{B}^{\mathrm{T}}\mathbb{C}_{NN}^{-1}\mathbb{B})^{-1}\mathbb{B}^{\mathrm{T}}\mathbb{C}_{NN}^{-1} \tag{8.107}$$

$$\hat{\boldsymbol{X}} = (\mathbb{B}^{\mathrm{T}}\mathbb{C}_{NN}^{-1}\mathbb{B})^{-1}\mathbb{B}^{\mathrm{T}}\mathbb{C}_{NN}^{-1}\boldsymbol{W} \tag{8.108}$$

$$\mathbb{C}_{EE} = (\mathbb{B}^{\mathrm{T}}\mathbb{C}_{NN}^{-1}\mathbb{B})^{-1} \tag{8.109}$$

The above three equations will be used in the derivation of the Kalman filter.

Kalman Filter for a Vector Case
The concept of the Kalman filter illustrated with a scalar case is stated for the general case of vector RV as follows. Assume that an a priori linear MMSE estimate with a linear observation model has already been computed with m a priori measurements and is to be updated with a new set of s additional measurements. Derive a recursive method which will combine the old estimate, the a priori *estimate*, with the new set of data to obtain a new, that is, updated, MMSE estimate.

Notations
The numbers of old and new observations are m and s, respectively. The old and new observation vector, observation error vector, observation matrix, and the

observation error covariance matrix have the corresponding dimensions. The dimension of the total observation vector and that of the corresponding observation error vector is both $(m + s)$. The dimension of the total observation matrix is $(m + s) \times n$ and the dimension of the total observation error covariance matrix, $(m + s) \times (m + s)$. The dimension of the state variable to be estimated, which is considered to be n, does not change. Therefore, the dimension of the estimator and that of the estimation error both remain the same at n, and the dimension of the estimation error covariance matrix remains the same at $(n \times n)$.

To distinguish the a priori and a posteriori estimates and data sets, we use the superscripts "tot," "old," and "new" to denote the total, old, and new data and estimates. The following notations are used in this book:

$$\hat{\boldsymbol{X}}^{\text{old}} = \mathbb{A}^{\text{old}}\boldsymbol{W}^{\text{old}} \qquad \hat{\boldsymbol{X}}^{\text{new}} = \mathbb{A}^{\text{new}}\boldsymbol{W}^{\text{new}} \qquad \hat{\boldsymbol{X}}^{\text{tot}} = \mathbb{A}^{\text{tot}}\boldsymbol{W}^{\text{tot}}$$

$$\boldsymbol{W}^{\text{old}} = \mathbb{B}^{\text{old}}\boldsymbol{X} + \boldsymbol{N}^{\text{old}} \qquad \boldsymbol{W}^{\text{new}} = \mathbb{B}^{\text{new}}\boldsymbol{X} + \boldsymbol{N}^{\text{new}} \qquad \boldsymbol{W}^{\text{tot}} = \mathbb{B}^{\text{tot}}\boldsymbol{X} + \boldsymbol{N}^{\text{tot}}$$

$$\boldsymbol{E}^{\text{old}} = \hat{\boldsymbol{X}}^{\text{old}} - \boldsymbol{X} \qquad \boldsymbol{E}^{\text{new}} = \hat{\boldsymbol{X}}^{\text{new}} - \boldsymbol{X} \qquad \boldsymbol{E}^{\text{tot}} = \hat{\boldsymbol{X}}^{\text{tot}} - \boldsymbol{X}$$

where the vectors and the matrices are defined as follows:

$$\hat{\boldsymbol{X}}^{\text{old}} = \begin{bmatrix} \hat{X}_1^{\text{old}} \\ \cdot \\ \cdot \\ \hat{X}_n^{\text{old}} \end{bmatrix} \qquad \hat{\boldsymbol{X}}^{\text{new}} = \begin{bmatrix} \hat{X}_1^{\text{new}} \\ \cdot \\ \cdot \\ \hat{X}_n^{\text{new}} \end{bmatrix} \qquad \hat{\boldsymbol{X}}^{\text{tot}} = \begin{bmatrix} \hat{X}_1^{\text{tot}} \\ \cdot \\ \cdot \\ \hat{X}_n^{\text{tot}} \end{bmatrix}$$

$$\boldsymbol{W}^{\text{old}} = \begin{bmatrix} W_1^{\text{old}} \\ \cdot \\ \cdot \\ W_m^{\text{old}} \end{bmatrix} \qquad \boldsymbol{W}^{\text{new}} = \begin{bmatrix} W_1^{\text{new}} \\ \cdot \\ \cdot \\ W_s^{\text{new}} \end{bmatrix} \qquad \boldsymbol{W}^{\text{tot}} = \begin{bmatrix} \boldsymbol{W}^{\text{old}} \\ -- \\ \boldsymbol{W}^{\text{new}} \end{bmatrix} = \begin{bmatrix} W_1^{\text{old}} \\ \cdot \\ W_m^{\text{old}} \\ -- \\ W_1^{\text{new}} \\ \cdot \\ W_s^{\text{new}} \end{bmatrix}$$

$$\boldsymbol{N}^{\text{old}} = \begin{bmatrix} N_1^{\text{old}} \\ \cdot \\ \cdot \\ N_m^{\text{old}} \end{bmatrix} \qquad \boldsymbol{N}^{\text{new}} = \begin{bmatrix} N_1^{\text{new}} \\ \cdot \\ \cdot \\ N_s^{\text{new}} \end{bmatrix} \qquad \boldsymbol{N}^{\text{tot}} = \begin{bmatrix} \boldsymbol{N}^{\text{old}} \\ -- \\ \boldsymbol{N}^{\text{new}} \end{bmatrix} = \begin{bmatrix} N_1^{\text{old}} \\ \cdot \\ N_s^{\text{old}} \\ -- \\ N_1^{\text{new}} \\ \cdot \\ N_s^{\text{new}} \end{bmatrix}$$

$$\boldsymbol{E}^{\text{old}} = \begin{bmatrix} E_1^{\text{old}} \\ \cdot \\ \cdot \\ E_n^{\text{old}} \end{bmatrix} \qquad \boldsymbol{E}^{\text{new}} = \begin{bmatrix} E_1^{\text{new}} \\ \cdot \\ \cdot \\ E_n^{\text{new}} \end{bmatrix} \qquad \boldsymbol{E}^{\text{tot}} = \begin{bmatrix} E_1^{\text{tot}} \\ \cdot \\ \cdot \\ E_n^{\text{tot}} \end{bmatrix}$$

$$
\mathbb{B}^{\text{old}} = \begin{bmatrix} b_{11}^{\text{old}} & \cdots & b_{1n}^{\text{old}} \\ \vdots & \ddots & \vdots \\ b_{m1}^{\text{old}} & \cdots & b_{mn}^{\text{old}} \end{bmatrix} \qquad \mathbb{B}^{\text{new}} = \begin{bmatrix} b_{11}^{\text{old}} & \cdots & b_{1n}^{\text{old}} \\ \vdots & \ddots & \vdots \\ b_{s1}^{\text{old}} & \cdots & b_{sn}^{\text{old}} \end{bmatrix}
$$

$$
\mathbb{B}^{\text{tot}} = \begin{bmatrix} \mathbb{B}^{\text{old}} \\ --- \\ \mathbb{B}^{\text{new}} \end{bmatrix} = \begin{bmatrix} b_{11}^{\text{old}} & . & b_{1n}^{\text{old}} \\ . & . & . \\ b_{m1}^{\text{old}} & . & b_{mn}^{\text{old}} \\ -- & -- & -- \\ b_{11}^{\text{new}} & . & b_{1n}^{\text{new}} \\ . & . & . \\ b_{s1}^{new} & . & b_{sn}^{\text{new}} \end{bmatrix}
$$

Observation Error Covariance Matrix

Using (2.25a), (2.30f), and the multiplication rule of block matrices, we obtain the following partitioned matrix

$$
\begin{aligned}
\mathbb{C}_{NN}^{\text{tot}} &= E\left(\boldsymbol{N}^{\text{tot}} \boldsymbol{N}^{\text{tot}^{\text{T}}}\right) \\
&= E\left\{ \begin{bmatrix} \boldsymbol{N}^{\text{old}} \\ -- \\ \boldsymbol{N}^{\text{new}} \end{bmatrix} \begin{bmatrix} \boldsymbol{N}^{\text{old}} \\ -- \\ \boldsymbol{N}^{\text{new}} \end{bmatrix}^{\text{T}} \right\} \\
&= E\left\{ \begin{bmatrix} \boldsymbol{N}^{\text{old}} \\ -- \\ \boldsymbol{N}^{\text{new}} \end{bmatrix} \left[\boldsymbol{N}^{\text{old}^{\text{T}}} | \boldsymbol{N}^{\text{new}^{\text{T}}} \right] \right\} \\
&= E\left\{ \begin{bmatrix} \boldsymbol{N}^{\text{old}} \boldsymbol{N}^{\text{old}^{\text{T}}} & | & \boldsymbol{N}^{\text{old}} \boldsymbol{N}^{\text{new}^{\text{T}}} \\ ----- & + & ----- \\ \boldsymbol{N}^{\text{new}} \boldsymbol{N}^{\text{old}^{\text{T}}} & | & \boldsymbol{N}^{\text{new}} \boldsymbol{N}^{\text{new}^{\text{T}}} \end{bmatrix} \right\} \\
&= \begin{bmatrix} E(\boldsymbol{N}^{\text{old}} \boldsymbol{N}^{\text{old}^{\text{T}}}) & | & E\left(\boldsymbol{N}^{\text{old}} \boldsymbol{N}^{\text{new}^{\text{T}}}\right) \\ ------ & + & ------ \\ E(\boldsymbol{N}^{\text{new}} \boldsymbol{N}^{\text{old}^{\text{T}}}) & | & E\left(\boldsymbol{N}^{\text{new}} \boldsymbol{N}^{\text{new}^{\text{T}}}\right) \end{bmatrix} \\
&= \begin{bmatrix} \mathbb{C}_{NN}^{\text{old}} & | & \mathbb{C}_{NN}^{\text{old,new}} \\ --- & + & --- \\ \mathbb{C}_{NN}^{\text{new,old}} & | & \mathbb{C}_{NN}^{\text{new}} \end{bmatrix}
\end{aligned} \tag{8.110}
$$

Assuming that the old and new observation errors are uncorrelated, we have

$$
\mathbb{C}_{NN}^{\text{old,new}} = \mathbb{C}_{NN}^{\text{new,old}} = 0
$$

$$
\mathbb{C}_{NN}^{\text{tot}} = \begin{bmatrix} \mathbb{C}_{NN}^{\text{old}} & | & 0 \\ -- & -+- & -- \\ 0 & | & \mathbb{C}_{NN}^{\text{new}} \end{bmatrix} \tag{8.111}
$$

Linear Mean Square Error Estimation

Using (8.108) and (8.109), consider the general solution of the a priori minimum mean square error estimation of $\boldsymbol{X}$ as follows:

$$\hat{\boldsymbol{X}}^{\text{old}} = \left(\mathbb{B}^{\text{old}^{\text{T}}}\mathbb{C}_{\boldsymbol{NN}}^{\text{old}^{-1}}\mathbb{B}^{\text{old}}\right)^{-1}\mathbb{B}^{\text{old}^{\text{T}}}\mathbb{C}_{\boldsymbol{NN}}^{\text{old}^{-1}}\boldsymbol{W}^{\text{old}} \tag{8.112}$$

$$\mathbb{C}_{\boldsymbol{EE}}^{\text{old}} = \left(\mathbb{B}^{\text{old}^{\text{T}}}\mathbb{C}_{\boldsymbol{NN}}^{\text{old}^{-1}}\mathbb{B}^{\text{old}}\right)^{-1} \tag{8.113}$$

Substituting (8.113) into (8.112), we obtain the following equation:

$$\hat{\boldsymbol{X}}^{\text{old}} = \mathbb{C}_{\boldsymbol{EE}}^{\text{old}}\mathbb{B}^{\text{old}^{\text{T}}}\mathbb{C}_{\boldsymbol{NN}}^{\text{old}^{-1}}\boldsymbol{W}^{\text{old}} \tag{8.114}$$

Left-multiplying both sides of the above equation by $\mathbb{C}_{\boldsymbol{EE}}^{\text{old}^{-1}}$ yields

$$\begin{aligned}\mathbb{C}_{\boldsymbol{EE}}^{\text{old}^{-1}}\hat{\boldsymbol{X}}^{\text{old}} &= \mathbb{C}_{\boldsymbol{EE}}^{\text{old}^{-1}}\mathbb{C}_{\boldsymbol{EE}}^{\text{old}}\mathbb{B}^{\text{old}^{\text{T}}}\mathbb{C}_{\boldsymbol{NN}}^{\text{old}^{-1}}\boldsymbol{W}^{\text{old}} = \mathbb{I}\mathbb{B}^{\text{old}^{\text{T}}}\mathbb{C}_{\boldsymbol{NN}}^{\text{old}^{-1}}\boldsymbol{W}^{\text{old}} \\ &= \mathbb{B}^{\text{old}^{\text{T}}}\mathbb{C}_{\boldsymbol{NN}}^{\text{old}^{-1}}\boldsymbol{W}^{\text{old}}\end{aligned} \tag{8.115}$$

Minimum Mean Square Error Solution for the Total Data

Using the general solution given by (8.108) and (8.109), the MMSE estimate for the total number of observations is given by the following equation:

$$\hat{\boldsymbol{X}}^{\text{tot}} = \left(\mathbb{B}^{\text{tot}^{\text{T}}}\mathbb{C}_{\boldsymbol{NN}}^{\text{tot}^{-1}}\mathbb{B}^{\text{tot}}\right)^{-1}\mathbb{B}^{\text{tot}^{\text{T}}}\mathbb{C}_{\boldsymbol{NN}}^{\text{tot}^{-1}}\boldsymbol{W}^{\text{tot}} = \mathbb{C}_{\boldsymbol{EE}}^{\text{tot}}\left(\mathbb{B}^{\text{tot}^{\text{T}}}\mathbb{C}_{\boldsymbol{NN}}^{\text{tot}^{-1}}\boldsymbol{W}^{\text{tot}}\right) \tag{8.116}$$

where

$$\mathbb{C}_{\boldsymbol{EE}}^{\text{tot}} = \left(\mathbb{B}^{\text{tot}^{\text{T}}}\mathbb{C}_{\boldsymbol{NN}}^{\text{tot}^{-1}}\mathbb{B}^{\text{tot}}\right)^{-1} \tag{8.117}$$

Substituting (8.110) into the above equation, we obtain the following estimation error covariance matrix for the total data:

$$\mathbb{C}_{\boldsymbol{EE}}^{\text{tot}} = \left\{\begin{bmatrix}\mathbb{B}^{\text{old}} \\ -- \\ \mathbb{B}^{\text{new}}\end{bmatrix}^{\text{T}}\begin{bmatrix}\mathbb{C}_{\boldsymbol{NN}}^{\text{old}} & | & 0 \\ -- & -+- & -- \\ 0 & | & \mathbb{C}_{\boldsymbol{NN}}^{\text{new}}\end{bmatrix}^{-1}\begin{bmatrix}\mathbb{B}^{\text{old}} \\ -- \\ \mathbb{B}^{\text{new}}\end{bmatrix}\right\}^{-1} \tag{8.118}$$

Using the matrix identity (2.68b) and the multiplication rule of block matrices of (8.118), we obtain the following equation:

$$\begin{aligned}\mathbb{C}_{EE}^{\text{tot}} &= \left\{ \begin{bmatrix} \mathbb{B}^{\text{old}} \\ -- \\ \mathbb{B}^{\text{new}} \end{bmatrix}^{\text{T}} \begin{bmatrix} \mathbb{C}_{NN}^{\text{old}^{-1}} & | & 0 \\ --- & + & --- \\ 0 & | & \mathbb{C}_{NN}^{\text{new}^{-1}} \end{bmatrix} \begin{bmatrix} \mathbb{B}^{\text{old}} \\ -- \\ \mathbb{B}^{\text{new}} \end{bmatrix} \right\}^{-1} \\ &= \left\{ \begin{bmatrix} \mathbb{B}^{\text{old}^{\text{T}}} | \mathbb{B}^{\text{new}^{\text{T}}} \end{bmatrix} \begin{bmatrix} \mathbb{C}_{NN}^{\text{old}^{-1}} \mathbb{B}^{\text{old}} \\ ------- \\ \mathbb{C}_{NN}^{\text{new}^{-1}} \mathbb{B}^{\text{new}} \end{bmatrix} \right\}^{-1} \\ &= \left(\mathbb{B}^{\text{old}^{\text{T}}} \mathbb{C}_{NN}^{\text{old}^{-1}} \mathbb{B}^{\text{old}} + \mathbb{B}^{\text{new}^{\text{T}}} \mathbb{C}_{NN}^{\text{new}^{-1}} \mathbb{B}^{\text{new}} \right)^{-1} \end{aligned} \tag{8.119}$$

Inversing (8.113), we have the following equation:

$$\mathbb{B}^{\text{old}^{\text{T}}} \mathbb{C}_{NN}^{\text{old}^{-1}} \mathbb{B}^{\text{old}} = \mathbb{C}_{EE}^{\text{old}^{-1}} \tag{8.120}$$

Substituting (8.120) into (8.119), we obtain the following equation:

$$\mathbb{C}_{EE}^{\text{tot}} = (\mathbb{C}_{EE}^{\text{old}^{-1}} + \mathbb{B}^{\text{new}^{\text{T}}} \mathbb{C}_{NN}^{\text{new}^{-1}} \mathbb{B}^{\text{new}})^{-1} \tag{8.121}$$

Substituting (8.111) and using the matrix identity (2.68b), we expand the term in the parentheses in (8.116) as follows:

$$\begin{aligned}\mathbb{B}^{\text{tot}^{\text{T}}} \mathbb{C}_{NN}^{\text{tot}^{-1}} \boldsymbol{W}^{\text{tot}} &= \begin{bmatrix} \mathbb{B}^{\text{old}^{\text{T}}} | \mathbb{B}^{\text{new}^{\text{T}}} \end{bmatrix} \begin{bmatrix} \mathbb{C}_{NN}^{\text{old}} & | & 0 \\ -- & -+- & -- \\ 0 & | & \mathbb{C}_{NN}^{\text{new}} \end{bmatrix}^{-1} \begin{bmatrix} \boldsymbol{W}^{\text{old}} \\ --- \\ \boldsymbol{W}^{\text{new}} \end{bmatrix} \\ &= \begin{bmatrix} \mathbb{B}^{\text{old}^{\text{T}}} & | & \mathbb{B}^{\text{new}^{\text{T}}} \end{bmatrix} \begin{bmatrix} \mathbb{C}_{NN}^{\text{old}^{-1}} & | & 0 \\ --- & -+- & --- \\ 0 & | & \mathbb{C}_{NN}^{\text{new}^{-1}} \end{bmatrix} \begin{bmatrix} \boldsymbol{W}^{\text{old}} \\ --- \\ \boldsymbol{W}^{\text{new}} \end{bmatrix} \\ &= \begin{bmatrix} \mathbb{B}^{\text{old}^{\text{T}}} & | & \mathbb{B}^{\text{new}^{\text{T}}} \end{bmatrix} \begin{bmatrix} \mathbb{C}_{NN}^{\text{old}^{-1}} \boldsymbol{W}^{\text{old}} \\ ------- \\ \mathbb{C}_{NN}^{\text{new}^{-1}} \boldsymbol{W}^{\text{new}} \end{bmatrix} \\ &= \mathbb{B}^{\text{old}^{\text{T}}} \mathbb{C}_{NN}^{\text{old}^{-1}} \boldsymbol{W}^{\text{old}} + \mathbb{B}^{\text{new}^{\text{T}}} \mathbb{C}_{NN}^{\text{new}^{-1}} \boldsymbol{W}^{\text{new}} \end{aligned} \tag{8.122}$$

Substituting (8.115) into (8.122), we obtain the following equation:

$$\mathbb{B}^{\text{tot}^{\text{T}}} \mathbb{C}_{NN}^{\text{tot}^{-1}} \boldsymbol{W}^{\text{tot}} = \mathbb{C}_{EE}^{\text{old}^{-1}} \hat{\boldsymbol{X}}^{\text{old}} + \mathbb{B}^{\text{new}^{\text{T}}} \mathbb{C}_{NN}^{\text{new}^{-1}} \boldsymbol{W}^{\text{new}} \tag{8.123}$$

Substituting (8.123) and (8.121) into (8.116), we obtain the following equation:

$$\hat{\boldsymbol{X}}^{\text{tot}} = \mathbb{C}_{EE}^{\text{tot}}\left(\mathbb{C}_{EE}^{\text{old}^{-1}}\hat{\boldsymbol{X}}^{\text{old}} + \mathbb{B}^{\text{new}\,\text{T}}\mathbb{C}_{NN}^{\text{new}^{-1}}\boldsymbol{W}^{\text{new}}\right) \tag{8.124}$$

where

$$\mathbb{C}_{EE}^{\text{tot}} = \left(\mathbb{C}_{EE}^{\text{old}^{-1}} + \mathbb{B}^{\text{new}\,\text{T}}\mathbb{C}_{NN}^{\text{new}^{-1}}\mathbb{B}^{\text{new}}\right)^{-1} \tag{8.125}$$

The first equation above shows the a posteriori estimate based on the total data in terms of the a priori estimate and the new data and the second equation, the associated estimation error covariance matrix. We will now transform the above two equations into the Kalman form.

Derivation of the Kalman Form

Expand (8.124) as follows:

$$\begin{aligned}\hat{\boldsymbol{X}}^{\text{tot}} &= \mathbb{C}_{EE}^{\text{tot}}\mathbb{C}_{EE}^{\text{old}^{-1}}\hat{\boldsymbol{X}}^{\text{old}} + \mathbb{C}_{EE}^{\text{tot}}\mathbb{B}^{\text{new}\,\text{T}}\mathbb{C}_{NN}^{\text{new}^{-1}}\boldsymbol{W}^{\text{new}} \\ &= \mathbb{C}_{EE}^{\text{tot}}\mathbb{C}_{EE}^{\text{old}^{-1}}\hat{\boldsymbol{X}}^{\text{old}} + K\boldsymbol{W}^{\text{new}}\end{aligned} \tag{8.126}$$

where

$$K \triangleq \mathbb{C}_{EE}^{\text{tot}}\mathbb{B}^{\text{new}\,\text{T}}\mathbb{C}_{NN}^{\text{new}^{-1}} \tag{8.127}$$

Substituting (8.125) into the above equation, we obtain

$$K = \left(\mathbb{C}_{EE}^{\text{old}^{-1}} + \mathbb{B}^{\text{new}\,\text{T}}\mathbb{C}_{NN}^{\text{new}^{-1}}\mathbb{B}^{\text{new}}\right)^{-1}\mathbb{B}^{\text{new}\,\text{T}}\mathbb{C}_{NN}^{\text{new}^{-1}} \tag{8.128}$$

Apply the earlier matrix identity given by (8.101) to the above equation in the reverse direction, that is,

$$\left(\mathbb{C}_{XX}^{-1} + \mathbb{B}^{\text{T}}\mathbb{C}_{NN}^{-1}\mathbb{B}\right)^{-1}\mathbb{B}^{\text{T}}\mathbb{C}_{NN}^{-1} = \mathbb{C}_{XX}\mathbb{B}^{\text{T}}\left(\mathbb{C}_{NN} + \mathbb{B}\mathbb{C}_{XX}\mathbb{B}^{\text{T}}\right)^{-1}$$

with the following substitutions

$$\mathbb{C}_{XX}^{-1} \to \mathbb{C}_{EE}^{\text{old}^{-1}} \qquad \mathbb{B}^{\text{T}} \to \mathbb{B}^{\text{new}\,\text{T}} \qquad \mathbb{C}_{NN}^{-1} \to \mathbb{C}_{NN}^{\text{new}^{-1}} \qquad \mathbb{B} \to \mathbb{B}^{\text{new}}$$

With these substitutions, (8.128) is changed to the following form:

$$K = \mathbb{C}_{EE}^{\text{old}}\mathbb{B}^{\text{new}\,\text{T}}\left(\mathbb{C}_{NN}^{\text{new}} + \mathbb{B}^{\text{new}}\mathbb{C}_{EE}^{\text{old}}\mathbb{B}^{\text{new}\,\text{T}}\right)^{-1} \tag{8.129}$$

Now, apply the matrix identity (8.102) to (8.125) in the reverse direction, that is,

$$(\mathbb{C}_{XX}^{-1} + \mathbb{B}^{\mathrm{T}}\mathbb{C}_{NN}^{-1}\mathbb{B})^{-1} = \mathbb{C}_{XX} - \mathbb{C}_{XX}\mathbb{B}^{\mathrm{T}}(\mathbb{C}_{NN} + \mathbb{B}\mathbb{C}_{XX}\mathbb{B}^{\mathrm{T}})^{-1}(\mathbb{C}_{XX}\mathbb{B}^{\mathrm{T}})^{\mathrm{T}}$$

with the following substitutions

$$\mathbb{C}_{XX}^{-1} \to \mathbb{C}_{EE}^{\mathrm{old}^{-1}} \qquad \mathbb{B}^{\mathrm{T}} \to \mathbb{B}^{\mathrm{new}\,\mathrm{T}} \qquad \mathbb{C}_{NN}^{-1} \to \mathbb{C}_{NN}^{\mathrm{new}^{-1}} \qquad \mathbb{B} \to \mathbb{B}^{\mathrm{new}}$$

With these substitutions, (8.125) is changed to the following form:

$$\begin{aligned}\mathbb{C}_{EE}^{\mathrm{tot}} &= \mathbb{C}_{EE}^{\mathrm{old}} - \mathbb{C}_{EE}^{\mathrm{old}}\mathbb{B}^{\mathrm{new}^{\mathrm{T}}}(\mathbb{C}_{NN}^{\mathrm{new}} + \mathbb{B}^{\mathrm{new}}\mathbb{C}_{EE}^{\mathrm{old}}\mathbb{B}^{\mathrm{new}^{\mathrm{T}}})^{-1}(\mathbb{C}_{EE}^{\mathrm{old}}\mathbb{B}^{\mathrm{new}^{\mathrm{T}}})^{\mathrm{T}} \\ &= \mathbb{C}_{EE}^{\mathrm{old}} - \mathbb{C}_{EE}^{\mathrm{old}}\mathbb{B}^{\mathrm{new}^{\mathrm{T}}}(\mathbb{C}_{NN}^{\mathrm{new}} + \mathbb{B}^{\mathrm{new}}\mathbb{C}_{EE}^{\mathrm{old}}\mathbb{B}^{\mathrm{new}^{\mathrm{T}}})^{-1}\mathbb{B}^{\mathrm{new}}\mathbb{C}_{EE}^{\mathrm{old}}\end{aligned} \tag{8.130}$$

Substituting (8.129) into (8.130), we obtain the following equation:

$$\mathbb{C}_{EE}^{\mathrm{tot}} = \mathbb{C}_{EE}^{\mathrm{old}} - K\mathbb{B}^{\mathrm{new}}\mathbb{C}_{EE}^{\mathrm{old}} \tag{8.131}$$

Substituting (8.131) into (8.126), we obtain the following equation:

$$\hat{X}^{\mathrm{tot}} = (\mathbb{C}_{EE}^{\mathrm{old}} - K\mathbb{B}^{\mathrm{new}}\mathbb{C}_{EE}^{\mathrm{old}})\mathbb{C}_{EE}^{\mathrm{old}^{-1}}\hat{X}^{\mathrm{old}} + KW^{\mathrm{new}} \tag{8.132}$$

Expanding the terms in the above equation, we obtain the following equation:

$$\begin{aligned}\hat{X}^{\mathrm{tot}} &= \mathbb{C}_{EE}^{\mathrm{old}}\mathbb{C}_{EE}^{\mathrm{old}^{-1}}\hat{X}^{\mathrm{tot}} - K\mathbb{B}^{\mathrm{new}}\mathbb{C}_{EE}^{\mathrm{old}}\mathbb{C}_{EE}^{\mathrm{old}^{-1}}\hat{X}^{\mathrm{old}} + KW^{\mathrm{new}} \\ &= \hat{X}^{\mathrm{old}} - K\mathbb{B}^{\mathrm{new}}\hat{X}^{\mathrm{old}} + KW^{\mathrm{new}} \\ &= \hat{X}^{\mathrm{old}} + K(W^{\mathrm{new}} - \mathbb{B}^{\mathrm{new}}\hat{X}^{\mathrm{old}})\end{aligned} \tag{8.133}$$

Summary of the Kalman Form

For the number of a priori data m, number of new observations s, and number of total (collapsed) observations $m + s$, the a priori estimate $\hat{X}^{\mathrm{old}}$ is updated by the following Kalman filtering algorithm to obtain a new estimate $\hat{X}^{\mathrm{tot}}$.

By (8.133),

$$\hat{X}^{\mathrm{tot}} = \hat{X}^{\mathrm{old}} + K\big(W^{\mathrm{new}} - \mathbb{B}^{\mathrm{new}}\hat{X}^{\mathrm{old}}\big)$$

By (8.129),

$$K = \mathbb{C}_{EE}^{\mathrm{old}}\mathbb{B}^{\mathrm{new}\,\mathrm{T}}(\mathbb{C}_{NN}^{\mathrm{new}} + \mathbb{B}^{\mathrm{new}}\mathbb{C}_{EE}^{\mathrm{old}}\mathbb{B}^{\mathrm{new}\,\mathrm{T}})^{-1}$$

By (8.131),

$$\mathbb{C}_{EE}^{\mathrm{tot}} = \mathbb{C}_{EE}^{\mathrm{old}} - K\mathbb{B}^{\mathrm{new}}\mathbb{C}_{EE}^{\mathrm{old}}$$

Comparison of the Kalman Form with the Original Form

The Original Form

By (8.124),

$$\hat{X}^{\mathrm{tot}} = \mathbb{C}_{EE}^{\mathrm{tot}}\left(\mathbb{C}_{EE}^{\mathrm{old}^{-1}}\hat{X}^{\mathrm{old}} + \mathbb{B}^{\mathrm{new}\,\mathrm{T}}\mathbb{C}_{NN}^{\mathrm{new}^{-1}}W^{\mathrm{new}}\right)$$

where, by (8.125),

$$\mathbb{C}_{EE}^{\mathrm{tot}} = \left(\mathbb{C}_{EE}^{\mathrm{old}^{-1}} + \mathbb{B}^{\mathrm{new}\,\mathrm{T}}\mathbb{C}_{NN}^{\mathrm{new}^{-1}}\mathbb{B}^{\mathrm{new}}\right)^{-1}$$

Both the original form and the Kalman form produce the same MMSE estimates. One form is preferred to the other depending on computational requirements.

Matrix inversion is computationally taxing. Therefore, depending on the dimension of the matrices to be inverted, one form is preferred to another. For example, if $m > s$, the Kalman form is preferred because the Kalman form only requires inverting $(s \times s)$ matrix, whereas the original form requires $(m \times m)$ and $(s \times s)$ matrix inversions.

The Kalman filter is used in dynamic situations where estimates are continuously updated as new measurements become available. Note that, after each iteration of estimation, the number of old measurements is increased by the number of new measurements just used. That is, the number of a priori data m keeps growing while the number of new measurements s stays small, that is, $m >> s$, and thus the Kalman form is computationally more efficient.

8.4 Queuing Theory Basics

In this section, we discuss the final example of the applications of random variables and stochastic processes, queueing theory. We first describe several real-life examples of queuing. We then discuss the common attributes of these queueing examples and define the queueing system in terms of the key parameters such as the arrival rate, the service rate, and the utilization factor. Finally, we discuss one special case of the queueing system, the M/M/1 queue.

8.4.1 Queueing Examples

Telephone call setup Queueing theory originated from the mathematical modeling of telephone call processing, which is briefly described below. Figure 8.7 depicts an end-to-end telephone call connection. A customer telephone line connects the customer's telephone set to the local switching office serving the customer, which is called the central office. The telephone lines are individually dedicated to

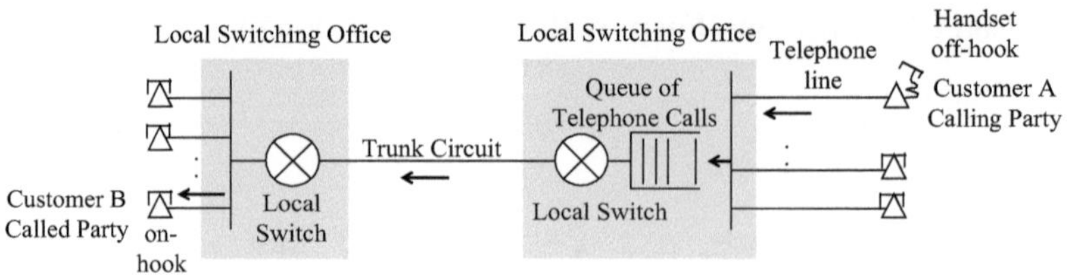

Fig. 8.7 Telephone call setup

customers and are not shared between the customers. In the figure, Customer A initiates a call to Customer B. Customer A's call setup request is placed in a queue at the customer's central office, along with other call requests. When Customer A's turn in the queue comes up, the local switch finds an idle circuit to the destination switch and routes the call on the circuit. If all the trunk circuits between the switches are occupied by other calls so that no idle circuit is available to route Customer A's call setup request, Customer A's call attempt is blocked.

The circuits between the switches are called the trunks. The trunks are not dedicated to any particular customers. While a call is in progress, the trunk carrying the call is used only by the current call and is not shared by any other call. However, when the call ends, the trunk is then available to be used by other calls. In this respect, unlike the customer telephone lines, the trunks are shared facility. The probability of call blocking is inversely proportional to the number of trunk circuits provided between the switches. The queueing model is used to determine, among others, the number of trunk circuits needed between the switches to meet a given objective of call blocking probability.

Customers at a Fast Food Restaurant

Customers arrive at a fast food restaurant, join a queue and wait for their turns, and, when their turns come, get service and leave the restaurant. As shown in Fig. 8.8, the restaurant has a pool of workers employed to provide the service. The level of service quality measured by, for example, customer waiting time, can be controlled by the size of the worker pool.

Customer Queue at a Bank

In Fig. 8.9, the bank customers join a single queue and advance toward the head of queue (HoQ). The customer at the HoQ goes to the first clerk who becomes free to serve the next customer, gets the service, and leaves the bank.

Cars at a Toll Gate

Cars arrive at a toll gate plaza and join a queue at a randomly selected toll booth. The cars advance in the queue and, when they reach the Head of Queue (HoQ), pay the toll and leave the booth. The amount of time a car must wait in the queue before it can pay the toll and leave the booth is a function of the amount of traffic, the number of toll booths, and the speed of the toll booth attendant in processing the toll (Fig. 8.10).

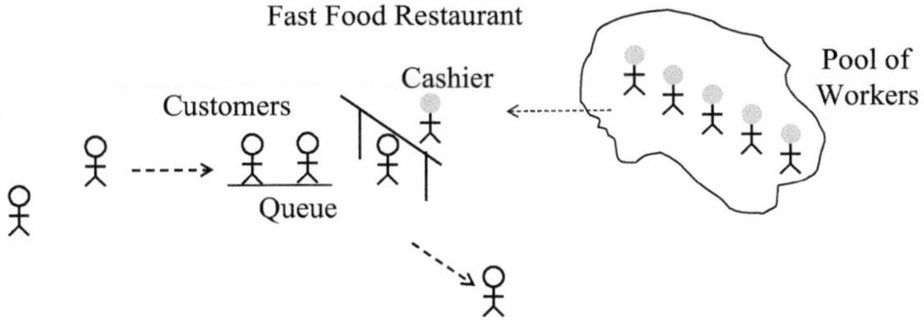

Fig. 8.8 Queueing at a fast food restaurant

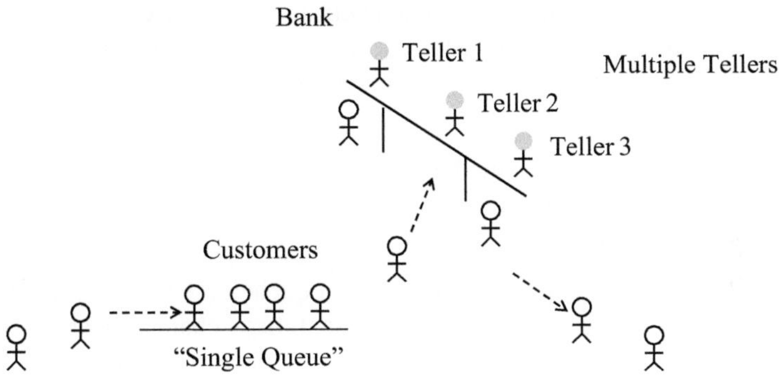

Fig. 8.9 Queueing at a bank

Fig. 8.10 Queueing at a toll gate

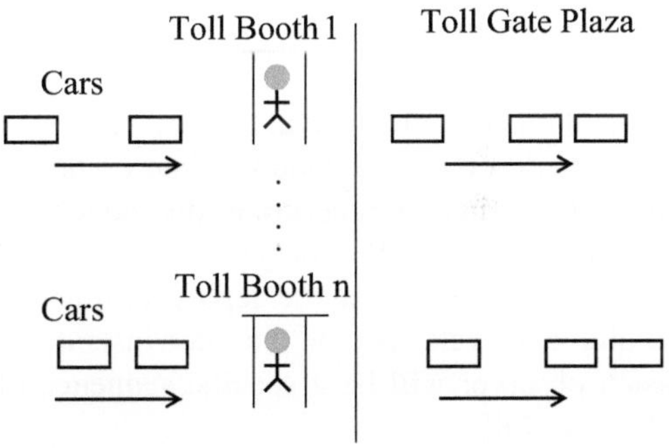

8.4.2 General Description of Queueing

In the four examples considered above, the central office, the fast food restaurant, the bank, and the toll booth, where the queues are formed, are called service stations, which is a generic term for the places or "entities" where customers go to obtain some kind of service. A queueing system may be described in general terms as follows. Customers "arrive" at a service station to obtain service, join a

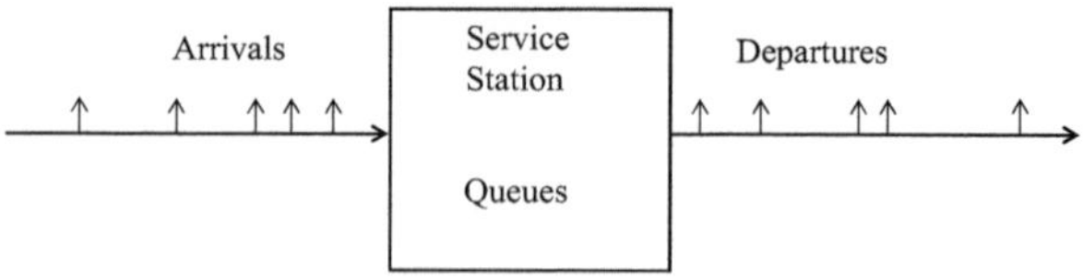

Fig. 8.11 A simplified diagram of a queueing system

queue at the tail of queue (ToQ), and advance through the queue towards the head of queue (HoQ). When a customer reaches the HoQ of the queue that he/she has joined, the customer is assigned to an available server for the service.

Different types of queueing model may be considered depending on whether the queueing system assumes a single queue or multiple queues and a single server or multiple servers. The queueing discipline determines how the customers arriving at a service station are classified into different types of queues and how the servers are assigned to the customers at the HoQs. The typical types of questions addressed by a queueing model are as follows:

How long would the customer line be?
How long would the customer have to wait before being served?
How quick would the service be?

Figure 8.11 shows a simplified diagram of a queueing system. A mathematical model of a queueing system is based on the probabilistic characterization of, among others, the customer arrival pattern and the service pattern by the service station.

8.4.3 Point Process

A random customer arrival pattern may be described by a point process. The point process describes the random time points at which the event of interest, e.g., an arrival, occurs. To describe the concept, suppose that observers have been dispatched to n different bus depots around a big city to record the instants of arrival of passengers at the depots starting at a reference time point. Assume that no multiple arrivals occur at the same instant. As illustrated in Fig. 8.12, the data from each observer will be a specific sequence of ascending time points for each bus depot as follows:

$$\begin{aligned} &\left\{t_1^j, t_2^j, \ldots, t_i^j, \ldots\right\}, \ t_1^j < t_2^j < \ldots < t_i^j < \ldots \\ &t_i^j = \text{time point of the } i\text{th occurence on the } j\text{th sample path} \end{aligned} \tag{8.134}$$

The ith time point across the collection of sample paths, $j = 1, 2, \ldots, n$, is an RV $T(i)$ taking on the values on the real line. The set of the RVs indexed in order of the time points i, $T(i)$s, is called the point process

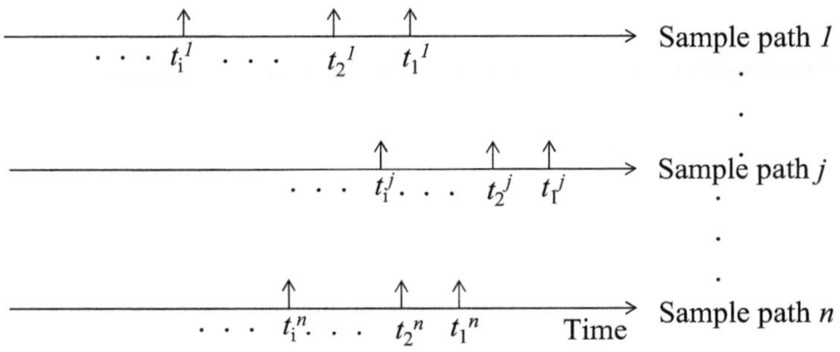

Fig. 8.12 Sample paths of arrival time points

$$\{T(1), T(2), \ldots, T(i), \ldots\}, \Omega_{\mathrm{T}} = R \tag{8.135}$$

Unlike the processes discussed so far, which are defined by time as the argument, the point process is defined by the ordering index i as the argument as discussed in Sect 6.1. $T(i)$ is an RV for a fixed i. For example, $T(1)$ is the RV representing the time points of the first occurrence of the event, e.g., the first arrival time points. Similarly, $T(i)$ is the RV of the ith arrival time points.

The following are some of the examples of the point process:

Sequence of the instants at which passengers arrive at a bus depot
Sequence of the instants at which cars arrive at a toll booth
Sequence of the instants at which babies are born
Sequence of the instants at which light bulbs fail
Sequence of the instants at which customers walk into a bar
Sequence of the instants at which customers leave a bar
Sequence of the instants at which telephone calls arrive at a central switching office

8.4.4 Statistical Characterization of the Point Process by the Counting Process

One of the methods of statistically characterizing the point process defined in Sect. 4.3 is to use the counting process, which is constructed as follows. Figure 8.13 illustrates the counting process. Select non-overlapping time intervals with a start and an end time points and define the RV $N(k)$ as the number of time points of event occurrence falling in the kth interval

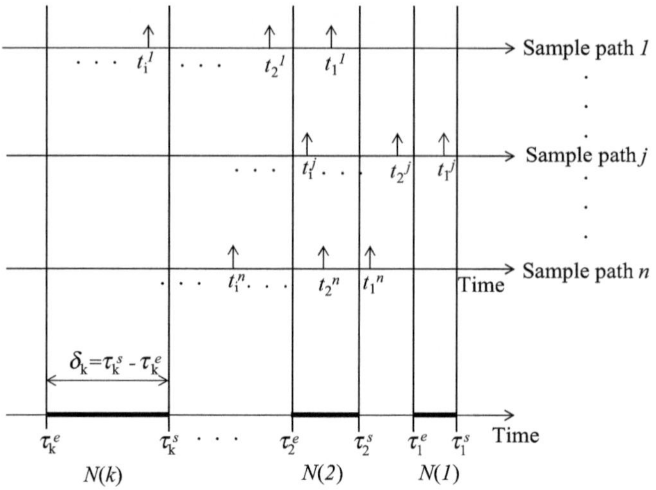

Fig. 8.13 Sample paths of counting process

$$\begin{aligned}\tau_i^s &= \text{starttimepointoftheithinterval}\\ \tau_i^e &= \text{endtimepointoftheithinterval}\\ \delta_k &= \text{intervallengthofthekinterval} = \tau_k^s - \tau_k^e\\ N(k) &= \text{numberofeventpointsoccurringinthekinterval}\end{aligned} \tag{8.136}$$

A collection of RVs $N(k)$, $k = 1, 2, \ldots$, is called the counting process

$$\{N(1), N(2), \ldots N(k), \ldots, N(n)\} \tag{8.137}$$

The counting process can be statistically characterized by determining the nth-order joint distributions for the n RVs defined by (8.137) for all possible values of n and all possible values of the start and end time points of the intervals.

8.4.5 Poisson Distribution

Modify Example 5.4.3 in Sect. 5.4.2 in the following manner. Suppose that the total observation period T is increased with the subinterval period τ fixed. Then,

$$\text{As } T \to \infty, p = \frac{\tau}{T} \to 0 \text{ and } n \to \infty \tag{8.138}$$

In (8.138), suppose that we increase T at a certain rate. Since τ is a fixed value, the probability p would decrease at the same rate as T. If the number of arrivals n and the observation period T increase at the same rate in such a way that the ratio

n to T converges to a constant λ, then the product np converges to a constant as follows:

$$\lim_{T\to\infty,\ n\to\infty} \frac{n}{T} = \lambda \tag{8.139}$$

In queueing theory, λ defined by (8.139) is referred to as the arrival rate.

Substituting (8.139) into the following equation, we see that, if (8.139) is met, the product np converges to the constant $\lambda\tau$

$$\lim_{T\to\infty, n\to\infty} np = n \times \frac{\tau}{T} = \tau \times \frac{n}{T} = \lambda\,\tau \tag{8.140}$$

The Poisson approximation to the binomial distribution is given by the following expression:

$$P[\{X = k\}] = \binom{n}{k} p^k (1-p)^{n-k} \approx \frac{(np)^k}{k!} e^{-np} \tag{8.141}$$

Substituting (8.140) into (8.141) and taking the limit as $T \to \infty$ and $n \to 0$ with np kept at a constant value of $\lambda\tau$, we can see that the above approximation approaches the exact expression as follows:

$$P[\{X = k\}] = \binom{n}{k} p^k (1-p)^{n-k} \to e^{-\lambda\tau} \frac{(\lambda\tau)^k}{k!} \tag{8.142}$$

To sum up, let the RV X be the number of random arrivals, the constant arrival rate, λ, and the event that k arrivals occur in a time interval of length τ, $\{X = k\}$. The probability of this event is given by the Poisson distribution as follows:

$$p_k = P[\{X = k\}] = e^{-\lambda\tau} \frac{(\lambda\tau)^k}{k!} \quad \text{for } k = 1, 2, \ldots \tag{8.143}$$

By replacing in (8.143)

$$\alpha = \lambda\tau \tag{8.144}$$

we can rewrite (8.143) as follows:

$$p_k = P[\{X = k\}] = e^{-\alpha} \frac{\alpha^k}{k!} \quad \text{for } k = 1, 2, \ldots \tag{8.145}$$

Given p_k by the above equation, the Poisson *pdf* is given by the following equation:

$$f_X(x) = \sum_k p_k \delta(x-k) \tag{8.146}$$

By integrating the *pdf* given by the above equation, we obtain the Poisson CDF as follows:

$$F_X(x) = \sum_k p_k \quad \text{for } k \le x \le k+1 \tag{8.147}$$

The mean and the variance of a Poisson RV are obtained as follows. The Taylor series expansion of the exponential function is given by

$$e^{\alpha} = \sum_{k=0}^{\infty} \frac{\alpha^k}{k!} \tag{8.148}$$

Taking the derivative of both sides of (8.148) with respect to α, we obtain the following equation:

$$e^{\alpha} = \sum_{k=0}^{\infty} \frac{k\alpha^{k-1}}{k!} \tag{8.149}$$

Multiplying both sides of the above equation by α, we obtain the following equation:

$$\alpha e^{\alpha} = \sum_{k=0}^{\infty} \frac{k\alpha^{k}}{k!} \tag{8.150}$$

Taking the derivative of both sides of (8.149) with respect to α, we obtain the following equation:

$$\begin{aligned} e^{\alpha} &= \frac{\mathrm{d}}{\mathrm{d}a}\left[\sum_{k=0}^{\infty} \frac{k\alpha^{k-1}}{k!}\right] \\ &= \sum_{k=0}^{\infty} \frac{k(k-1)\alpha^{k-2}}{k!} \\ &= \frac{1}{2}\sum_{k=0}^{\infty} \frac{k(k-1)\alpha^{k}}{k!} \\ &= \frac{1}{2}\sum_{k=0}^{\infty} \frac{k^2\alpha^{k}}{k!} - \frac{1}{2}\sum_{k=0}^{\infty} \frac{k\alpha^{k}}{k!} \end{aligned} \tag{8.151}$$

Rearranging the terms of (8.151), we obtain the following equation:

$$\sum_{k=0}^{\infty} \frac{k^2\alpha^k}{k!} = \alpha^2\left[e^{\alpha} + \frac{1}{\alpha^2}\sum_{k=0}^{\infty} \frac{k\alpha^k}{k!}\right] \tag{8.152}$$

Substituting (8.150) into (8.152), we have

$$\sum_{k=0}^{\infty} \frac{k^2\alpha^k}{k!} = \alpha^2\left[e^\alpha + \frac{1}{2}\alpha e^\alpha\right] = e^\alpha\left(\alpha^2 + \alpha\right) \tag{8.153}$$

Substituting (8.145) and (8.146) into the following equation and using the property of the Dirac delta function, we obtain the following equation:

$$E(X) = \int_{-\infty}^{+\infty} x f_X(x)\mathrm{d}x = e^{-\alpha} \sum_{k=0}^{\infty} \frac{k\alpha^k}{k!} \tag{8.154}$$

Substituting (8.150) into (8.154), we obtain the following equation:

$$E(X) = e^{-\alpha}\alpha e^\alpha = \alpha \tag{8.155}$$

Substituting (8.153) and (8.155) into the following, we have

$$E\left(X^2\right) = e^{-\alpha} \sum_{k=0}^{\infty} \frac{k^2\alpha^k}{k!} = e^{-\alpha}\, e^\alpha\left(\alpha^2 + \alpha\right) = \alpha^2 + \alpha \tag{8.156}$$

Substituting (8.155) and (8.156) into the following, we obtain the following equation:

$$\mathrm{Var}(X) = E\left(X^2\right) - [E(X)]^2 = \alpha^2 + \alpha - \alpha^2 = \alpha \tag{8.157}$$

8.4.6 Poisson Process

For the counting process defined by (8.137), postulate the following assumptions:

1. The RVs, $N(k)$s, are mutually independent for any k and for any start and end times of the intervals, τ_k^s and τ_k^e. This means that the RV $N(k)$ is not influenced by what happens in the previous intervals, e.g., $N(k-1)$, and does not influence what will happen in the future intervals, e.g., $N(k+1)$.
2. The statistical properties of the RVs, $N(k)$s, remain unchanged if $N(k)$s are shifted by the same amount keeping the interval lengths δ_k unchanged. In other words, if the start and end points of the intervals are all shifted by the same amount so that the interval lengths remain unchanged, the statistical properties of $N(k)$s remain unchanged. This means that the statistical properties of the RVs, $N(k)$s, depend on the interval lengths δ_k and do not depend on the absolute values of the start and end points of the intervals.
3. $P[\{N(k) = 0\}] \neq 0$ for any k and for any interval length. This means that there is a nonzero probability that nothing will happen in any interval of any length.
4. The probability that more than one events will occur simultaneously is zero.

5. There exists a constant rate of arrival $\lambda = n/T$ defined by (8.139).
6. There is an infinite reservoir of arrivals, that is, $n \to \infty$, as $T \to \infty$. This assumption needs to be carefully examined before being taken and will be discussed further later.

With the above assumptions, referring to Fig. 8.14, define the Poisson process $X(t)$ as the number of random events occurring in the interval from time 0 to t. For convenience, let us call the events the arrivals. For a fixed value t, $X(t)$ is an RV. Under these assumptions, (8.143) gives the following probability distribution

$$P[\{X(t) = m\}] = p_m = e^{-\lambda t} \frac{(\lambda t)^m}{k!} \qquad \text{for } m = 1, 2, \ldots \tag{8.158}$$

The unit of λ and t must be consistent in time. For example, if $\lambda = 120/\text{h}$ in the above equation, t needs to be expressed in hours. If t is expressed in minutes, λ must be converted to 120/60 min, which yields 2/min.

By (8.155) and (8.157), the mean and the variance of the Poisson process $X(t)$ are given by

$$E\{X(t)\} = \lambda t \qquad \text{Var}\{X(t)\} = \lambda t \tag{8.159}$$

Under the above assumptions, we have, by (8.143),

$$P[\{N(k) = m\}] = e^{-\lambda \delta_k} \frac{(\lambda \delta_k)^m}{m!} \qquad \text{for } k = 1, 2, \ldots \tag{8.160}$$

where δ_k is the interval length of kth interval defined by (8.136). By the first assumption discussed at the beginning of this section, the probabilities of the number of random arrivals in two non-overlapping intervals of length δ_m and δ_n, $N(m)$ and $N(n)$, are independent. The probabilities of future arrivals are not affected by the previous arrivals. This is true only with an infinite reservoir of arrivals. The Poisson arrivals are often referred to as the "pure random arrivals." The applicability of the Poisson arrival model must be evaluated for each application under consideration for reasonableness.

Example 8.4.1

Suppose that packets arrive at a packet switch at the rate of $\lambda = 6 \times 10^6/\text{min}$. Assuming a Poisson arrival, what is the probability that three packets will arrive in a 10-ms interval?

Let

$$k = 3;\ \lambda = {}^{6\times10^6}\!/_{m} = {}^{1\times10^5}\!/_{s} = {}^{0.1}\!/_{ms};$$
$$t = 10\ \text{ms};\ \lambda t = 0.1 \times 10 = 1$$

$$P\{3 \text{ in } 10\,\text{ms}\} = e^{-1}\frac{(1)^3}{3!} = 0.06 \quad \left(\text{Use } e^{-1} \cong 0.37\right)$$

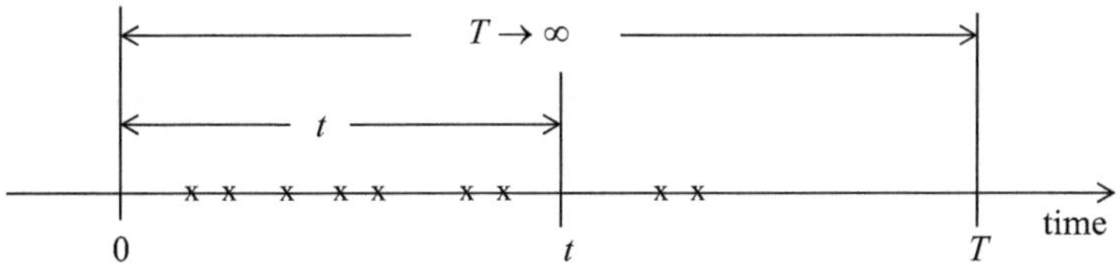

Fig. 8.14 Definition of Poisson process

The interarrival times of the Poisson arrivals are exponentially distributed. The CDF and the *pdf* of the interarrival times are given by

$$F(t) = 1 - e^{-\lambda t} \qquad f(t) = \lambda e^{-\lambda t} \tag{8.161}$$

The mean interarrival time is $1/\lambda$.

Example 8.4.2
Consider people arriving at a bus station in a large city with millions of people. In this situation, the assumption of infinite reservoir of arrivals, assumption 6 of the Poisson process discussed at the beginning of this section, may be reasonable. Now, consider people arriving at a bus station in a small community, say, with five commuters. In this case, the Poisson model would not apply. For example, if five people have already arrived so far, the probability of additional arrivals would obviously be zero, and assumption 6 would not be valid.

8.4.7 Key Parameters of a Queueing System

Arrival Rate
A long-term arrival rate is defined by (8.139). The long-term arrival rate can be determined by taking either the time average or the ensemble average depending on the conditions the arrival process satisfies. If the arrival process is stationary, the arrival rate can be determined by taking the time average. If the process is mean ergodic, the arrival rate can also be determined by taking the ensemble average.

Service Rate
Let the total number of customers served during the period of length T be m. The long-term service rate, denoted by γ, is the number of customers served in a unit time and is defined by the following equation:

$$\gamma = \lim_{T \to \infty, m \to \infty} \frac{m}{T} \tag{8.162}$$

The service time, denoted by τ_S, is the time taken to serve one customer and is given by the inverse of the service rate as follows:

$$\tau_S = \frac{1}{\gamma} = \frac{T}{m} \tag{8.163}$$

Utilization Factor

The utilization factor, denoted by ρ, is a measure of how fully the resource of the queueing system is used to meet customer needs and is defined by the ratio of the arrival rate to the service rate as follows:

$$\rho = \frac{\lambda}{\gamma} \tag{8.164}$$

Substituting (8.139) and (8.162) into (8.164), we see that the utilization factor can also be expressed as the ratio of the number of customers arriving at the service station to the number of people that the service station can serve during the same time period, that is, the service station's capacity to serve, as follows:

$$\rho = \frac{n}{m} \tag{8.165}$$

For example, a utilization factor of 70% means that the customers arrive at the service station at the rate of 70% of the service station's capacity. As will be seen later in the analysis of the *M*/*M*/1 queue, the utilization factor must be less than 1 to have a steady-state queue; otherwise, the queue length grows indefinitely.

8.4.8 Little's Theorem

Let the queue length and the delay or "waiting time" through the queue be RVs *N* and *D*, respectively, with the means as follows:

$$\mu_N = E(N) \qquad \mu_D = E(D)$$

The following relation holds true:

$$\mu_N = \lambda \times \mu_D \tag{8.166}$$

where λ is the arrival rate.

Heuristic Interpretation of Little's Theorem

A proof of Little's theorem is beyond the scope of this book. The heuristic interpretation of the theorem is depicted in Fig. 8.15.

In Fig. 8.15, Customer A joins the queue at the ToQ at time t_1 and advances to the HoQ at t_2. The delay through the queue $D = t_2 - t_1$. As Customer A moves forward in the queue, new customers join the queue behind Customer A at the rate of the arrival rate λ. The amount of time Customer A takes to reach the HoQ is D. Therefore, the number of people who join the queue behind Customer A by the time Customer A reaches the HoQ, which is the queue length N, is given by $N = \lambda \times D$ as stated by Little's theorem.

8.4.9 M/M/1 Queue

Under certain conditions of random arrivals and departures, that is, services, such as the Markov chain, a queuing system can be modeled as a special class called a "birth-death" process. An important class of queue referred to as the $M/M/1$ queue is the "birth-death" process.

The $M/M/1$ queue is a single server queue. The $M/M/1$ queue is based on the following assumptions. The customer arrivals are assumed to be a Poisson process with an arrival rate λ. The service time is assumed to be exponentially distributed with a service rate γ with the following CDF and *pdf*.

$$F(t) = 1 - e^{-\gamma t} \qquad f(t) = \gamma\, e^{-\gamma t} \tag{8.167}$$

Based on the analysis of the birth-death process, it can be shown that, in the steady state of the $M/M/1$ queue, the probability that the number of customers in the queue, N, is equal to n, denoted by p_n, is given by the following equation:

$$p_n = P(N = n) = (1 - \rho)\rho^n \qquad n = 0, 1, 2, \ldots \tag{8.168}$$

where ρ is the utilization factor defined by (8.162). The steady-state solution given by above equation is possible if the utilization factor is less than 1.

$$\rho = \frac{\lambda}{\gamma} < 1 \quad \text{or} \quad \lambda < \gamma \tag{8.169}$$

The above condition states that, in order for the M/M/1 queue to attain a steady state, the customers arriving at the service station must be served at a faster rate than the rate by which they arrive.

Setting $n = 0$ in (8.168), we obtain the probability that the queue is empty as follows:

$$p_0 = P(N = 0) = 1 - \rho \tag{8.170}$$

For example, if $\rho = 0.7$, by the above equation, the probability that there is no customer to serve at the service station, that is, empty queue, is 0.3, which is the complement of the utilization factor. This means that 30% of the time the service station is idle.

Consider the probability that the queue length N is less than k as follows:

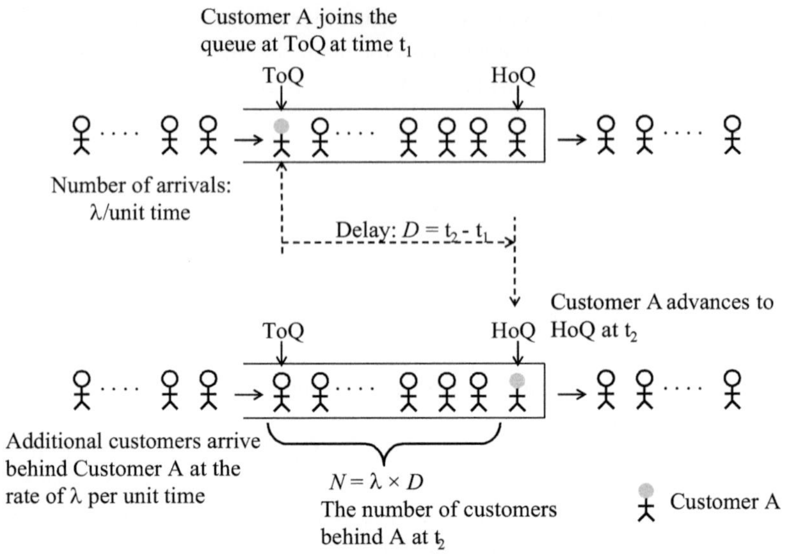

Fig. 8.15 Heuristic interpretation of Little's theorem

$$P(N \leq k) = \sum_{n=0}^{k} (1-\rho)\rho^n = (1-\rho)\sum_{n=0}^{k} \rho^n = (1-\rho)S \tag{8.171}$$

where

$$S = \sum_{n=0}^{k} \rho^n \tag{8.172}$$

To find the closed-form expression of S, consider the following operation:

$$\begin{aligned} S - \rho S &= \sum_{n=0}^{k} \rho^n - \rho \sum_{n=0}^{k} \rho^n \\ &= \left(1 + \rho + \rho^2 + \cdots + \rho^k\right) - \left(\rho + \rho^2 + \cdots + \rho^k + \rho^{k+1}\right) \\ &= 1 - \rho^{k+1} \end{aligned}$$

Hence

$$S = \frac{1 - \rho^{k+1}}{1 - \rho} \tag{8.173}$$

Substituting (8.173) into (8.171), we obtain

$$P(N \leq k) = (1-\rho)\frac{1-\rho^{k+1}}{1-\rho} = 1-\rho^{k+1}, \quad k = 0, 1, 2, \ldots \tag{8.174}$$

The complement of the above probability is given by the following equation:

$$P(N > k) = 1 - P(N \leq k) = \rho^{k+1}, \quad k = 0, 1, 2, \ldots \tag{8.175}$$

The mean, or expected value, of the queue length is, by definition, given by the following equation:

$$\mu_N = E(N) = \sum_{n=0}^{\infty} n p_n \tag{8.176}$$

Substituting (8.168) into the above equation, we obtain the following expression for the mean queue length:

$$\mu_N = (1-\rho)\sum_{n=0}^{\infty} n\rho^n \tag{8.177}$$

From (8.172) and (8.173), we have the following expression:

$$S = \sum_{n=0}^{k} \rho^n = \frac{1-\rho^{k+1}}{1-\rho}$$

and taking the limit $k \to \infty$ yields the following expression

$$\sum_{n=0}^{\infty} \rho^n = \frac{1}{1-\rho}$$

Taking the derivative of both sides of the above equation with respect to ρ, we obtain the following equation:

$$\sum_{n=0}^{\infty} n\rho^{n-1} = \frac{1}{(1-\rho)^2}$$

or, rewriting the left-hand side of the above as follows,

$$\rho^{-1}\sum_{n=0}^{\infty} n\rho^n$$

we obtain the following equation:

Fig. 8.16 Mean queue length, μ_N, as a function of utilization factor, ρ

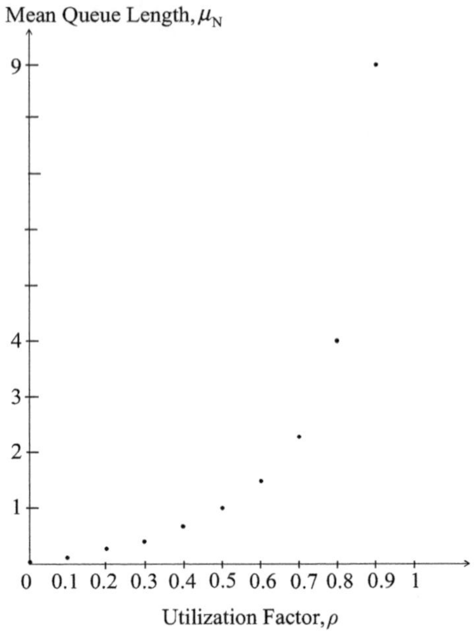

$$\sum_{n=0}^{\infty} n\rho^n = \frac{\rho}{(1-\rho)^2} \tag{8.178}$$

Substituting (8.178) into (8.177), we obtain the following equation for the mean queue length:

$$\mu_N = (1-\rho)\frac{\rho}{(1-\rho)^2} = \frac{\rho}{1-\rho} \tag{8.179}$$

Figure 8.16 plots the mean queue length, μ_N, as a function of the utilization factor, ρ, given by the above equation.

The graph shows a "knee" at $\rho = 0.7$ at which the mean queue length increases more steeply as a function of ρ as follows:

ρ	μ_N
0.7	2.3
0.8	4.0
0.9	9.0
1.0	∞

Because of this, 70% utilization factor is sometimes considered an optimum operating point of a service station. For example, the manager can control the utilization factor by hiring more or less workers to staff the service station to keep the utilization factor near 70%. For example, if the service station is currently

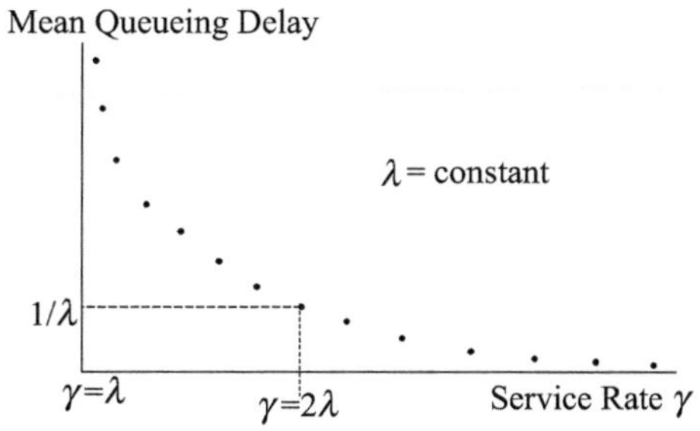

Fig. 8.17 ean queueing delay as a function of the service rate

operating at the utilization factor of 85%, the performance of the service station can be dramatically improved by bringing the utilization factor down to 70% by adding more workers.

By a method like that used to derive the mean queue length, the variance of the queue length is obtained to be as follows:

$$\sigma_N^2 = \frac{\rho}{(1-\rho)^2} \tag{8.180}$$

From Little's theorem given by (8.166), we obtain the following relationship:

$$\mu_D = \frac{\mu_N}{\lambda} \tag{8.181}$$

Substituting (8.179) into the above equation, we obtain the following equation for the mean delay through the queue

$$\mu_D = \frac{\rho}{\lambda(1-\rho)}$$

Substituting (8.164) into the above equation, we obtain

$$\mu_D = \frac{\frac{\lambda}{\gamma}}{\left(1-\frac{\lambda}{\gamma}\right)} = \frac{1}{\gamma-\lambda} \tag{8.182}$$

Figure 8.17 plots the mean queueing delay as a function of the service rate.

About the Author

Dr. Kun I. Park has extensive experience in telecommunications. Through his career, he worked at Bell Laboratories, Lucent and MITRE as Member of Technical Staff and Technical Supervisor. He also worked as Adjunct Professor, with the rank of Full Professor, at the Stevens Institute of Technology, Hoboken, New Jersey, USA for over 20 years, where he taught, among other courses, probability and stochastic processes. He received his Ph.D. in Electrical Engineering from the University of Pennsylvania, Philadelphia, Pennsylvania, USA.

K.I. Park, *Fundamentals of Probability and Stochastic Processes with Applications to Communications*, https://doi.org/10.1007/978-3-319-68075-0

Bibliography

1. R.V. Churchill, *Introduction to Complex Variables and Applications*, 1st edn. (McGraw-Hill Book Company, Inc., New York, 1948)
2. G.B. Thomas Jr., *Calculus and Analytic Geometry* (Addison-Wesley Publishing Company, Inc., Cambridge MA, 1953)
3. P. Lancaster, M. Tismenetsky, *The Theory of Matrices*, 2nd edn. (Academic Press, Inc., Orlando, 1985)
4. G. Arfken, *Mathematical Methods for Physicists*, 3rd edn. (Academic Press, Inc., San Diego/London, 1985)
5. D. Middleton, *An Introduction to Statistical Communication Theory* (McGraw-Hill Book Company, Inc., New York, 1960)
6. W. Feller, *An Introduction to Probability Theory and Its Applications*, vol I, 3rd edn. (Wiley, New York, 1950)
7. A.O. Allen, *Probability, Statistics, and Queueing Theory with Computer Science Applications*, 2nd edn. (Academic Press, Inc., Boston, 1990)
8. A. Papoulis, *Probability, Random Variables, and Stochastic Processes*, 3rd edn. (McGrow-Hill, Inc., Auckland, 1991)
9. J.G. Proakis, M. Salehi, *Communication Systems Engineering* (London, Prentice Hall, 1994)

K.I. Park, *Fundamentals of Probability and Stochastic Processes with Applications to Communications*, https://doi.org/10.1007/978-3-319-68075-0

Index

K.I. Park, *Fundamentals of Probability and Stochastic Processes with Applications to Communications*, https://doi.org/10.1007/978-3-319-68075-0

MIX
Papier aus verantwortungsvollen Quellen
Paper from responsible sources
FSC® C105338

If you have any concerns about our products,
you can contact us on
ProductSafety@springernature.com

In case Publisher is established outside the EU,
the EU authorized representative is:
Springer Nature Customer Service Center GmbH
Europaplatz 3, 69115 Heidelberg, Germany

Printed by Libri Plureos GmbH
in Hamburg, Germany